D1783890

# Environmental Ergonomics:
## Sustaining Human Performance in Harsh Environments

# Environmental Ergonomics

## Sustaining Human Performance in Harsh Environments

Edited by

**Igor B. Mekjavic**
**Eric W. Banister**
**James B. Morrison**

*School of Kinesiology, Simon Fraser University,*
*Burnaby, Canada*

*Taylor & Francis*
*Philadelphia · New York · London*
*1988*

**USA**    Taylor & Francis Inc., 242 Cherry St, Philadelphia,
PA 19106–1906

**UK**    Taylor & Francis Ltd, 4 John St, London, WC1N 2ET

**British Library Cataloguing in Publication Data**
Environmental ergonomics: sustaining
   human performance in harsh environments.
   1. Human engineering   2. Human physiology
   I. Mekjavic, Igor B.   II. Banister, Eric W.
   III. Morrison, James B.
   612'.00246208      TA166

   ISBN 0-85066-400-4


**Library of Congress Cataloging in Publication Data is
Available.**

*Cover design by Russell Beach*
Typeset in 11/12pt Bembo by Katerprint Typesetting
Services, Oxford
Printed in Great Britain by Taylor & Francis (Printers) Ltd,
Basingstoke, Hants.

# Preface

Through the ages, our continuous endeavour to harness the Earth's natural resources has led us to live and work in a variety of harsh environmental conditions. It is only through seemingly limitless human ingenuity that we have been able to survive in environments posing thermal stresses, such as in equatorial and polar regions, to perform unhindered at altitude and on the ocean floor, and most recently to conquer space. The problem of sustaining unhindered human performance in different environments requires a multi-disciplinary approach, thus continued progress relies on collaborative effort between researchers in the Life Sciences and Engineering Sciences.

To date, although much collaboration exists within specific scientific fields worldwide, the results of such studies are often reported at separate professional meetings in a specific branch of the Engineering or Life Sciences. There are only rare opportunities for scientists working in the field of Ergonomics, specifically investigating the protection of workers from the environmental variables, to meet in similar scientific forums. It was partially this need to exchange scientific experiences in the field of Environmental Ergonomics that stimulated a group of scientists from several countries to establish biennial meetings in Environmental Ergonomics. This book contains a selection of papers presented at the 2nd International Environmental Ergonomics Conference, held at Whistler, British Columbia (Canada), on 21–25 July 1986 and sponsored by the School of Kinesiology, Simon Fraser University. The meeting established strong links between the participating scientists and a commitment to maintain such a forum in Environmental Ergonomics in the future. It is fitting therefore, to thank all conference participants, the organizations and institutions committed to this aim. We would like to thank them also for their support in directly assisting with the publication of this volume.

The chapters in this volume cover seven distinct topics. Part I deals with the physiological problems encountered during exposures to hot or cold environments. As clothing forms the most common protective barrier from the environment, Part II reviews current trends in the assessment of protective clothing. With the advent of offshore industry, the hazard of cold water immersion has instigated the development of specialized clothing to sustain life during prolonged cold water immersion. Recent developments in this area are discussed in Part III. Part IV introduces current advances in modelling the physiological responses to thermal stress. The physiological and technological limitations to work underwater are reviewed in Part V, and some effects of hypobaric environments are presented in Part VI. The volume concludes with a discussion of the effects of increased gravitational load on physiological mechanisms. Undoubtedly, with the projected development of an orbiting space station in the 1990s, many exciting developments in this field of Environmental Ergonomics may be anticipated.

The broad range of topics presented in this volume emphasizes the need for further collaboration in the several areas of specialization discussed here. The benefits of meetings at which scientists, members of industry and agencies regulating work practices share and exchange experiences are substantial.

We are greatly indebted to the Natural Sciences and Engineering Research Council of Canada and Simon Fraser University for the substantial assistance provided for the conference and this publication. We also appreciate the admirable concern for the safety and welfare of workers in arctic and offshore industrial installations demonstrated by Mustang Industries Inc. and Petro Canada in their corporate sponsorship of the conference.

We are indebted to Margot Coard, Pat Good and Karen Mittleman, who helped with numerous aspects of both the conference and this publication. In addition, the editorial assistance of David Grist of Taylor & Francis is much appreciated.

*Burnaby, B.C.*
*June 1987*

*Igor B. Mekjavic*
*Eric W. Banister*
*James B. Morrison*

# Contents

# Contributors

J. R. Allan — Environmental Sciences Division, RAF Institute of Aviation Medicine, Farnborough, Hants GU14 6SZ, England

Peter B. Bennett — F. G. Hall Laboratory, Duke University Medical Center, Durham, NC 27710, USA

Larry G. Berglund — John B. Pierce Foundation Laboratory, New Haven, CT 06519, USA

Hilding Bjurstedt — Dept of Environmental Medicine, Karolinska Institute, 104 01 Stockholm, Sweden

John Bligh — Institute of Animal Physiology, Babraham, Cambridge CB2 4AT, England

Monika Buse — Institute of Physiology, MA 4/59, Ruhr University, 4630 Bochum 1, FR Germany

G. W. Crockford — London School of Hygiene and Tropical Medicine, Malet St, London WC1, England

Gunnar O. Dahlbäck — Defence Materiel Administration, Testing Directorate, 580 13 Linköping, Sweden

A. Pharo Gagge — John B. Pierce Foundation Laboratory, New Haven, CT 06519, USA

Ralph F. Goldman — Multi-Tech Corp., 1 Strathmore Rd, Natick, MA 01760, USA

T. Guthe — Institute of Work Physiology, Oslo, Norway

G. Havenith — IZF/TNO Institute for Perception, 3769 ZG Soesterberg, The Netherlands

Philip Hayes — Environmental Sciences Division, RAF Institute of Aviation Medicine, Farnborough, Hants GU14 6SZ, England

T. R. Hennessy — Admiralty Research Establishment, Queens Rd, Teddington, Middx TW11 0LN, England

Kaoru Inoue       Dept of Anatomy, School of Medicine, Hokkaido University, Kita-ku, Sapporo 060, Japan

W. A. Lotens       IZF/TNO Institute for Perception, 3769 ZG Soesterberg, The Netherlands

Kazuya Matsuda       Dept of Ergonomics, Kyushu University of Design Sciences, 4-9-1 Shiobaru, Minami-ku, Fukuoka 815, Japan

Igor B. Mekjavic       School of Kinesiology, Simon Fraser University, Burnaby, B.C. V5A 1S6, Canada

James B. Morrison       School of Kinesiology, Simon Fraser University, Burnaby, B.C. V5A 1S6, Canada

Yasunobu Nishi       Dept of Architectural Engineering, Hokkaido Institute of Technology, Hokkaido, Sapporo 006, Japan

Sarah A. Nunneley       USAF School of Aerospace Medicine, Brooks AFB, TX 78235-5302, USA

Takahimu Oohori       Dept of Electrical Engineering, Hokkaido Institute of Technology, Hokkaido, Sapporo 006, Japan

K. Rodahl       Institute of Work Physiology, Oslo, Norway

Masahiko Sato       Dept of Ergonomics, Kyushu University of Design Sciences, 4-9-1 Shiobaru, Minami-ku, Fukuoka 815, Japan

Ed A. Smallhorn       Product Development Division, Nova Scotia Research Foundation, PO Box 790, Dartmouth, N.S. B2Y 3Z7, Canada

P. J. Sullivan       School of Kinesiology, Simon Fraser University, Burnaby, B.C. V5A 1S6, Canada

K. H. Umbach       Bekleidungsphysiologisches Institut e.V. Hohenstein, 7124 Bönningheim, FR Germany

Jürgen Werner       Institute of Physiology, MA 4/59, Ruhr University, 4630 Bochum 1, FR Germany

Eugene H. Wissler       University of Texas at Austin, Austin, TX 78712, USA

Akira Yasukuochi       Dept of Industrial Physiology, National Institute of Industrial Health, Nagao, Tama-ku, Kawasaki 214, Japan

# PART I
# DRY THERMAL ENVIRONMENTS

# 1. Human Cold Exposure and the Circumstances of Hypothermia

## John Bligh

## 1. An evolutionary perspective

The principal selective pressures which culminated in mammalian endothermic homeothermy are presumed to have been the advantages in predator–prey interactions of the continuous ability to move at speed without a prior need to raise muscle and other tissue temperatures. The essentially independent, but functionally integrated, metabolic and regulatory features of endothermic homeothermy probably evolved in a warm and humid climate. Here the energetic wastefulness of a permanently elevated metabolic rate was no handicap since it could be readily sustained by a continuously available and adequate energy supply. The total flow of heat down the thermal gradient from organism to environment was never so great as to overwhelm the compensatory physiological processes of thermoregulatory heat production and conservation.

Subsequently various ecological pressures—including reproductive success in the early mammals, whose rapidly increasing numbers would have threatened to outstrip the immediately available energy supply—compelled migration north and south into progressively less advantageous regions, where both the energy supply and ambient temperature were subject to wide seasonal fluctuations. The seasonal concurrence of the lowest ambient temperatures and the lowest availability of energy—food—threatened the very survival of the endothermic homeothermic animals in high latitudes and constituted a further selection pressure which resulted in appropriate adaptations. Some of the latter were structural, such as the development of sub- and super-cutaneous thermal insulation. Others were functional, and these may be classified as being either autonomic or behavioural. Autonomic functions are essentially reflex ones which do not involve higher brain centres, and which do not involve elaborate body movements; shivering and sweating are examples. Behavioural functions do involve the higher brain structures, and so involve elaborate body movements; huddling, nest-building and food caching are examples.

It is conjectured that pre-man evolved as a fruit-gathering primate in a tropical forest habitat, and later moved on to open equatorial plains where survival depended upon successful competition with fast, powerful and clawed carnivores for a share of the available energy supply. This was achieved largely by behavioural means: the development of a prowess in hunting strategies and tactics based upon memory, anticipation, planning, inventiveness and team-work; the development of the fore-limbs for using hand-held and impelled weapons; and in the development of a capacity to sustain rapid movement across open country. These attributes, supposedly acquired by pre-man or early-man as a means of hunting competitively, constitute the basic features of contemporary man, and it thus would seem likely that evolution of the higher functions of the human brain has its origins in the selective pressures of competitive hunting, as maybe also does the loss of hair from much of the body, and the development of skin glands as copious secreters of an aqueous fluid during physical activity.

The loss of hair and development of sweat glands ensured that humans had become anatomically and physiologically adapted to a tropical environment, but were quite unsuited to endure the seasonal coldness and famine of the higher latitudes. The impediment of a naked skin, however, clearly has not precluded human colonization of subarctic terrains, and hunting excursions into the true arctic. The reason for this being, of course, the ability of humans to use the foresight, inventiveness and manual dexterity, acquired as a competitive hunter in the tropics, to engineer comfortable, or at least tolerable, mini- and micro-climates for themselves, by means of suitable housing and clothing.

Thus humans can survive in high latitudes, not by any structural or functional capacity of their bodies to withstand cold stress, but by virtue of their engineering prowess in minimizing the stress upon their bodies of the seasonal variations in ambient temperature and food supply: the one by housing and clothing, and the other by planned production, harvesting, storage and distribution of foodstuffs.

Armed with the capability to create and control the environment by engineering prowess, which is an elaborate form of behaviour, there is now nowhere on the Earth's terrestrial surface that humans cannot venture into, and even colonize. It it, indeed, this same genetic consequence of their successful competition as an open-plain hunter which now enables humans to venture into environments such as outer space and oceanic depths, to which they are even less structurally and physiologically suited. So long as the engineering is adequate, and suitable mini- and micro-climates are sustained, there is no environmental stress acting upon the body, and therefore, there is no physiological strain.

## 2. The background to hypothermia

Notwithstanding the implied assertion that humans have the behavioural capability to engineer a tolerable immediate environment for themselves anywhere on, above or below the Earth's surface, the fact remains that all such exploits are attended by risks of failure and fatality. Obviously, in all cases of disaster there must have been a failure in the adequacy of the engineering, which means, in biological terms, a failure of the organism (man) to succeed behaviourally, in isolating itself adequately from the potential stress of a hostile environment. This, however, is an over-simplification. Practical considerations, both physical and physiological, often compel a compromise in engineering design with the attendant risk that if a calculated degree of environmental stress is exceeded, physiological strain will then be unavoidable. Compromises which arctic and antarctic residents and travellers may be compelled to make, the calculation of what is a reasonable risk, and the circumstances in which reasonable risk may become a life-threatening reality will be discussed. There is also the important, but often overlooked fact that all humans are not physiologically equal, which means that while a calculated degree of physiological strain may be tolerable for the majority, it may constitute an intolerable strain for others. Consideration must therefore also be given to the pathophysiological conditions pre-disposing some individuals to suffer hypothermia in circumstances that would be regarded as free of hazard for the majority.

### Thermoregulation in the cold: the limiting factor

The basic relationships between the thermoregulatory functions of an endothermic homeothermic mammal and the thermal component of its environment are expressed in Figure 1.1, modified from Mount (1979). Within a relatively narrow range of ambient temperature ($T_a$), known as the thermoneutral zone (TNZ), the principal autonomic thermoregulatory effector functions of heat production (HP) by shivering and heat loss (HL) by evaporation of sweat are inoperative. Within this zone the balance between the rates of HP and HL, ensuring that a particular level of body temperature ($T_b$) is maintained, is achieved to some extent by changes in the skin temperature. This is effected by changes in peripheral vasomotor tone, but mostly by a whole variety of behavioural means ranging from postural adjustments which vary the ratio of exposed surface area to mass, to the choice of clothing and the

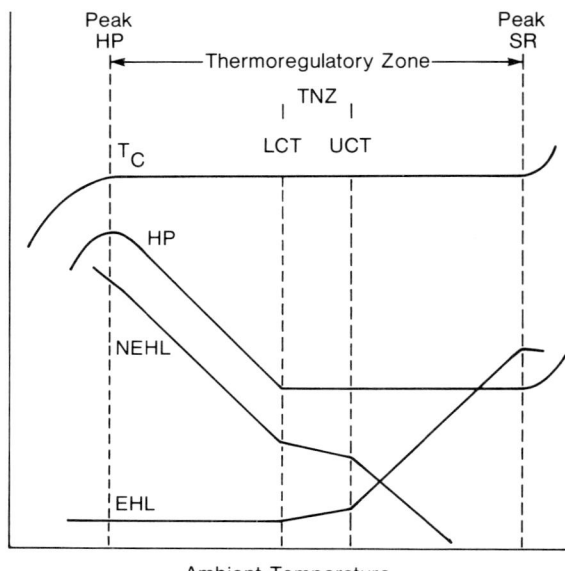

Ambient Temperature

*Figure 1.1. A non-quantitative expression of the relationships between ambient temperature ($T_a$) and body core temperature ($T_c$), heat production (HP), nonevaporative heat loss (NEHL) and evaporative heat loss (EHL). Within the thermoneutral zone of $T_a$ (TNZ) neither thermoregulatory HP nor EHL are active. Only when $T_a$ is below the lower critical temperature (LCT) is there thermoregulatory HP, the intensity of which is proportional to the extent by which $T_a$ is below the LCT. At some low $T_a$, HP becomes maximal (peak HP), and when $T_a$ falls below this point, HL exceeds HP, and $T_c$ declines (i.e., hypothermia then occurs). Only when $T_a$ is above the upper critical temperature (UCT) does EHL by sweating occur, at a rate proportional to the extent by which $T_a$ exceeds the UCT. At some high $T_a$, the rate of sweating becomes maximal (peak EHL), and at higher $T_a$s, HP exceeds HL and $T_c$ rises (i.e., hyperthermia then occurs). The zone of thermoregulation thus extends from the $T_a$ at which peak HP occurs to that at which peak EHL occurs.*

construction of thermally less stressful indoor environments in which to live, especially when inactive. By such means, activated largely by conscious perception of thermal comfort or discomfort, humans can keep their body temperature inside the zone of thermoneutrality within which neither active shivering nor sweating occur. No absolute values may be given to the ambient temperatures defining the zone boundaries, since they are continuously shifting as one seeks, by behavioural means, to stay within the TNZ whatever the ambient conditions may be.

When, for one reason or another, it is not possible to equate the rate of HP within the body with that of HL by one or more of many behavioural means, then autonomic processes of thermoregulation are activated in the maintenance of a stable $T_c$. The ambient temperature at which heat production by shivering occurs is called the lower critical temperature (LCT), and marks the lower limit of the TNZ in a particular set of circumstances. The intensity of the HP by shivering increases progressively as $T_a$ falls below the LCT, until a $T_a$ is reached at which the rate of HP becomes maximal, the $T_a$ of peak HP. If $T_a$ falls below that at which peak HP occurs, HL then exceeds HP and $T_c$ begins to decline. When $T_a$ rises to the upper limit of the TNZ, known as the upper critical temperature (UCT), at which point heat production and heat loss can no longer be maintained by behavioural means, evaporative heat loss by sweating commences. The rate of sweating increases progressively as $T_a$ rises above the UCT until it reaches its peak rate. If $T_a$ rises still further HP can no longer be matched by HL, and body temperature starts to rise. When evaporation of secreted sweat is prevented, either by clothing or by the high relative humidity of the air, the threshold $T_a$ at which $T_c$ starts to rise is lowered. Thus the full range of ambient temperature variations within which homeothermy may be sustained is set by the $T_a$s at which HP and HL, by autonomic means of shivering and sweating respectively, reach their peak rates.

Since our concern is with the occurrence of hypothermia, we shall concentrate our attention on the circumstances by which individuals may fail to engineer sufficient lowering of their LCT by behavioural means below the prevailing ambient temperature, or at least to a point where peak HP by shivering does not occur. An important point to note here is that shivering at a peak rate is time limited, and cannot be sustained. This may be due to fuel exhaustion, metabolite accumulation or some other component of 'muscle fatigue'.

As is illustrated diagrammatically in Figure 1.2, there is virtually no limit to the depression of the LCT which may be achieved by thermal insulation. This means that there is no problem in keeping body core temperature within the TNZ by behavioural means at any $T_a$ that may be encountered on the surface of the Earth, provided always that the materials needed for its engineering are available. As is evident from Figure 1.2, as extra-cutaneous thermal insulation is progressively increased, not only is the LCT progressively lowered, but the slope of the increase in thermoregulatory HP is also reduced as ambient temperature falls progressively further below the LCT. This means that the $T_a$ at which the rate of HP will peak is depressed to an even greater extent than is the LCT as thermal insulation is increased.

The LCT and UCT are also shifted by variations in the rate of non-thermoregulatory heat production, that is to say, in the rate of resting heat production or metabolism and in heat production associated with the performance of external work. This point is illustrated by Figure 1.3. With the addition of the endogenous HP of work, both the LCT and the UCT for any given degree of thermal insulation are shifted to the

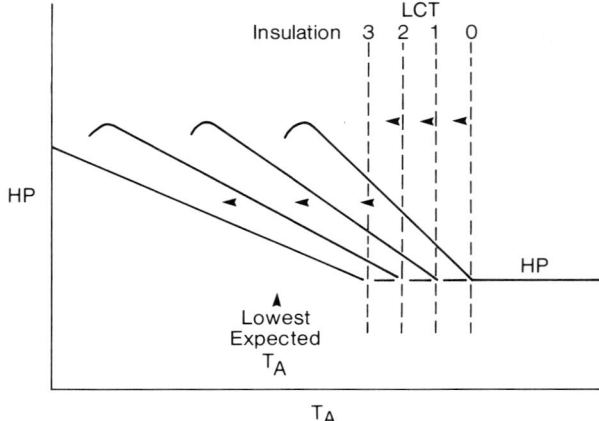

Figure 1.2.   *The effect of increases in the thermal resistivity of the super-cutaneous insulative layer of clothing, indicated by the horizontal arrows on the lower critical temperature (LCT), on the rate of increase of HP below LCT, and on the $T_a$ at which peak HP occurs. The vertical arrow indicates the lowest expected $T_a$ (based on local weather knowledge). By relating this arrow to the increases in HP as $T_a$ falls it may be seen, in this non-quantitative representation, that none of the degrees of thermal resistivity (indicated by the numbers 1, 2 and 3 on vertical dashed lines) would be sufficient to eliminate the need for thermoregulatory HP at the lowest anticipated $T_a$. Thermal insulation represented by the number 2 would, however, prevent the demand upon thermoregulatory HP exceeding 50% of peak HP.*

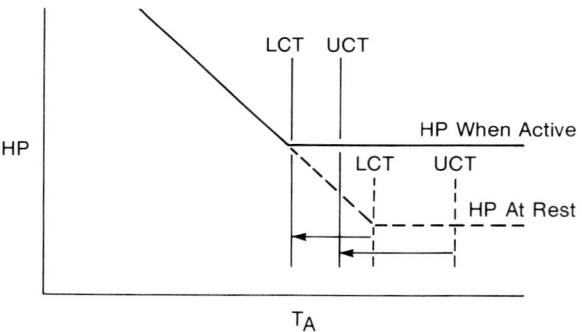

Figure 1.3.   *A diagrammatic representation of the shifts in the upper critical temperature (UCT) and lower critical temperature (LCT) which occur with the transition from a state of rest to that of work or exercise, or vice versa, if no compensatory adjustment is made in the thermal insulation of the clothing worn. In this instance, the level of thermal insulation needed to keep a person within the thermoneutral zone (TNZ) at rest would cause them to be above the UCT during work. Sweating would then occur beneath the clothing. Likewise, the level of thermal insulation necessary to keep a subject within the TNZ during activity will be insufficient to keep the subject within the TNZ at rest, and thermoregulatory HP by shivering will then become necessary.*

left. If there is no compensatory reduction in the thermal insulation, the UCT may be exceeded, and sweating will occur beneath the clothing. This will probably be unable to evaporate and fluid accumulation in the clothing will cause a persistent decrease in its thermal resistivity. This may be particularly disadvantageous, depending on the ambient thermal conditions when the period of work is over. If a working or exercising subject is carrying only an amount of thermal insulation needed to stay within the TNZ while active, or the thermal insulation of the fabric is decreased by wetness, then upon a return to resting at the same $T_a$, the TNZ will be shifted to the right. The $T_a$ will then be below the subject's TNZ, and shivering will occur in order to sustain $T_c$.

From these considerations it is evident that there can be only two basic reasons for the occurrence of hypothermia: failure to provide that degree of thermal insulation required at the prevailing $T_a$ to keep the rate of heat flow from the body less than the body's maximum capacity for heat production, and/or a failure to sustain the body's fuel supply, with a consequent reduction in the body's maximum capacity for heat production. Neither circumstances need occur with adequate provisioning. Why then does man, a species of mammal that has long relied upon imagination, foresight, planning and teamwork for his survival, sometimes fail to provide himself with the thermal insulation and energy supply necessary for survival in the cold? There is, of course, no simple answer, because the circumstances of each incident of hypothermia are always particular to the situation. Misfortune, miscalculation, mis-placed confidence, logistic failure and individual pathophysiological susceptibility are the broad categories of citable cause.

Where, for example, hypothermia has occurred because of the mechanical failure of a vehicle and/or its heating system, the verdict could be misfortune. On the other hand a failure to carry emergency thermal insulation and food may be considered as a miscalculation or logistic failure. In biological terms, a lack of imagination, foresight and planning rendered the individual unfit to survive.

Military personnel required to operate in arctic terrains may not be able to operate efficiently while carrying all the insulative clothing required when at rest, and may therefore rely upon logistic support to provide additional protective clothing during periods of rest. The unpredictability of military operations, however, means that from time to time logistic support will fail. Here no blame can be attached to an individual who becomes hypothermic.

Competitive long-distance skiers may also be unwilling to carry the additional thermal insulation required to keep them within thermo-neutrality while active. Reliance for safety is often placed upon the individual competitor's perceived certainty of reaching the next shelter before resting. Should some accident or other circumstances cause a period of rest away from shelter, then the skier may be quite unable to equate HL with HP, and hypothermia is inevitable. Is such an occurrence a mis-adventure, a misfortune or a miscalculation, or was it a calculated risk in which the possibility of disaster was known and accepted?

An important fact in occurrences such as those cited is that it is often not practical to carry the degree of thermal insulation which will protect the individual in the worst possible developing situation. There may be no initial misunderstanding about what could go wrong, or what the consequences of that would be. This raises the issue of what is a reasonable risk, and what is a thoroughly irresponsible one. There can be no hard and fast rule on what is reasonable, if only because means of communication and the degree of remoteness from help will be additional factors to be taken into consideration. A rough-and-ready guideline can, however, be suggested. The perpendicular arrow in Figure 1.2 is a notational indication of the lowest $T_a$ likely to be encountered in a particular area at a particular time of year. While it may be deemed impractical to carry that degree of protective clothing which would keep a person within his TNZ should the worst likely situation be encountered, it would clearly be foolhardy to carry only that degree of protection such that survival would depend on a sustained maximum rate of heat production. A reasonable compromise would seem to be the provision of that degree of thermal insulation which would require shivering thermogenesis at a rate of not more than 50% of peak HP rate to equate heat loss with heat production, since this rate of HP could probably be sustained indefinitely so long as energy supplies held out.

In terms of the non-quantitative expressions in Figure 1.2 it is apparent that the degree of thermal insulation which would set the LCT at level '2' would meet this specification. This concept of a reasonable calculated risk is convertible into approximate insulation values; but assumptions would still have to be made about the physiological normality of the individual and the reliability of the available meteorological data.

As has been indicated, however, this calculation is that of a tolerable, but not ideal, relationship between humans and an environment which brings them closer to the condition at which homeothermy will fail than they would behaviourally choose to be. Thus it is necessary to consider what unpredictable occurrences would confound the calculation and put people at risk. The sort of situations that might arise are indicated in Figure 1.4 and include such misfortunes as a reduction in thermal insulation of the clothing as might result from mechanical damage or water damage, exhaustion, loss or logistic failure of food

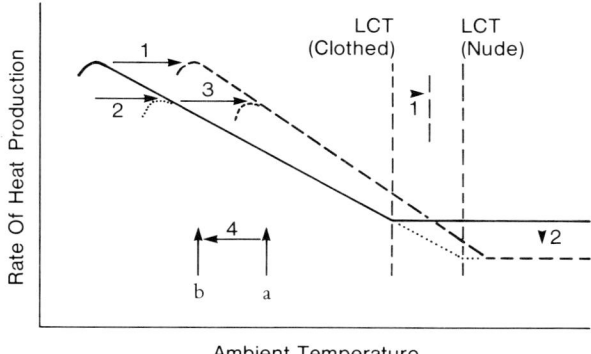

*Figure 1.4. Here it is assumed that two arbitrary units of insulation are adequate to cause a rise in HP to no more than 50% of the peak MR at the lowest anticipated $T_a$, indicated by the vertical arrow labelled a. The numbers 1 to 4, together with the solid arrows, indicate circumstances in which the estimated position of safety (HP of not more than 50% of peak MR) will be disturbed, and the risk of hypothermia increased. These are:*

*1. Reduction in the effective insulation by 50%, such as might occur by moisture penetration, mechanical damage, loss of gear while being carried (not worn), or by a logistic failure to provide the additional means of insulation required when at rest.*

*2. Reduction in the resting rate of HP, and/or in the summit rate of HP and/or in the slope of the increment in HP as $T_a$ declines below the LCT, as may be occasioned by inadequate rations, undiagnosed metabolic abnormalities, or drug effects.*

*3. The concurrence of circumstances 1 and 2 above.*

*4. An unexpectedly low $T_a$ (indicated by the vertical arrow labelled b).*

supplies leading to a reduction in either resting or maximal HP, and/or the occurrence of far more severe weather conditions than those forecast. Such unscheduled occurrences as these will raise the effective LCT and thus greatly increase the risk of exhaustion of the process of shivering thermogenesis, with the consequent onset of hypothermia. Since each eventuality may arise concurrently with others, additive consequences are also depicted in Figure 1.4.

Where compromise between a 'safe' level of thermal insulation and an 'acceptable risk' level is made, it is imperative that the personnel are cognizant of this situation, and the compromise. Events which would add to the risk should be identified and prepared for in the planning of any expedition in compromised circumstances.

## 3. Conditions which vary the capacity of individual persons to withstand cold exposure

In any proposed calculation of risks entailed in a compromise between over-caution and under-caution during exposure to low ambient temperatures it is assumed that the subjects are mentally fit and physically healthy. It may be assumed that officially-sponsored arctic and antarctic expeditioners have all undergone prior medical examination, and are capable of a high metabolic rate within the 'normal' range. One can be less sure of industrial workers on arctic projects, since the enthusiasm to get the job, or rather to earn the money, can lead to deliberate concealment of a pathological condition. Where the population in question is that of settled arctic residents, and covers a full age spectrum including the neonate and the elderly, there are many conditions of the body which will place individuals at serious risk of hypothermia in ambient conditions considered to be risk-free for the active and healthy members of the community.

The recognition of the pathophysiological conditions which may render particular persons more suceptible to hypothermia is, then, of paramount importance. Unfortunately—apart from the particular and extreme cases of the neonates and the elderly who, it is well recognized, are at risk from acute hypothermia in winter conditions, even in temperate regions—there seems to have been little discussion of the various human conditions which increase the risk of hypothermia, and which are not specifically applicable to the neonates and the elderly. It should be noted, however, that the known conditions that predispose the elderly to hypothermia, listed by Cooper and Ferguson (1983), are not exclusive to them. These are:

1. Changes in perception of environment temperature.
2. Changes in the ability to generate heat in the cold.
3. Failure of behavioural responses to cold.
4. Effects of prescription drugs, and drugs used to alter mood.
5. Simple cold exposure due to an accident.
6. Severe autonomic dysfunction.

Lipton (1984) has contributed a useful discussion of the pathological states which disturb the capacity to thermoregulate at high and low ambient temperatures.

Many of the conditions detrimental to the maintenance of homeothermy during cold exposure have a direct or indirect effect on the maximum capacity to produce heat, and these are expressible in terms of the arrows in Figure 1.4. There are, however, more subtle factors such as the interference of central nervous synaptic activities by drugs, including alcohol, and a possible effect of cold exposure on the working of the heart, which need to be considered and which cannot be adequately expressed by that format. It may, therefore, be of greater heuristic and mnemonic value to express the various pathological and drug-induced interferences with thermoregulatory performance in terms of their points of action on a simple representation of the system by which body temperature is regulated. This is given in Figure 1.5, and is derived from Cremer and Bligh (1969). The physiological system, expressed in this way, is clearly analogous to an engineered system of thermoregulation of, say, a climatic chamber, with comparable components. This does not mean, however, that the actual principle of regulation is the same in physiological and engineered processes of thermoregulation, thus vague descriptive terms such as 'shifts in the set-point', with overtones of engineering analogues such as adjustable reference signals, are studiously avoided.

'Warm' and 'cold' sensors, both centrally- and peripherally-located, transduce and transmit information about deep-tissue and cutaneous temperatures to the central nervous system (CNS), which somehow translates this received information into appropriate instructions to the thermoregulatory effectors of increased heat production and increased heat loss. These actions are mostly effected via efferent nervous pathways, with some hormonally-transmitted modulatory influences, particularly on the heat production effectors. Venous blood draining from tissues and organs actively involved in thermoregulatory heat loss is, obviously, at a lower temperature than the arterial blood, by an extent determined by the rate of heat loss. Likewise, the venous blood

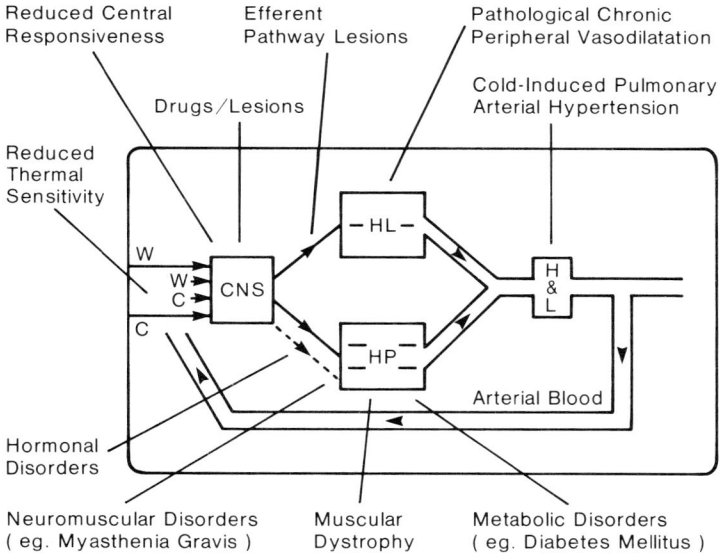

*Figure 1.5. Representation of the components and processes of mammalian thermoregulation (redrawn from Cremer and Bligh, 1969) upon which are superimposed indications of the pathological and pharmacological conditions which may interfere with thermoregulation, and thereby increase susceptibility of individuals to hypothermia in ambient conditions which a healthy person could tolerate easily.*

draining from tissues and organs involved in thermoregulatory heat production is at a higher temperature than that of the arterial blood. Thus venous return to the heart can be thermally streamed, but has become well mixed thermally by the time it has traversed the pulmonary circulation and the heart, within which negligible heat exchange occurs. Thus the temperature of the arterial blood leaving the left cardiac ventricle is the mean temperature of the venous blood entering the right heart, and is rapidly affected by any changes in the rates of heat production and loss in the body. Arterial blood is distributed throughout the body, and therefore provides feedback to the thermosensors, wherever in the body they may be located, of the thermal consequences of the thermoregulatory effector functions.

Superimposed on Figure 1.5 are indications of some of the pathological or pharmacological conditions and their points of action that either definitely, or possibly, impair thermoregulation. These are discussed briefly below.

## Peripheral vasomotor tone

While the majority of the abnormalities which are conducive to the occurrence of hypothermia result in impairment to heat production, autonomic control of the rate of heat loss through peripheral vasomotor tone plays a role in the control of body temperature within the zone of thermoneutrality. Persistent peripheral vasodilatation in the cold, by keeping skin temperature abnormally high, would be expected to lower the threshold $T_a$ of shivering. Collins *et al.* (1977) and MacMillan *et al.* (1967) have both noted impaired peripheral vasomotor tone in some elderly people, and Macmillan *et al.* (1967) correlated this condition with the occurrence of hypothermia. Since pathological loss of peripheral vasomotor tone control can occur at any age, this condition should not be overlooked.

## Sensors

Reduced sensitivity to cold, evidenced by impaired conscious awareness and behavioural responsiveness to cold exposure in the elderly (Horvath *et al.*, 1955; Collins *et al.*, 1977) may also lower the threshold ambient temperature for shivering, and thus predispose a person to hypothermia; and the condition of impaired thermal sensitivity may not be peculiar to the elderly. Since an elderly person may drift into a hypothermia state, apparently, without the warning a bout of intensive shivering would provide, it seems possible that core temperature thermosensitivity, as well as peripheral thermosensitivity, is impaired, although this has not been positively shown to be so.

## Afferent pathways from thermosensors

Little can be said of the consequences of lesions. Extensive destruction of the skin, as can occur in leprosy, might impair the immediate behavioural response to changes in ambient temperature, but since deep body thermosensitivity is apparently unimpaired, one would not expect wide-spread destruction of skin structures to result in a profound disturbance to thermoregulation, and none seem to have been reported. High-level spinal transections which sever the pathways to the brain from skin areas below the section also usually sever the efferent pathways to the musculature below the section. Thus any observed thermolability in paraplegic patients may be as readily attributed to efferent pathway severence as afferent pathway severence.

## Central nervous thermosensor–thermoregulatory effector interface

Because of the redundancy and plasticity in the CNS, the gradual destruction of the hypothalamic thermoregulatory interface, for example, by a tumour, may not result in severe thermolability (Lipton, 1984). There are, however, pathological conditions of the CNS, a symptom of which may be hyper- or hypothermia, or a loss of thermoregulation at high and low ambient temperatures (Lipton *et al.*, 1977) which, though rare occurrences, should not be overlooked.

A much more common cause of thermal disturbances of central origin relate to the central actions of drugs. While the full magnitude and complexity of the pathways, and the interactions between pathways at the central nervous interface between thermosensors and thermoregulatory effectors are still far from understood, there can be no doubt that there are multi-synaptic pathways from the sensors to the effectors, with interactions between them, and excitatory and inhibitory influences converging onto them. A basic neuronal pattern for which there is some evidence (Bligh, 1977) is given in Figure 1.6. The obvious existence of many synapses is what is important, however, not the details of the neuronal pattern. It is now known that some 50 transmitter substances play roles in interneuronal communication, and that many of these will induce thermoregulatory effects when applied pharmacologically to the regions of the CNS involved in thermoregulation. It is also known that the majority of centrally active drugs exert their actions at the synapses by mimicking or modulating transmitter actions. Thus, it is by no means surprising that many of the drugs prescribed for the alleviation of mental disturbances, or self-administered to produce changes in mood, cause interferences with the processes of thermoregulation.

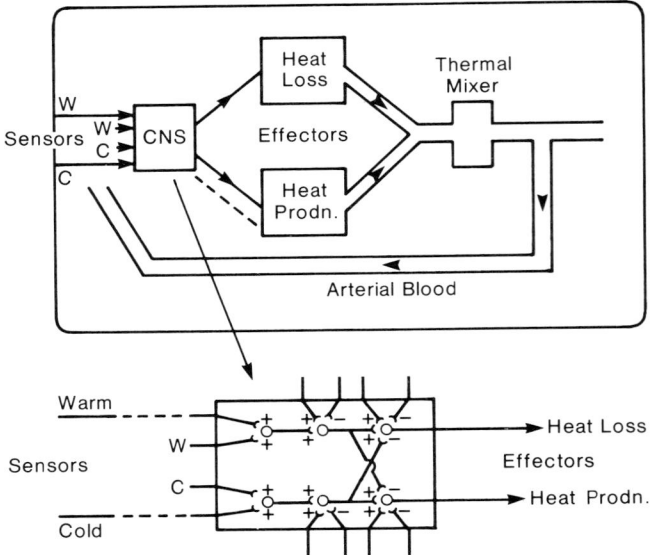

*Figure 1.6. Showing the possible neuronal basis of neurology of the central nervous interface between the input from thermosensors and the efferent output to thermoregulatory effectors.*

With the exception of alcohol, which obviously interferes with the functioning of the brain, and is frequently related to an occurrence of hypothermia, relatively little attention has been paid to the practicalities of hypothermia as a side-effect of drug therapy. The prescription dosages of the tranquillizer and antidepressant drugs, for instance which are known to interfere centrally with thermoregulation (Cooper and Ferguson, 1983; Lipton, 1984), are mostly based on animal studies and human trials conducted at 'room temperature' (about 20°C), and are not generally varied according to the working environment of the patient. A substance which interferes with the thermoregulatory responsiveness to cold exposure may well have no effect on thermoregulation in the thermoneutral zone of thermal comfort. Thus such trials have given no warning of the serious consequence the 'safe' and accepted dose may have upon someone whose job or lifestyle exposes him or her to very low ambient temperatures. This is a relatively uninvestigated matter which is currently of little clinical concern, even in relation to arctic residents of all ages.

## The efferent neural pathways

Mention has already been made of spinal cord injuries involving severence of the descending pathways from the CNS to the musculature. The susceptibility of paraplegic patients to hypothermia is, of course, well recognized. Thus, except through some misfortune or neglect, it is unlikely that such persons would be allowed to encounter extreme cold stress. Nevertheless, such a condition will accentuate other factors such as those related to old age, or to drug administration, particularly in inadequately heated houses, and there is need for attention to the domestic circumstances of paraplegics.

## The efferent humoral pathways

The capacity to generate heat in response to cold exposure, as well as when at rest at thermoneutrality, is to some extent influenced by hormonal actions upon the thermoregulatory effectors of heat production. An increase in the release of thyroid hormone during cold exposure has been demonstrated, as has the effect of the thyroid hormone upon basal metabolism and the increased susceptibility to hypothermia in the thyroid deficient mammal. Thus, the possibility that hypothyroidism, and perhaps other hormonal imbalances, could be factors in the occurrence of hypothermia should be considered.

## The neuromuscular junction

There has been little, if any, discussion of a role of neuromuscular disorders in the susceptibility to hypothermia, but obviously any impairment in the transmission of signals from motor nerves to muscles, as occurs in myasthenia gravis, could have such an effect. And as yet undiagnosed incipient conditions could be particularly hazardous, since there would then be no awareness of the need for care about cold exposure.

## Muscular disorders

Clearly, any condition involving muscular dystrophy will, by lowering the capacity to shiver, increase the susceptibility to hypothermia upon cold exposure. Likewise, a failure to sustain the oxygen and/or energy supply to the muscle, as may occur with impaired circulation of the blood, or with diabetes mellitus, respectively, will lower the metabolic response to cold exposure by shivering and thus predispose an individual to hypothermia.

## Pulmonary arterial hypertension

In sheep (Chauca and Bligh, 1976) and cattle (Will *et al.*, 1978), exposure to cold has been found to induce an increase in pulmonary arterial blood pressure ($P_{pa}$) which, at high altitude, is additive to the effect of hypoxia upon $P_{pa}$, and thus can precipitate the occurrence of acute high altitude sickness. Exposure of the facial area to cold air evidently induces a rise in $P_{pa}$ in man also (Burns *et al.*, 1984). This raises the possibility that by increasing the work that must be done by the right heart, in pumping blood through the pulmonary capillary bed, cold exposure could over-stress an ailing heart, with fatal consequences (Figure 1.7). There is, indeed, evidence of a higher incidence of fatal cardiac malfunctions in northern Scandinavia in winter than in summer (Leppälvoto, 1984; Näyhä, 1984) which could be related to an effect of cold exposure upon $P_{pa}$. The supposed beneficial effect of covering the nasobuccal areas of the face during cold exposure, particularly recommended for persons with histories of heart trouble in subarctic environments, possibly relates to an easing of the resistance

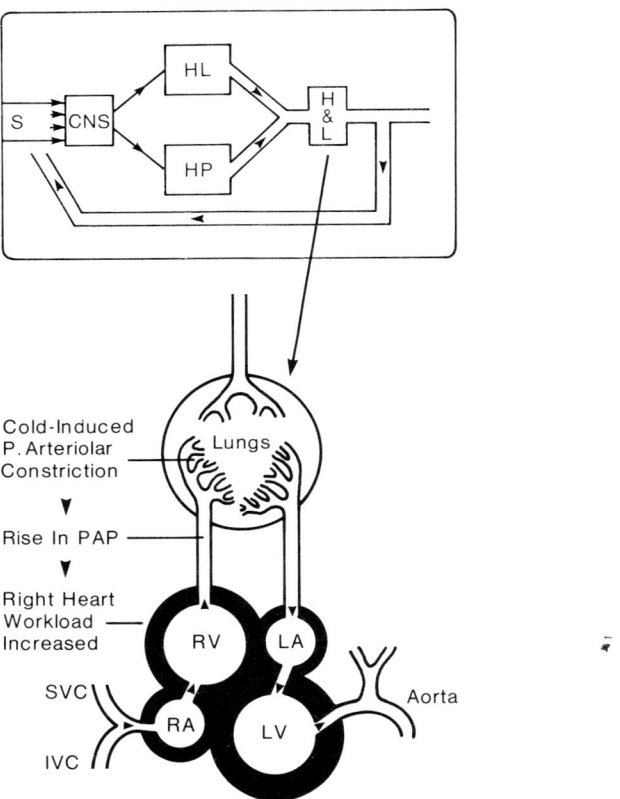

*Figure 1.7.    An illustration of the consequences of cold-induced pulmonary arterial hypertension upon the work load of the right heart. PAP = pulmonary arterial pressure; RA = right cardiac artrium; RV = right cardiac ventricle; LA = left cardiac atrium; LV = left cardiac ventricle.*

against which the right heart has to work in the cold. A more important aspect of this finding may not be the condition itself, and its possible consequence, but the reminder it gives that there may be other unrecognized or scarcely recognized factors and conditions which place some individuals at risk of hypothermia. This may be a secondary rather than a primary consequence of cold exposure.

## 4. *Discussion*

Concern about the occurrence of hypothermia is oddly polarized. There is considerably more interest in its treatment than in its cause and more concern about accidental hypothermia (which is primarily due to a failure to provide adequate thermal insulation) than about incidental hypothermia (which is primarily due to a functional disorder of the body) in arctic populations. The reverse is the situation in more temperate regions which only have unpredictable spells of cold weather. Whether this contrast of interest reflects true geographical differences in the occurrence of accidental and incidental hypothermia is not clear. Lack of clinical concern about incidental hypothermia, including that of neonates and the elderly, in the Alaskan sub-arctic interior was attributable in 1977 (personal observations, 1977–1985) to its alleged rarity. At that time, however, there was an almost total inability to detect such cases: home visits were virtually unknown; low-reading clinical thermometers were non-existent; when a normal clinical thermometer failed to register a low body temperature it elicited little curiosity; and the failure of many such patients to keep a subsequent appointment was regarded as a customer prerogative. At the same time, it must be borne in mind that since incidental hypothermia is always incidental to some other bodily condition, death can be attributable to the primary conditions rather than to the secondary condition of hypothermia, even though, had hypothermia been promptly diagnosed, the primary conditions need not have caused death at that time.

Attention has been drawn to a basic distinction between accidental and incidental hypothermia. It has been emphasized that healthy human beings can, and generally do, defend themselves adequately from the effects of cold exposure by engineering adequate thermal insulation for themselves in the form of protective clothing and housing; the basic cause of accidental hypothermia is a failure to anticipate and provide for such a contingency adequately. Some persons will, however, always be willing to take risks in hostile environments, whether for adventure or gain. The practicality of a particular situation may also limit the additional insulative equipment that may be readily carried, for use in rest periods during work at low ambient temperatures. Thus consideration here is given to where the distinction lies between a foolhardy and irresponsible risk of becoming hypothermic, and a reasonable and calculated risk. It is suggested that a compromise lies in the provision of an amount of thermal insulation which, at the lowest ambient temperature likely to be encountered, will require heat production of not more than 50% of the maximum. Even so, misfortune resulting in hypothermia will occur periodically. The nature of such misfortunes has been indicated, so that sojourners or travellers in cold places will recognize the immediate and long-term perils of the situation in which they are placed, and be able to take whatever protective action is possible, immediately and proficiently.

Calculation of the attendant risk of hypothermia, when available insulation is less than necessary to keep the subject within the thermoneutral zone, is based upon an assumption about the importance of the role played by physical fitness. This may be a reasonable, but not a wholly safe, assumption about arctic and antarctic adventurers, all of whom are usually physically examined and found to possess good cardiorespiratory function before an adventure. The physical fitness of industrial operatives in arctic conditions generally receives little attention; and that of the full age spectrum of the inhabitants of an arctic settlement even less. The many recognized and possible pathological conditions which can impair resistance to hypothermia are discussed in the context of a diagrammatical representation of the processes of thermoregulation, and the point is made that there may be yet other unrecognized conditions that predispose individuals to hypothermia in circumstances that are not considered dangerous to the majority of a population.

It is evident that of all these conditions, those central disturbances to thermoregulation resulting from prescribed and unprescribed drugs, including alcohol, are the ones in most urgent need of recognition and attention.

## References

Bligh, J., 1977, The central neurology of mammalian thermoregulation. *Neuroscience*, **4**, 1213–1236.

Burns, B., Belblum, J., McCauley, M., Chodoff, P. and Stene, J., 1984, Elevation in pulmonary arterial pressure and resistance following brief exposure of the naso-maxillary region to cold air (5°C) in humans. In *Hypoxia, Exercise and Altitude*, Eds J.R. Sutton, C.S. Houston and N.L. Jones (New York: A.R. Liss), p. 453.

Chauca, D. and Bligh, J., 1976, An additive effect of cold exposure and hypoxia on pulmonary artery pressure in sheep. *Res. Vet. Sci.*, **21**, 123–124.

Collins, K.J., Dore, C., Exton-Smith, A.N., Fox, R.H., McDonald, I.C. and Woodward, P.M., 1977, Accidental hypothermia and impaired temperature homeostasis in the elderly. *Brit. Med. J.*, **1**, 353–356.

Cooper, K.E. and Ferguson, A.V., 1983, Thermoregulation and hypothermia in the elderly. In *The Nature and Treatment of Hypothermia*, Eds R.S. Pozos and L.E. Wittmers (Minneapolis, MN: University of Minnesota Press), pp. 35–45.

Cremer, J.E. and Bligh, J., 1969, Body temperature and responses to drugs. *Brit. Med. Bull.*, **25**, 299–306.

Horvath, S.M., Radcliffe, C.E., Hatt, P.K. and Spurr, G.B., 1955, Metabolic responses of old people to a cold environment. *J. Appl. Physiol.*, **8**, 145–148.

Leppäluoto, J., 1984, Cold as a disabling factor in northern countries. *Arct. Med. Res.*, **37**, 10–12.

Lipton, J.M., 1984, Thermoregulation in pathological states. In *Heat Transfer in Medicine and Biology*, Vol. 1, Eds A. Shitzer and R.C. Eberhart (New York: Plenum), pp. 79–105.

Lipton, J.M., Kirkpatrick, J. and Rosenberg, R.N., 1977, Hypothermia and persisting capacity to develop fever: occurrence in a patient with sarcoidosis of the central nervous system. *Arch. Neurol.*, **34**, 498–504.

MacMillan, A.L., Corbett, J.L., Johnson, R.H., Smith, A.C., Spalding, J.M.K. and Wollner, L., 1967, Temperature regulation in survivors of accidental hypothermia. *Lancet*, **2**, 165–169.

McMurtry, I.F., Reeves, J.T., Will, D.H. and Grover, R.F., 1975, Hemodynamic and ventilatory effects of skin-cooling in cattle. *Experientia*, **31**, 1303–1304.

Mount, L.E., 1977, *Adaptation to Thermal Environment: Man and his Productive Animals* (London: Edward Arnold).

Näyhä, S., 1984, The cold season and deaths in Finland. *Arct. Med. Res.* **37**, 20–24.

Will, D.H., McMurtry, I.F., Reeves, J.T. and Grover, R.F., 1978, Cold-induced pulmonary hypertension in cattle. *J. Appl. Physiol.*, **45**, R469–R473.

# 2. Physiological Limitations of Human Performance in Hot Environments, with Particular Reference to Work in Heat-Exposed Industry

**K. Rodahl and T. Guthe**

## 1. Introduction

Low levels of heat stress may cause discomfort and fatigue; higher levels may impair performance; and exceeding the heat tolerance level may be a health hazard.

The purpose of this chapter is to review present knowledge on the effects of heat stress on performance, the assessment of occupational heat stress, the physiological monitoring of heat strain, available data on heat stress in hot industries, limits of temperature tolerance and health hazards of excessive heat exposure, and finally to discuss practical application of existing knowledge. In this discussion, we shall not consider sedentary work at ordinary indoor temperatures and the concept of comfort under such conditions.

## 2. Effects of heat on performance

The temperature of the environment is one of a variety of factors affecting human performance (Figure 2.1). Neuromuscular function and the energy-yielding processes which are the basis for human performance are, within limits, temperature dependent, reaching their optimum at temperatures above resting levels (Åstrand and Rodahl, 1986). At temperatures substantially higher than these optimal levels, both physical and mental performance may deteriorate due to a complicated interplay of physiological and patho-physiological processes. In

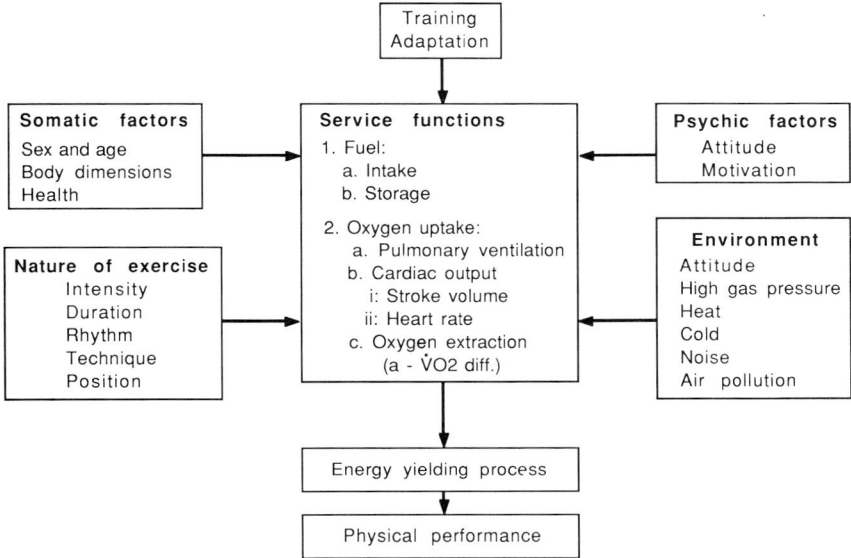

*Figure 2.1. Factors affecting performance (from Åstrand and Rodahl, 1986).*

excessive heat the available blood volume generally speaking has to serve a dual purpose: in addition to transporting oxygen it has to transport heat from the interior of the body to the skin where it can be dissipated in order to prevent overheating. This may limit the oxygen transporting capacity of the blood circulation, since phylogenetically temperature regulation appears to have priority over oxygen transport (Hardy, 1967). In addition, prolonged heat stress may lead to hypohydration, which in itself impairs performance, especially endurance. In addition, prolonged heat strain may impair mental and psychomotor functions, affecting performance.

## Physical work capacity

Heat exposure, in itself, represents an extra load on the blood circulation. Exhaustion occurs much sooner during heavy physical exercise in the heat because the blood, in addition to carrying oxygen, now also has to serve as a cooling fluid. Consequently, the heart rate has to increase, and the circulation time has to be shorter. This represents an extra burden on the heart. The stress of heat and the hydrostatic factors in prolonged standing work may be added to the stress of work itself.

Williams *et al.* (1967) observed no difference in maximal $O_2$ uptake in subjects working in the heat and at comfort temperatures. At submaximal work rates, however, they found that the major change in haemodynamics in the heat was an increase in heart rate and a fall in stroke volume. Neither cardiac output nor arteriovenous difference were significantly altered compared with comfortable conditions (Rowell, 1974). Williams *et al.* (1967) also demonstrated a larger lactate production in the subject who exercised in the heat as compared with one exercising in a neutral environment. This finding can be explained as a result of the reduced muscle blood flow.

Heat stress can also be a significant problem for pilots during low-level flights in hot climates, especially in fighter aircraft that impose high task loads and the need for application of repetitive manoeuvring forces (Nunneley and Flick, 1981). Pilots noted lowered G-tolerance and increased general fatigue on the hotter flights. Weight losses up to 2·3% were observed.

## Mental work capacity

An evaluation of the mental or intellectual performance during exposure to heat or cold is hampered by subjective variations and lack of suitable objective testing methods (Pepler, 1963). As a rule, a deterioration is observed when the room temperature exceeds 30–35°C if the individual is acclimatized to heat. For the unacclimatized, clothed individual, the upper limit for optimal function is about 25°C. The observed deterioration in performance capacity refers to the precise manipulation requiring dexterity and coordination, the ability to observe irregular, faint optical signs, the ability to remain alert during prolonged, monotonous tasks, and the ability to make quick decisions. During a 3-hour drilling operation, the best results were achieved at 29°C; but at a room temperature of 33°C the performance was reduced to 75%, at 35·5°C to 50% and at 37°C to 25%. A high level of motivation may to some extent counteract the detrimental effect of the climate.

Wyon *et al.* (1979) examined the effect of moderate heat stress (up to 29°C) on mental performance in seventeen year old boys and girls. They were subject to rising air temperature conditions, typical of occupied classrooms in the range 20–29°C. Sentence comprehension was significantly reduced by intermediate levels of heat stress in the third hour. A multiplication task was performed significantly more slowly in the heat by male subjects, showing a minimum at 28°C. Recognition memory showed a maximum at 26°C, decreasing significantly at temperatures below and above. The authors consider these results to support the hypothesis of reduced arousal in moderate heat stress in the absence of conscious effort.

Colquhoun (1970) observed that in the normal range of rectal temperatures, performance in a calculation test rose and fell with the rise and fall in body temperature.

## Water deficit

High sweat rates with excessive loss of body fluids may cause a deficit of body water (hypohydration or dehydration). Since the regulation of body temperature has priority over the regulation of body water, hypohydration may be a threat to life if the environment is very hot and water is not available.

Prolonged exposure to heat and/or prolonged exercise almost always causes hypohydration. In both situations a decrease in plasma volume has been noticed. Costill and Fink (1974) report a 16–18% reduction in plasma volume at hypohydration equivalent to a 4% decrease in body weight. There is a shrinkage of the red cells during hypohydration (therefore, changes in haematocrit are not a reliable measure of changes in plasma volume: see Costill and Fink, 1974; Harrison *et al.*, 1975). During exercise and/or exposure to heat stress there is a movement of protein from the interstitial spaces, particularly in skeletal muscles, to the vascular volume. The increase in plasma protein concentration will raise plasma osmotic pressure, thereby helping to maintain blood volume by reducing water loss and enhancing water gain (Senay, 1972; Harrison *et al.*, 1975). Irrespective of the cause of sweating, hypohydration is associated with a decrease in stroke volume during exercise and a concomitant increase in heart rate during submaximal exercise (Figure 2.2). It is remarkable, however, that during maximal exercise, oxygen uptake, cardiac output and stroke volume are not modified by a sweat loss of up to 5% of body weight. However, endurance, i.e., the time that a standardized maximal exercise load can be tolerated, is definitely reduced after dehydration (Figure 2.3). It has also been found that rectal temperatures are significantly higher in the dehydrated subject (Figure

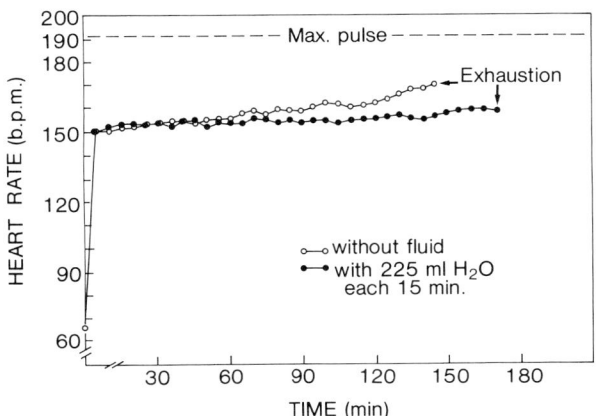

*Figure 2.2.   Mean heart rates in subjects running on the treadmill at 70% maximal $O_2$ uptake until exhaustion, with and without fluid (from Staff and Nilsson, 1971).*

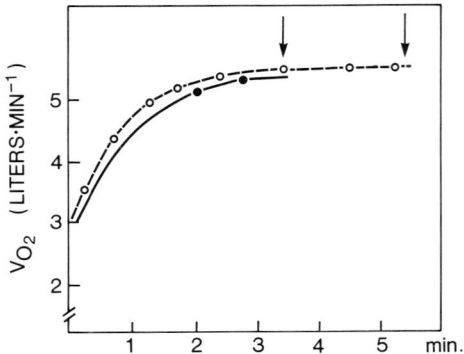

*Figure 2.3.    Oxygen uptake at an exercise load which could be tolerated for 5½ minutes during normal conditions (open symbols) but only for 3½ minutes after hypohydration (filled symbols). Arrows indicate maximal exercise time (from Saltin, 1964).*

2.4), and the rise is related to the weight loss incurred (Gisolfi and Copping, 1974). The exercise rise in core temperature with hypohydration is probably due to inadequate seating (Greenleaf and Castle, 1971).

Unpublished results from a study in a Norwegian cement factory, where the ambient temperature was fairly high even in winter, showed that the heart rate of the workers at a standardized submaximal work rate was significantly lower (more than 10 beats $min^{-1}$) at the end of the work shift when 2 litres of fluid was taken in the course of the shift, compared with the days when no fluid was taken between meals.

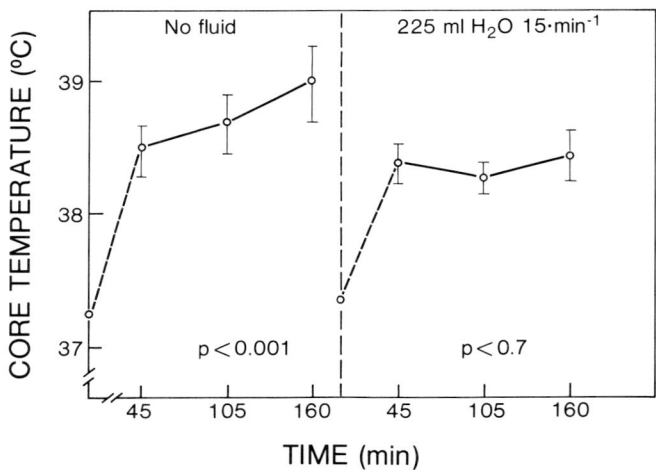

*Figure 2.4.    Mean rectal temperature in subjects running on the treadmill at 70% maximal $O_2$ uptake, with and without fluid. When no fluid was taken, rectal temperature rose significantly from the 45th to the 165th minute (from Staff and Nilsson, 1971).*

The explanation for the gradual decrease in physical performance as hypohydration develops is not available at present. It apparently cannot be primarily a modification of the aerobic energy yield, because the maximal aerobic power was not impaired in Saltin's experiments, despite a pronounced hypohydration (Saltin, 1964). The explanation should be sought at the cellular level, where changes may occur during hypohydration. The maximal isometric strength after a water deficit is reported to be unaffected (Saltin, 1964) or slightly decreased in connection with progressive hypohydration (Bosco *et al.*, 1968). In any case, a reduced water content within the muscle cell and a disturbed electrolyte balance can easily influence the muscle cell's ability to contract and its susceptibility to metabolites. The reduction in work performance is more marked if a water deficit is caused by extended heavy exercise than after exposure to hot environment without exercise being involved.

Dehydration causes a reduced tilt-table tolerance. A person who could normally tolerate prolonged 45°C head-up tilt with a heart rate of 90 beats $min^{-1}$ fainted within 7·5 minutes after a fluid loss corresponding to 3% of his body weight. Following a fluid loss corresponding to 6% of his body weight, he fainted within 1·5 minutes. His heart rate before fainting was 115 and 135, respectively. A low stroke volume was a characteristic finding (Adolph, 1947).

Dehydration also affects crew members of high performance aircraft. Nunneley and Stribley (1979) have shown that dehydration tends to lower G-tolerance and increase the variability of response to heat.

During prolonged and physically heavy training or during participation in certain competitive sports, the sweat rate may be very high. In some cases it may be as high as $2\, l\, h^{-1}$.

Well-trained subjects are less affected in their performance by hypohydration than untrained subjects (Buskirk *et al.*, 1958; Saltin, 1964). Acclimatization to heat does not seem to protect from the deteriorating effect of hypohydration.

The simplest method of determining whether the fluid intake has been adequate is by weighing the individual under standard conditions. Even a reduction in body weight of 1–2% may represent a deterioration in physical performance (Pitts *et al.*, 1944; Adolph, 1947; Ladell, 1955; Saltin, 1964; Gisolfi and Copping, 1974).

It should be emphasized that the degree of body hypohydration is overestimated from measurements of body weight when large amounts of glycogen have been metabolized, which will deliver a 'surplus' of water. The fluid loss during prolonged heat exposure should preferably be replaced by drinking 100–150 ml water several times per hour. The water temperature should be about 15°C. In the case of heat exposure lasting for several weeks, the ingestion of salt tablets is advisable,

5–15 g per day depending on diet, climate and degree of physical activity.

## Age

Although the experimental data are still limited, the available evidence suggests that heat tolerance is reduced in older individuals (Robinson, 1963; Leithead and Lind, 1964; Lind *et al.*, 1970). They start to sweat later than do young individuals. Following heat exposure, it takes longer for their body temperature to return to normal levels. Older people react with a higher peripheral blood flow, but their maximal capacity is probably lower. In one study, it was found that 70% of all individuals who suffered heat stroke were over 60 years of age (Minard and Copman, 1963). On the other hand, Davies (1979) observed no evidence for differences in thermoregulatory function which could be ascribed to sex or age, in his one-hour treadmill exercise experiments on subjects 18 to 65 years of age in a moderate environment. However, Davies (1981) has shown that thermal responses of children are quantitatively different from young adults, evaporative sweat loss being lower in children and skin temperature higher for the same environmental conditions than in adults.

## Sex

The available experimental evidence shows that women require lower evaporative cooling in both hot-wet and hot-dry environments (Shapiro *et al.*, 1981). Women have a lower tissue conductance in cold and a higher tissue conductance in heat than do men. This fact indicates a greater variation in the peripheral reaction to climatic stress in women. It appears that this fact is of no importance for the performance of work, however. Studies by Drinkwater *et al.* (1982) reveal that in healthy older women, aging does not diminish the functional capacity of the sweating mechanism to cope with heat stress while resting. From studies of active men versus active women during acclimatization to dry heat, Horstman and Christensen (1982) concluded that active women performed work of equal relative intensity in dry heat as well as active men. Ventilatory, metabolic and cardiovascular differences between sexes were minimal.

Frye and Kamon (1983) observed no differences in sweating efficiency between sexes in the dry heat, but the women maintained a significantly higher sweating efficiency than the men in the humid heat. In both environments, the men recruited a significantly lower percentage of their available sweat glands than did the women.

Aerobic capacity is an important factor to be considered when men and women are compared in the heat. When fitness levels are similar, the previously reported sex related differences in response to an acute heat exposure seem to disappear (Avellini *et al.*, 1980; Nielson, 1980).

## State of training

It appears that a trained individual is better able to adjust to heat than one who is untrained. Physical training will enhance the sweating mechanism at a given level of central sweating drive. The increased metabolic rate during training necessitates a high thermoregulatory demand, and apparently this demand will induce an increased peripheral sensitivity of the sweat glands to the central sweating drive. For an individual who increases his maximal aerobic power by training, a given rate of exercise will require a lower percentage of the maximal oxygen uptake. Concomitantly, the core temperature will decrease during a standardized work rate. The increased capacity for heat dissipation behind this adaptation is due to an enhanced sweating response (Shvartz *et al.*, 1979). In addition, heat acclimatization results in a further enhancement of the sweating response at a given level of central sweating drive by lowering the zero point of the central nervous system drive for sweating. In other words, physical training seems to increase the slope of the sweating rate–core temperature curves, i.e., the activity of the sweat glands increases at a given core temperature. In contrast, acclimatization to heat seems to lower the threshold core temperature at which sweating starts (it moves the curves 'to the left' without changing the slope; Nadel *et al.*, 1974). For an optimal heat acclimatization, simultaneous exposure to both heat and exercise is recommended. Physical training will also improve the circulatory potential. Thereby the trained individual may be able to maintain a cardiac output sufficient to meet metabolic requirements and the demand for peripheral bloodflow for a longer period of time than untrained people (Drinkwater *et al.*, 1976).

## Acclimatization

After several days' exposure to a hot environment, one is able to tolerate the heat much better than when first exposed. This is associated with an increased sweat production, a lower skin temperature, and a reduced heart rate, as illustrated in Figure 2.5 (Robinson *et al.*, 1943; Kuno, 1956; Bass, 1963; Leithead and Lind, 1964; Wyndham, 1973; Rowell, 1974; Libert *et al.*, 1983). The increased sweat rate provides for a more effective cooling of the skin through the evaporative heat loss, and the resultant lowered skin temperature provides a better cooling of the blood flowing through the skin. Thus, the body can afford to cut

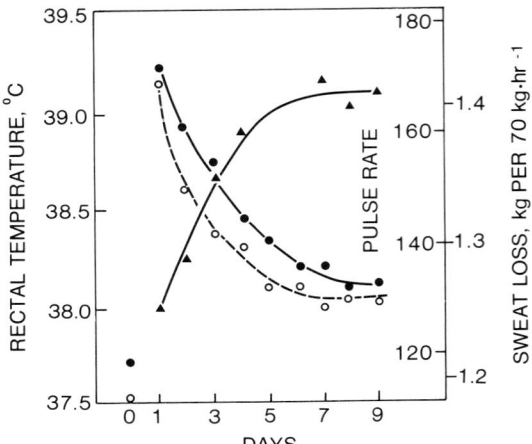

*Figure 2.5.   Mean rectal temperature (●), heart rate (○) and sweat loss (▲) in a group of men during a 9-day acclimatization to heat. On day 0 they exercised for 100 minutes at a rate of 1·2 MJ (300 kcal h⁻¹) in a cool climate (48·9°C dry-bulb and 26·7°C wet-bulb temperature) (from Åstrand and Rodahl, 1986).*

down on the skin blood flow. The fundamental basis for the mechanism of the acclimatization to heat is the fact that the sweat glands have the capacity to produce more sweat than they do under ordinary circumstances.

In the armed forces, and even in certain industries, the problem of an efficient and rapid method of acclimatizing a large number of people may be a practical problem. Daily exposure to a hot environment for about an hour will, after a week, result in some acclimatization. However, studies by Wyndham and co-workers (1973) to establish the minimal number of days required for acclimatization showed that a person cannot be acclimatized adequately for a normal shift of 6–8 hours in less than 4 hours per day and in less than 8–9 days. They applied a step test with a gradual increase in rate of exercise up to an oxygen uptake of 1·4 l min⁻¹ combined with heat stress conditions (air temperature 31·7°C, air saturated with water vapour). This programme is applied to recruits for gold mines in South Africa.

## 3. Assessment of heat stress

### Assessment of the environment

During the past several decades, considerable efforts have been devoted to developing new methods for the assessment of environmental heat exposure at different workplaces and refining some of the old heat stress indices, such as the Effective Temperature and the Wet Bulb Globe Temperature (WBGT). By such indices one tries to predict the heat load on the individual.

They are usually based on data from physical measurements of air temperature (dry bulb), radiant temperature (globe), relative humidity (wet bulb) and air velocity. A measurement of the individual's metabolic rate is also often included, i.e., the energy demand of the work is measured or estimated (Kerslake, 1972; Åstrand *et al.*, 1975). The WBGT index is rather complicated, and the recording of it is time-consuming. In the case of a predominantly dry, radiant heat environment, the much simpler and faster reacting Wet Globe Thermometer or Botsball, devised by Botsford (1971), records similar values and the results are interchangeable with a simple formula (Ciriello and Snook, 1977).

A number of investigators have attempted to evaluate the different indices and compared them under a variety of conditions. Brief and Confer (1971) measured five heat stress indices simultaneously under varying environmental conditions in an environmental test room. They found that the WBGT and the WGT were best suited to hot work situations where radiant heat energy is present, and that the Effective Temperature was the preferred index for hot, moist conditions.

Botsball readings were compared with the WBGT at 30 different workplaces in industrial plants (Beshir *et al.*, 1982). The measurements were taken twice daily over a one-year period for a total of 13 489 observations. The relationship in °C between the WGT and WGBT was found to be: WBGT $= 1\cdot01$ WGT $\pm 2\cdot6$, and the correlation coefficient was very high ($r = 0\cdot956$). They emphasize the advantages with the Botsball thermometer in terms of convenient size, ruggedness and ease of use, and the fact that it may easily be located near the worker and thus permit an accurate evaluation of the working environment. On the other hand the rapid water loss which required frequent refilling represented a major disadvantage of the WGT.

Ciriello and Snook (1977) investigated the relationship between the Botsball temperature and the WBGT under 210 different environmental conditions in an environmental chamber. The results indicate that as long as wind speed, humidity and radiant heat can be estimated fairly accurately, the WBGT can be predicted from WGT within $\pm 0\cdot4$°C. Only when adequate knowledge of the mentioned environmental conditions are not available did the predictability fall to $\pm 2\cdot9$°C. It should be kept in mind that in the case of practical application in industry, a difference of some 3°C may only be of academic significance.

A number of different heat stress indices were compared by Pulket *et al.* (1980) in a hot-humid environment. They concluded that, because of its ease of use, the WGT or the CET (Corrected Effective Temperature) index may be the best choice for preliminary industrial surveys in hot–humid environments.

The empirical relationships between several heat stress indices were examined by Mutchler and Vecchio (1977), including WBGT and WGT derived from a series of studies in 14 representative, hot industries. Essentially they verified the findings of others that the differences between the indices are not dramatic. They point out that the predicted values for WBGT from selected values of WGT vary only slightly among investigators, the maximum difference being 2·2°C.

On the basis of the measurements of heat stress indices in 200 working places, Locati *et al.* (1974) concluded that indices such as the WGT and the WBGT could be correlated with sufficient accuracy.

Plunkett and Carter (1974) having described some of the problems encountered when trying to apply the WBGT index to assess heat stress in three Alcoa smelting plants. They showed that all workplaces investigated had 2-hour time-weighted WBGT values above the recommended levels, with standard deviations of considerable magnitude, without this being associated with significant risks of heat disorders. They pointed out that Alcoa had about one reported heat illness per one million man-hours worked, and that many of the cases reported as heat illness were not truly cases of such disorders. Their investigations showed that the mass of precise data required to obtain a meaningful time-weighted average WBGT, when applied to aluminium smelting activities where workers are exposed to conditions which vary from place to place and from day to day, makes this procedure impractical as an index to determine work practices. They were unable to determine any significant correlation that would allow them to use one, or a few, monitoring stations to represent an entire pot-room.

Malchaire (1976) has made a critical evaluation of the readings attained with different natural wet bulb and wet globe thermometers, with particular emphasis on design specifications. He concludes that it is a major drawback of the WGBT and WGT indices that they are so dependent on the design characteristics of the measuring instruments.

In 1969, Taylor *et al.* described an instrument for direct measurement of the Wet Bulb Globe Temperature index, with a computer based on the mercury-in-glass thermometer. More recently, the Brüel and Kjær company has produced a commercially available WBGT recorder. Rapp and Aubertin (1985) have also developed a computer program for the calculation of mean radiation temperature in workplaces. Mutchler *et al.* (1976) have even developed a model for the prediction of WBGT in a hot environment, based on outside weather conditions.

Kamon and Ryan (1981) have suggested an Effective Heat Strain Index (EHSI) using a hand calculator with memory for on-the-site evaluation of prevailing ambient conditions. The inputs to the pro-grammed calculator include dry-bulb, wet-bulb and globe tempera-tures, an estimate of metabolism and air movements. The index is based on a program for calculating the total heat balance and the efficiency of sweating.

Ramsey and Chai (1983) have examined the inherent variability in heat-stress decision rules. They conclude that rather than being lulled into a sense of accuracy which may result from obtaining highly accurate thermal measurements, it might be more appropriate to utilize such measurements as general indicators of thermal stress, and then to provide reasonable work practices for reducing the risk of heat disor-ders. They suggest that the use of a simplified set of decision rules can basically serve the same purpose as a seemingly more precise approach.

## Assessment of physical response to heat stress (heat strain)

In this discussion we are primarily dealing with hot industrial work-places where heat is the major stress. We not considering the state of comfort in an indoor sedentary working environment.

As pointed out by Mitchell (1973), the criterion by which an indus-trial heat stress index must be judged is its usefulness in actual industrial situations. According to Kerslake (1972) it is simply not possible to produce a heat stress index which is both accurate and universally applicable.

In the words of Gagge and Nishi (1976), there is no single physical index of the thermal environment that is universally useful for judging both comfort and varying levels of heat strain: "The only true tempera-ture index is one based on the heat balance equation between man and his thermal environment". In other words, the environment is one thing, but the human reaction to that environment is something else, and in fact something of the greatest importance. For this reason, greater emphasis should be placed on the assessment of human response to the heat stress encountered by workers during the performance of their everyday work at their actual place of work. Without such information any discussion of safe upper limits of exposure would seem meaningless. Furthermore, it should be made quite clear that what is needed is field studies of individuals doing their everyday work, not simulated laboratory studies, which, however realistic, never entirely mimic real life. Results from laboratory investigations involving steady-state work and thermal loadings cannot always be transferred to

actual working conditions in real life, where the occurrence of thermal
and work stress are characteristically intermittent and are subject to
worker-controlled pacing, making it possible for workers to operate
without ill effects at work and heat levels which in the laboratory might
produce symptoms and signs of heat disorders.

Fuller and Smith (1981) maintain that the use of physiological
measurements, such as the heart rate and oral temperature, for evaluating heat stress in an industrial plant is simpler, less expensive, less
restrictive and more protective than the evaluation of hot jobs by
environmental measurements and subjective evaluation of metabolic
rate (the technique used with the WBGT criteria).

In excessive heat, the ability to perform and survive is a matter of
maintaining thermal balance at a level compatible with body function.
This means that heat production in the long run must equal heat loss.
Thus, the ideal way to monitor this state of affairs would be to apply
the heat balance equation:

$$S = M \pm R \pm C - E$$

where $M$ = metabolic heat production; $R$ = radiant heat exchange;
$C$ = convective (and conductive) heat exchange; and $E$ = evaporative
heat loss.

There is no practical way, however, to record all the elements of this
equation in everyday life in any number of people. We are, therefore,
forced to consider simplified ways of obtaining an estimate of thermal
balance, causing changes in body heat content, i.e., a net gain or loss of
heat.

*Skin temperature*

Skin temperature can be measured relatively easily, either by thermo-
couples, thermistors or disposable probes. It may be used as an
indication of general heat stress or thermal balance. In our experience,
however, it can indeed be used as an indication of how close the worker
is to the heat source (see Figure 2.6, showing how the skin temperature
in a tapper in an aluminium plant rises abruptly each time the tapper
approaches the pot). This illustrates the importance of proper shielding
of the temperature sensor under conditions of intense radiating heat in
order to obtain real values of skin temperature. Otherwise it should be
noted that the skin temperature recorded on the inside of the thigh
closely resembles mean skin temperature under most ordinary
circumstances (Ramanathan, 1964; see Figure 2.7).

Skin temperature is the best single temperature index of warm and
cold sensations, both in warm and cold environments. Skin wettedness
is the best single index of warm discomfort (Gagge *et al.*, 1974).

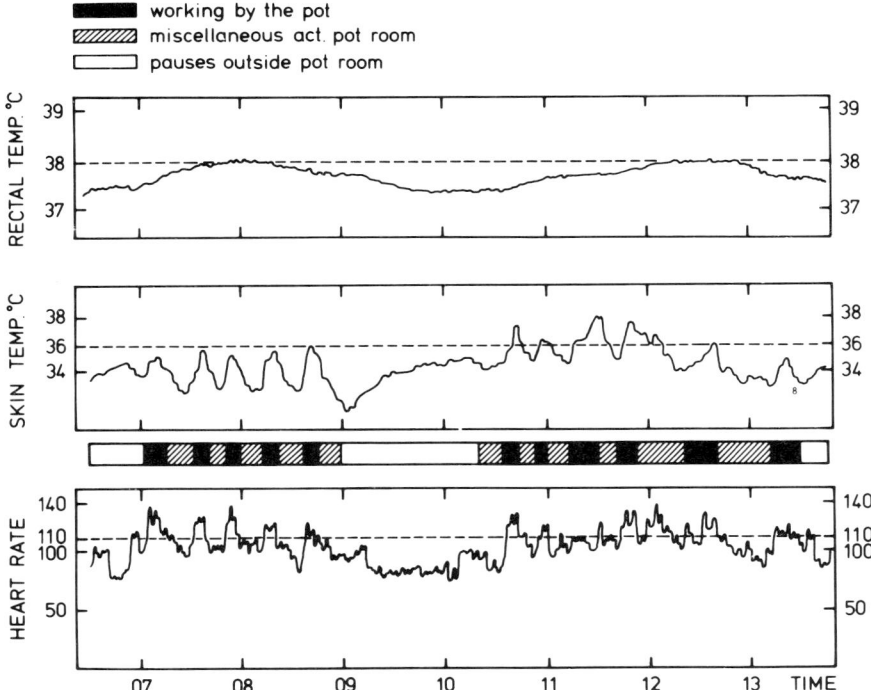

Figure 2.6.   Skin temperature of a tapper in an aluminium plant, indicating radiant heat exposure each time he is close to the pot tapping the finished metal.

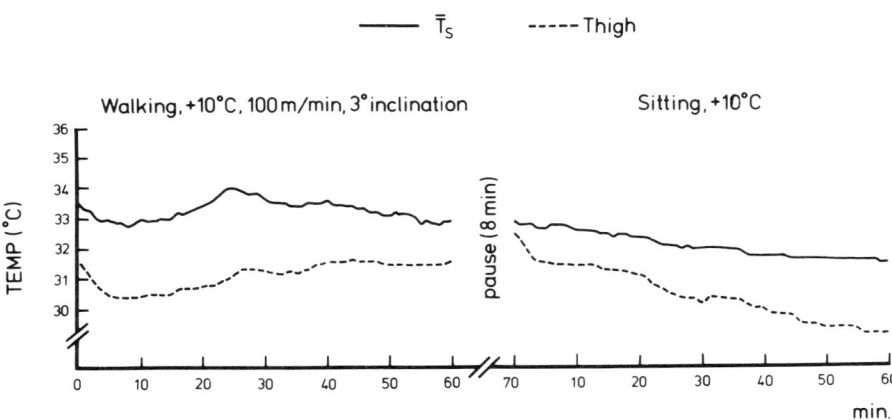

Figure 2.7.   Mean skin temperature and skin temperature on the inside of the thigh in a subject walking on a treadmill at 10°C, lightly dressed, at a speed of 100 m min$^{-1}$, inclination 3°.

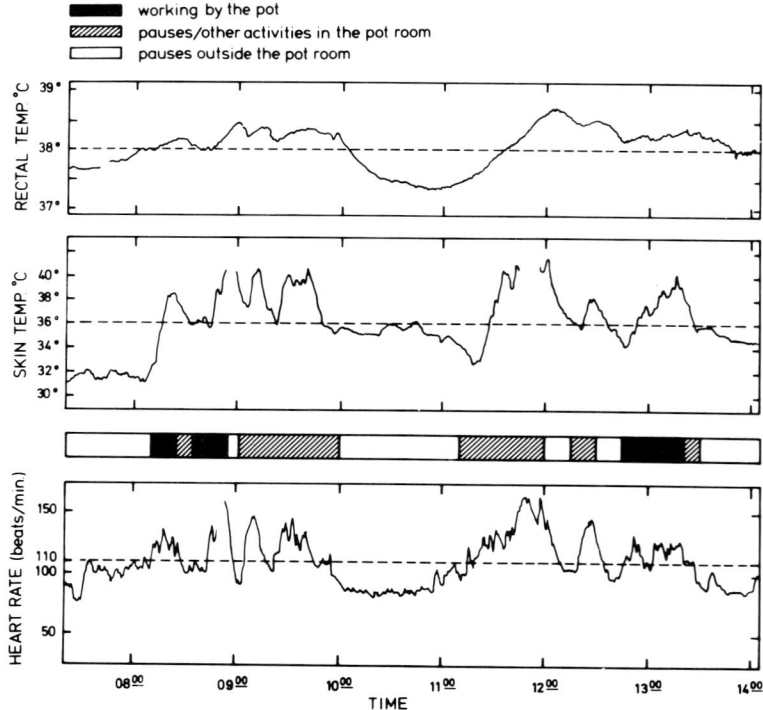

*Figure 2.8. Body temperature and heart rate in a pot tender, working in an aluminium plant.*

## Rectal temperature

Rectal temperature mirrors core temperature. In the long run it may indeed reflect body heat gain or body heat loss. Under such conditions rectal temperature may be used as an index of heat stress, as long as sufficient consideration is given to the fact that it is a slow parameter. In most cases of industrial heat exposure it takes some 45 minutes for the rectal temperature to reach a plateau. This emphasizes the importance of continuous recording of the rectal temperature, either by portable tape recorders such as the Medilog, or portable electronic loggers such as the Vitalog. This, however, requires cooperative subjects and patience and persuasion on the part of the investigator. But, when properly done, the continuous recording of the rectal temperature in workers in the course of their normal work and leisure is indeed a most revealing parameter in terms of thermal strain (Figure 2.8).

## Oral temperature

Oral temperature is probably a poor substitute for rectal temperature, and cannot always be recorded continuously under field conditions.

Under certain ideal conditions, however, oral temperature may be used as an indication of central body temperature. Under conditions of varying work loads, Mairiaux *et al.* (1983) observed that oral temperature was not so well related to oesophageal temperature as was rectal temperature.

Morganes *et al.* (1981) have shown that ear canal temperature has only limited usefulness in field studies, because it is affected by external conditions such as wind, unless the subjects can wear a protective helmet or similar device throughout the period of data collection.

*Heart rate*

Heart rate is an index of a combination of factors, including work load, energy expenditure and heat stress. It is also an index of the autonomous nervous system's reaction to emotional stress. On the whole it is a useful index of work stress in general, but is of less value as a direct indicator of heat strain unless the other influencing factors are adequately controlled.

In our experience, heart rate may more or less follow rectal temperature in a general pattern, but it may also in some cases rather closely parallel the changes in skin temperature (see Figure 2.6 illustrating a tapper in an aluminium plant).

Vogt *et al.* (1973) have suggested a method of using continuously recorded heart rate as a method for estimating thermal stress as well as physical work stress in actual work situations. The proposed procedure is based on laboratory experiments, and requires subjecting each individual concerned to standard tests in a climatic chamber in order to establish their circulatory responses to work and heat. There is no doubt that it is possible to distinguish between heart rate increased due to work and due to heat, especially under controlled conditions. In our experience, however, the practical application of such a method under field conditions in real work situations would appear complicated considering the need for climatic chamber pre-tests. Another approach would be to subject the individuals to a series of standardized cycle ergometer work loads at the site of the actual workplace at the prevailing ambient temperature, in addition to recording the resting heart rate, and the heart rate during the performance of their actual work (see Figure 2.9).

*Sweat rate*

Sweat rate, as determined by changes in body weight, is a reflection of the amount of fluid lost as a part of the evaporative heat loss, when carefully controlled (records kept of weight of food and fluid intake, and weight of stools and urine output). It is therefore a meaningful

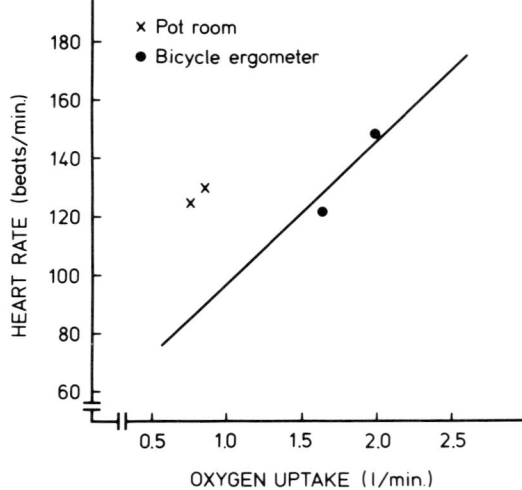

*Figure 2.9.   The heart rate of the same subject is higher at the same work load in the heat (×) than in a cool environment (●).*

index of heat stress. Furthermore, it is relatively easy to measure, since it merely requires an accurate scale, capable of recording changes in body weight with a precision of some 50 g.

## 4.  Heat stress in hot industries

### A brief review of some of the available literature covering specific studies in different types of hot industries

The basic aspects of thermal physiology and temperature regulation have been dealt with in considerable detail elsewhere (Åstrand and Rodahl, 1986). In this review we shall therefore limit ourselves to the more applied aspects with particular reference to published results of studies carried out in hot industrial workplaces, and especially heat exposure of the workers.

A review of more than 500 scientific publications published in the past 10 years dealing with heat stress and heat exposure, showed that over 75% concerned studies of methodology or measurements of environmental heat by different procedures, evaluation of heat exposure limits and considerations regarding such limits by regulatory agencies, and basic research on the physiology of thermoregulation and the effects of heat on the organism. Only about 25% of the publications dealt with studies of heat stress in specific industries or occupational situations. It is evident that heat stress problems have not been most frequently studied in the traditional hot industries, i.e., production and casting of steel, iron, ferroalloys, aluminium, magnesium, silicon carbide, etc., industries using processes in which energy is transferred into heat. Deep mining (gold, coal) comes second, representing industries which do not produce heat, but where the operations are performed in an excessively hot environment. Surprisingly, only a few studies have been concerned with the combination of internal and external ambient heat, i.e., a heat-liberating industry located in a tropical climate.

A review of the available recent Eastern European literature on heat stress indicates that much of it involves surveys of existing conditions with a view to improving efficiency and the working environment, largely by changing procedures and practical improvements (Stolbun *et al.*, 1970; Babaev and Khodzhibaev, 1972; Dorofeyeva, 1973; Krosnoschekov and Chervyakov, 1973; Tymanski and Zaborski, 1975; Detsik *et al.*, 1976; Bakaev and Bobokhodzhaev, 1977; Paltsev, 1982; Shishlyannikova *et al.*, 1983).

One of the earlier studies of the iron and steel industry was made by Christensen *et al.* (1952), who carried out an extensive physiological study of workers at Uddeholm Iron Works at Nykroppa, Sweden in 1951–52. They observed peak energy expenditure during work of up to 16 kcal min$^{-1}$ (3·35 l $O_2$), mean energy expenditure during the shift of 5–12 kcal min$^{-1}$, mean heart rate of 110–165 beats min$^{-1}$, and maximal core temperature of over 38°C in 10 out of 12 cases, in two cases in excess of 39°C. Fluid loss in excess of 6 litres was recorded. They also made the remarkable observation that the heart rate of the worker doing the same work in the plant at an air temperature of 32–44°C and outside the building at an air temperature of 19–20°C differed some 60 beats min$^{-1}$ (166 as against 104).

In an attempt to arrive at a physiological evaluation of the resulting strains in the industrial worker in a steel mill, Minard and Goldsmith (1971) measured ambient heat load (including WBGT), level of physical activity, core temperature and heart rate of workers in the open hearth department of a Pittsburgh steel mill. Heart rate responses were continuously monitored, using a miniature tape recorder. The results were analysed by a digital computer. Core temperature was measured intermittently by radiotelemetry using a radiosonic capsule. When a worker was engaged in tapping the furnace, heart rate increased progressively in an irregular manner with intermittent peaks associated with increasing task level and increasing heat exposure, reaching peak rates of 168 beats min$^{-1}$. Core temperature lagged behind heart rate, reaching a maximum of 38°C. They observed that the heart rate responses varied widely in different workers performing the same job, and the more physically fit workers had lower heart rates during the shift than did those who were less fit.

Gresenz *et al.* (1973) made an attempt to classify the different work operations in the iron and steel industry on the basis of heart rate and oxygen uptake during work, considering ambient temperature and humidity.

Folprechtova *et al.* (1973) measured ambient temperature, energy expenditure, heart rate, sweat rate and body temperature in 20 forge-men during three consecutive days in the summer and in the winter. They showed, as expected, that the sweat rate was significantly higher in the summer than in the winter, although the energy expenditure was significantly lower in the summer. Hunting *et al.* (1974) examined the physical work load in a dropforge by continuous registration of heart rate, as well as measurement of the radiant heat load. They observed a mean working heart rate of some 40 beats min$^{-1}$ higher than the resting heart rate.

Dorofeyeva *et al.* (1974) studied the fluid and electrolyte balance in heat exposed workers in metallurgic plants, and observed the familiar changes of heat exposed subjects. They pointed out the importance of adequate fluid intake and the need for a suitable beverage for the maintenance of fluid and electrolyte balance.

Vogt *et al.* (1977) show that it is possible to improve working conditions at blast-furnaces by reducing the physiological load in terms of reduced thermal stress due to radiation, and reduced motor stress, without reducing productivity.

Reischi *et al.* (1977) have developed a radiotelemetry system which was used for the evaluation of heat stress in a steel factory. The system was used as a station monitor for determining environmental tempera-ture conditions and as a personal monitor to obtain data on the physiological response of the worker in terms of heart rate, skin and rectal temperature.

In India, 35 heat-exposed foundry workers were monitored in terms of physiological strain, as assessed by heart rate, recovery heart rate, oral temperature, sweat loss and fluid balance (Parikh *et al.*, 1977). No significant physiological heat strain was observed, in spite of the fact that the workers were exposed to considerable heat stress. They explained this by the fact that the foundry workers were performing light work only. They recorded WBGT observations up to 48°C in the summer, and 41°C in the winter. During the summer the WBGT exceeded 30°C for 94% of the time. The authors emphasize the importance of adequate water intake during the work shift to minimize the strain, pointing out that the average water deficit at the end of the day approached 2% of the body weight.

Klimmer *et al.* (1984) examined 27 workers, aged about 40 years, in a

German iron and steel work. The mean energy expenditure corresponded to about $1 \, 1 \, O_2 \, min^{-1}$, varying between 0·5 and 1·7 $1 \, min^{-1}$. Mean heart rates varied between 80 and 180 beats $min^{-1}$. In one case the heart exceeded 120 beats $min^{-1}$ for almost 70% of the time.

Heat stress in the aluminium industry has been investigated by Dinman *et al.* (1974) who studied 34 workers at three Alcoa aluminium smelting plants. In addition to measuring WBGT at the site of each specific job location, they measured heart rates by palpation before and after the work shifts. Heart rate samplings were also taken immediately after specific job-associated exertions. Oral temperatures were recorded at the beginning and end of each shift, and after each job component occurring in the course of the day. Four-hour sweat rates were determined by body weight changes, taking fluid intake and urinary output into consideration. On the basis of their data they concluded that there were no significant body temperature and pulse rate elevations or weight losses over the eight-hour shift, despite ambient conditions greater than 85°F (~35°C) WBGT, under conditions of light and moderate work.

Mining operations may be associated with considerable heat stress. For a review of the applied physiology problems associated with the gold mining industry, the reader is referred to Wyndham (1974), Martinson (1977) and Strydom (1980).

In the case of coal mining, Stebel (1981) has reviewed the influence of microclimate on the worker in underground German coal mines, and discussed methods of improvement. Hausman and Petit (1978) have pointed out that neither rationalization of ventilation nor air conditioning can solve all the problems in coal mines where there are severe climatic conditions. They point out that in the case of an effective temperature of less than 35°C and a low velocity air flow, merely stirring the air by ordinary ventilation is enough to improve conditions considerably. However, in the case of temperatures above 35°C either the air which is blown on the worker must be cooled, or the worker must wear cooled clothing. They remind us of the fact that greater emphasis must be made on the cooling of the head. They maintain that the simplest and most effective solution is to issue the worker with a hooded jacket of double sponge fabric soaked in cold water. Resetjuk *et al.* (1976) have reported studies in the Donetz coal mines, in which the spontaneous work–rest schedule was replaced by a schedule based on physiological principles, i.e., shorter, more frequent breaks, rational organization of work, etc.

The functional efficiency of heading-drivers engaged in non-mechanized mine operations with air temperatures up to 31°C was studied by Vanin (1972). He pointed out that their normal performance efficiency could not be maintained under the conditions of high air temperature, and emphasized the importance of acclimatization and the need for air conditioning in deep mines under these conditions. Gaebein (1983) measured heart rate, body temperature and sweat loss in six drivers working in a dry-warm climate in an underground potash mine in East Germany. He concluded that when adhering to a proper work–rest regimen, the heat stress was not excessive.

Considerable heat exposure of workers operating close to the furnaces in glass factories was demonstrated by Abeysekera (1981) in two mechanized glass factories in Sri Lanka, and by Boulos and Boulos (1980) in Saudi Arabia. In a study by Sköldström *et al.* (1980) in a Swedish glass factory, rectal temperatures exceeding 38°C and heart rates up to 150 beats min$^{-1}$ were observed in some of the workers. While in most cases the heat exposure was rather moderate, the physiological heat strain was at times excessive, although the WBGT did not at any time exceed the recommended maximal value. It was therefore considered that under these conditions the WBGT was a poor indication of the actual heat exposure.

Zöller *et al.* (1981) studied 39 acclimatized workers at a ball-bearing forge. They observed a slight decline in systolic blood pressure and an increase in heart rate during the shift, and a drop in blood pH and an increase in serum lactate and triglycerides.

The fact that thermal radiation may be a problem for engine-room staff on ships at sea was demonstrated by Mæhlum *et al.* (1980), Jebens *et al.* (1980) and Tymanski *et al.* (1975).

## Our own studies of industrial heat stress

During the past several years the Norwegian Institute of Work Physiology has been involved in a series of extensive studies of heat stress in several types of heat-exposed industries, especially the electrochemical industry. This has included the production of steel, ferroalloys, aluminium, magnesium, silicon carbide and paper, in addition to studies of sailors on board ships operating in tropical zones. In most of these studies, measurements of the environmental heat load were made, and the thermoregulatory response of the individual worker assessed. This is a brief review of some of the more pertinent aspects of these studies.

Two series of studies, one in the winter and one in the summer, were made by members of the Norwegian Institute of Work Physiology at the ferroalloy plant Fiskaa Verk in Kristiansand (Magnus *et al.*, 1980; Egeland *et al.*, 1981). The purpose was to make a survey of the degree of heat stress encountered in this type of industry, and to investigate its effects on the workers, especially in the light of the findings of Koetzel *et al.* (1973) suggesting a connection between heat exposure and the development of hypertension in workers in the smelting industry.

The first series of studies took place in 1978, involving 12 workers. Botsball readings revealed ambient temperatures above the recommended ceilings at all examined work areas in the pot-room (26–38°C WGT).

The continuously recorded heart rate by a miniature portable recorder revealed considerable circulatory stress, not only in heat exposed workers (in one case a heart rate of 130 for 20% of the observation period), but also in manual workers engaged in the breaking up of finished metal with sledge-hammers in a cool environment (in one case a heart rate of 125 beats min$^{-1}$ for 45% of the observation period).

Skin and rectal temperatures were also recorded continuously. The rectal temperature exceeded 38°C in all but two subjects. In one of the tappers the rectal temperature remained above 38°C for 87 minutes, and in a maintenance worker for 93 minutes. Rectal temperatures in excess of 38°C were also observed in one of the manual workers not exposed to ambient heat, indicating that the rise in rectal temperature was due to muscular work and not to ambient heat stress (Figure 2.10). Rectal temperatures in excess of 38°C for extended periods were also observed

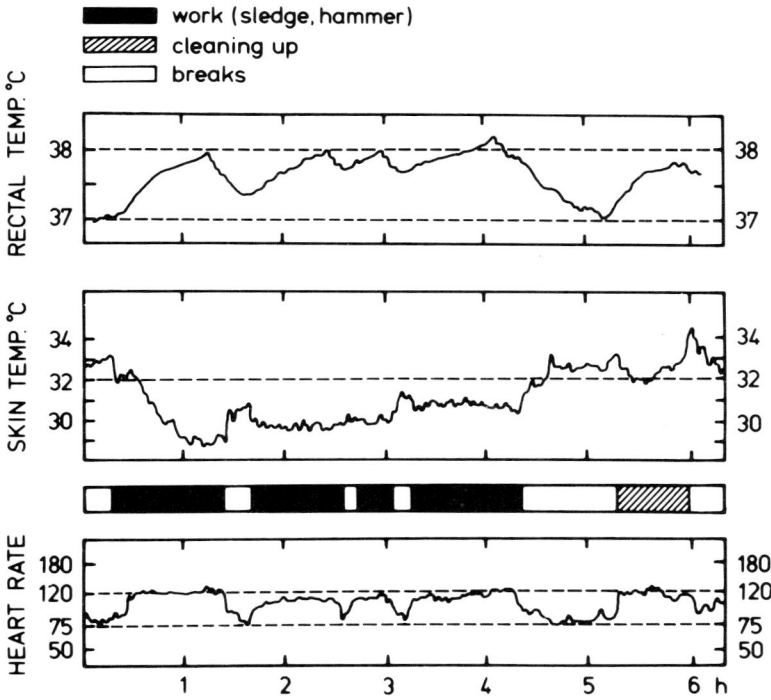

*Figure 2.10. Heart rate and body temperature in a manual worker engaged in the breaking up of finished metal with sledge hammers in a cool environment in a ferroalloy plant.*

in maintenance workers engaged in repair work by the pot, with heart rates up to 180 beats $min^{-1}$.

Sweat loss averaged about 2 l during the working shift, the highest value recorded being 3·6 l in the case of one of the tappers. The hourly sweat rate, on the other hand, could in some cases be rather high, in one case as high as 0·508 l $h^{-1}$. It was observed that only about half of the fluid lost through sweating was replaced by fluid intake during the working shifts, leading to hypohydration. The degree of dehydration expressed in percentage of the body weight was 1·15% on the average. This is close to the level at which objective signs usually occur in the form of increased heart rate and impaired endurance (Åstrand and Rodahl, 1986).

The blood studies revealed plasma renin concentrations significantly higher during heat exposure compared to values before the beginning of the working shift ($P < 0.05$). This was interpreted as an indication of reduced kidney blood flow, due to heat or physical work or both, since this elevation also occurred in the manual workers not exposed to heat but engaged in strenuous physical work. The serum concentrations of creatinine, urea and uric acid showed a significant increase in the course of the working shift, most pronounced in the case of creatinine ($P < 0.001$).

The urinary analyses showed reduced Na and K excretions during the working shift, compared to values off work. Creatinine clearance, which is a measure of the glomerular filtration in the kidneys, was on average lower during the working shift than was the case off work. The greatest difference in creatinine clearance between work and leisure was observed in the manual workers engaged in exceptionally hard physical work, who also had the highest norepinephrine eliminations. The mean urinary epinephrine elimination was on the whole quite similar during the working shift and during the time off work.

The 'logical reasoning test' as an index of mental fatigue or perform-ance was administered to all the subjects during the study, before and after the working shift, and showed no difference of statistical significance.

The blood pressure, both systolic and diastolic, was measured in a standardized manner before, during and after the working shift in 10 subjects, and showed no significant difference. Nor did the measure-ment of blood pressure in 50 heat-exposed and 50 non-heat exposed workers in this plant. Subsequent, extensive studies of the blood pressure and the incidence of hypertension in workers of this plant, in whom the blood pressure was followed over several years, verified this finding: that exposure to heat stress is not associated with the develop-ment of high blood pressure, as previously claimed by Kloetzel *et al.* (1973).

A subsequent series of studies at this plant (Egeland *et al.*, 1981) verified or substantiated the findings of the first study. In addition, it emphasized the importance of acclimatization and indicated the occurrence of some degree of haemodilution during the working shift in the heat exposed workers. Specifically, it pointed out that no ill effects were observed as a consequence of the heat exposure encountered at this plant, and that all the physiological parameters measured were within the normal range. It appeared that one person was susceptible to orthostatic hypotension and was apt to faint when heat exposed, and was thus unsuitable for this type of work. This also applies to persons suffering from cardiovascular disease or chronic kidney disease. Otherwise no evidence of health hazards were detected for these kinds of industrial operations. The seasonal differences in the observations were minimal.

Extensive studies of Norwegian aluminium plant workers by continuously recorded rectal and skin temperatures and heart rates with the aid of miniature recorders, by Botsball temperature recordings, and by fluid balance recordings revealed ambient temperatures considerably above the recommended upper limits at a number of job locations (Rodahl, 1981; Jansen *et al.*, 1982). The studies were made twice a year for two years. There was surprisingly little difference between winter and summer temperature in the plant, i.e., only about 10°C.

The workers in this particular aluminium plant were exposed to the heat of the pot-room for 44% of the total shift, on average, and 34% of the total shift was spent working. The rest of the time was spent in the canteen drinking coffee, talking, reading, playing cards or moving about. Similar findings have been made in other ferroalloy industries. As is often the case, the workers preferred to work rather intensely in order to get the job done as quickly as possible, to be able to rest that much longer. The result may be extended periods of excessive heat stress instead of more frequent brief periods of exposure interspaced with brief cooling-off periods outside the pot-room. A burner cleaner may serve as an example (Figure 2.11). The time he spent actually cleaning the burners amounted to three one-hour periods. The rest of the time was spent outside the pot-room in a comfortable cool canteen. The time spent by the pot, on the whole, varied from 30 to 190 minutes per shift.

In 15 out of 22 workers the rectal temperature exceeded 38°C (see, for example, Figure 2.12). In one case the rectal temperature exceeded 38°C for 63% of the observation period, i.e., over half of the entire shift period. In our experience, the rectal temperature is a good index of heat stress.

When exposed to the pot-room temperature, the subject's rectal temperature rose gradually, levelling off in 45–60 minutes, whereas the

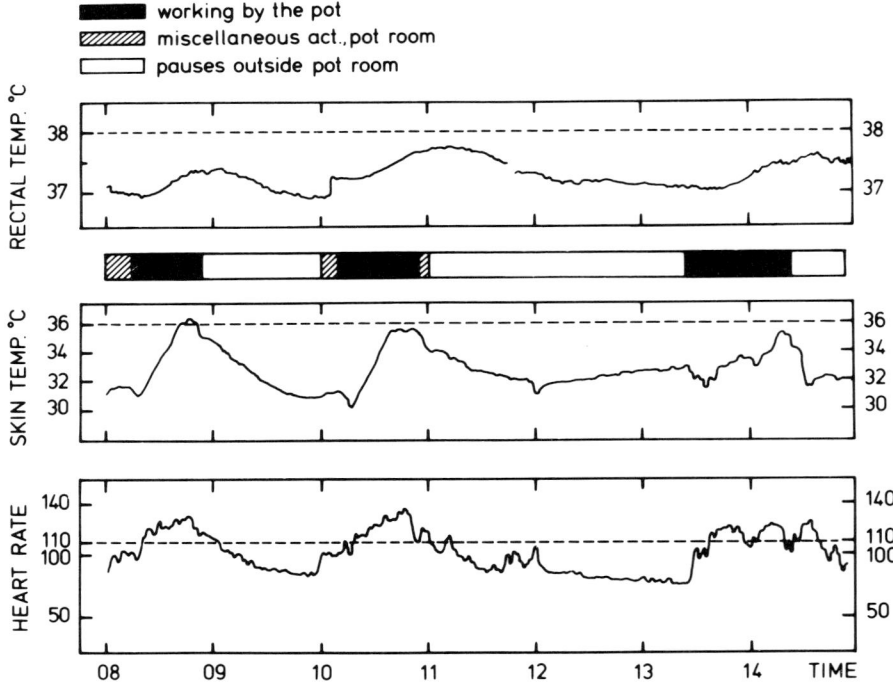

Figure 2.11.    Work–rest pattern in a burner-cleaner in an aluminium plant.

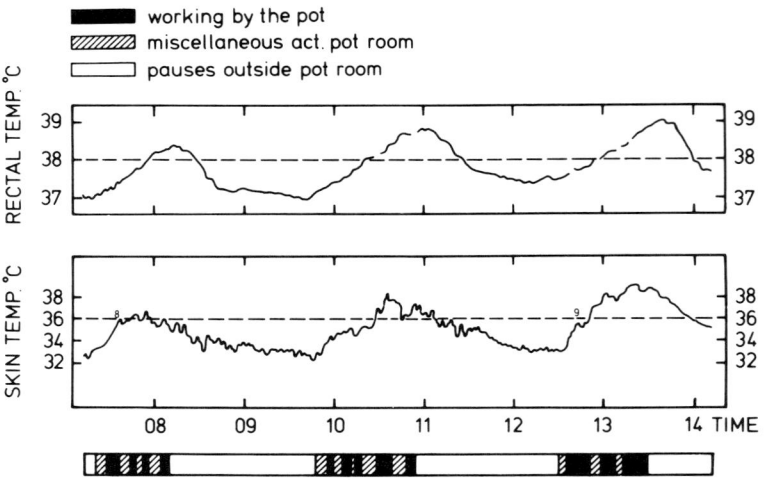

Figure 2.12.    A jack raiser in an aluminium plant; an example of rectal temperature exceeding 38°C.

skin temperature reacted much faster and is a good indicator of how close the subject is to the heat source. An example of this is given in Figure 2.6 showing a tapper making 11 trips during the shift to fill his truck with melted aluminium. The figure shows 11 peaks in skin temperature, indicating when the tapper was close to the pot. In our experience, skin temperatures over 36°C are usually associated with profuse sweating, which is also in accordance with experimental data (Nielsen, 1969).

In almost all subjects studied, the heart rate exceeded 110 beats min$^{-1}$ for considerable periods of time, in some cases up to 30% of the time. In several cases it exceeded 130 beats min$^{-1}$, which in most people represents about half of the heart rate reserve (halfway between resting and maximal heart rate). Figure 2.8 illustrates the relative changes in rectal temperature, skin temperature and heart rate in a pot tender during a work shift.

In several cases the weight loss of the workers exceeded 1% of the body weight in the course of the shift, indicating a considerable negative fluid balance, or hypohydration. The highest sweat rates which we observed were about 0·8 l h$^{-1}$. One worker consumed as much as 6 l of water in the course of his working shift. Thus, excessive fluid loss due to sweating is a major problem in workers regularly working in our heat exposed industries.

Our blood studies showed a lower haemoglobin value at the end of the shift compared to the values before the shift (15·4 ± 0·2, as against 14·7 ± 0·3), indicating haemodilution. Similar findings were made in our studies in the ferroalloy industry. We also found that those who had the greatest sweat losses also had the greatest drop in sodium concentrations of the blood, but in no case did we find values which were outside the normal range. In every case the sodium loss was next day. We therefore concluded that there was no advantage in taking salt tablets during work.

Urinary catecholamine elimination was also examined in our studies, but was found to be of no value as an indication of stress under these conditions. Individual differences were greater than fluctuations with stress exposure in the same individual.

## 5. Limits of temperature tolerance

It appears to be generally agreed that there is no single index or recordable parameter which simply reflects the strain to which a worker is exposed during his work in heat exposed industry. It appears that too much emphasis has been placed on the academic accuracy at the expense of practicability. This is probably due to lack of practical

experience of the actual prevailing conditions in heat exposed industries
on the part of those who have developed the indices, essentially based
on laboratory experiments. This may have caused them to overlook the
fact that work in these cases on the whole is performed under rather
varied conditions, with varying degrees of physical work load, heat
stress and work periods. Industrial work of the kind in question is very
seldom comparable with the schedules used in laboratory experiments.
For this reason the suggested so-called safe upper limits of heat exposure under conditions of continuous work, work 75%, 50% or 25% of
the time may be artificial and therefore not applicable in real life
situations. Furthermore, the workers move frequently from place to
place, with greatly varying ambient temperatures. In addition, the
thermal stress varies with time, from hour to hour, and from night to
day (see Figure 2.13). Finally, different members of the same work
team may be exposed differently even though they are performing the
same type of work. Differences in the type of clothing used may also
affect the workers differently, even though they are engaged in the
same task. Individual differences and degree of acclimatization must
also be taken into account, as well as the level of fitness in relation to the
physical work load to be endured and the physical work load *per se* and
its effect on the workers' heat production. Age and sex also have to be
taken into consideration.

Permissible upper limits for safe heat exposure have been based on
the assumption that the rectal temperature of the worker should not

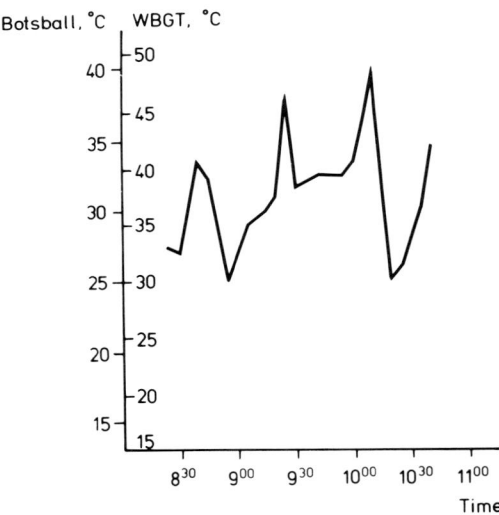

*Figure 2.13.   Showing the fluctuation in the ambient temperature in one location in the pot-
room in an aluminium plant.*

exceed 38°C. At ordinary room temperature this is usually the temperature reached in continuous cycle ergometer work in the laboratory at a work rate representing approximately 50% of the individual's maximal oxygen uptake (Åstrand *et al.*, 1975). But when working in the heat, part of the temperature rise is due to the heat gain of the body due to the ambient temperature, which is added to the internal heat mainly produced by the working muscles. As is evident from our field observations, core temperatures exceeding 38°C are rather frequent, without appearing to have any harmful effect. In fact, one of our subjects came to work on a bicycle and had a rectal temperature of about 38°C upon arrival, due to the work of bicycling. The same applies to the sophisticated heat indices, which hardly provide a better assessment of the stress of heat than that felt subjectively by the worker himself. This does not mean that a reliable thermal index is of no interest, for there is certainly a need for an objective assessment of the thermal environment, especially in order to assess objectively whether or not any change in fact represents an improvement of the environment.

Heat stress may be more or less uniform throughout the work shift, such as in a pot-room, or it may fluctuate markedly with brief periods of extensive heat exposure, as may be the case in a ferroalloy plant during tapping, or in a pepper mill during certain maintenance procedures. In the latter case, mean values for heat exposure may be questionable. Prolonged moderately high heat exposures may be as bad as or worse than brief intense heat exposures. In addition, humidity may be a decisive factor for tolerance.

Ramsey *et al.* (1975) studied the performance of subjects engaged in four sedentary tasks in warm and hot environments for work periods of up to 2 hours. The results were compared with recommended limits according to occupational health and safety regulations. They concluded that this limit is not a single line, but rather a range of temperature–time combinations. They pointed out that it is indeed possible to compensate adequately during short exposures, permitting a higher temperature limit for brief work bouts.

## The merits of some of the most commonly used indices

### Predicted 4-hour sweat rate (PSR)

The PSR is an old physiological heat stress index based on the prediction of the amount of sweat which would be produced in the course of 4 hours, knowing the air movement, dry and wet temperature and globe temperature in the case of radiant heat. In addition, work load and clothing has to be taken into account.

Initially, an upper limit of 4·5 l was considered, but over the years a number of values have been proposed. While the magnitude of the sweating certainly is a parameter of interest, the problem with this index is its complexity, which limits its applicability. At any rate, the recording of the changes in body weight in a number of representative members of the working staff in question would certainly give a reasonably good idea of the sweat-producing effect of the working environment.

In our studies of workers in Norwegian heat exposed industries we very seldom observed values as high as 4·5 l of sweat in 4 hours. We did observe, however, that the avoidance of prolonged, profuse sweating is of paramount importance in terms of preserving homeostasis and a proper state of hydration.

### Belding and Hatch's Heat Stress Index (HSI)

This is an index which is based on analysis of the heat exchange between the body and the environment under conditions of maintained thermal balance. It is an expression of the evaporative heat loss required for the maintenance of thermal balance, compared to the maximally possible evaporative heat loss:

$$\text{HSI} = \frac{E_{\text{req}}}{E_{\text{max}}} \, 100$$

In terms of practical applicability, this index also has its limitations, especially since it is based on a constant skin temperature of 35°C, which from our own observations is often not the case.

### Effective temperature (ET)

This is an index based on subjective comparisons of different ambient environments primarily intended to be used for the assessment of comfort in individuals engaged in sedentary or light physical work. It would not be very applicable under the extreme conditions of work and heat encountered in heat exposed industries.

### Corrected Effective Temperature (CET)

CET is a modification of the ET, the dry air temperature being replaced by globe temperature.

Pafnote and Vaida (1981) recorded climatic parameters and physiological strain in terms of oral and skin temperatures, heart rate and sweat loss over a 4-hour period in 140 subjects at various hot metallurgic workplaces. On this basis they claim the safe limits for light work to be 28°C CET and 26°C for average work.

## Wet Bulb Globe Temperature (WBGT)

The WBGT is an index which was introduced in 1957 (Yaglou and Minard, 1957) as an improvement of the ET, taking radiation into account. The index is based on a simple formula, which has been the subject of several modifications over the years. This may be confusing since it reflects different parameters and different weightings of some of these parameters (Botsford, 1971). Nonetheless, it is a meaningful way to describe the total heat stress of the environment, under specific circumstances. A major disadvantage with the WBGT is the time it takes for the instrument to equilibrate (up to 30 minutes).

Based on the WBGT the American Conference of Governmental Industrial Hygienists (ACGIH, 1976) have suggested tolerance limits for heat stress exposure for industrial workers. This includes continuous work and intermittent work (75, 50 and 25% work, 25, 50 and 75% rest), and work load (light and moderately heavy). The recommendations are based on the assumption that the rectal temperature of the worker should not be allowed to exceed 38°C.

There are a number of obvious objections to these recommendations. In the first place it is based too much on theory and assumption rather than on physiological observations at actual workplaces. The basic assumption of a rectal temperature ceiling of 38°C is unfounded in the first place. The values stipulated for intermittent work are inadequately based on actual observations. Furthermore, industrial work is in reality seldom performed in this manner or according to the proposed schedule between rest and work. Finally, there is no direct evidence showing exact environmental limits causing health effects. This depends as much on how the work is done, the length of each exposure and cooling-off period, protection, fluid intake, etc. In the final analysis the exposed worker himself is usually able to tell how he feels, and how hot it is, more so than any index is able to do.

Parikh *et al.* (1976) undertook a field evaluation of the WBGT index under the conditions of severe heat stress experienced by workers in Indian industries. They recorded heart rate and environmental temperature (WBGT) in exposed workers in nine small foundries in three seasons. They found that while the WBGT far exceeded the WHO proposed limits, the average heart rate during the shift rarely exceeded 110 beats min$^{-1}$.

Ljungberg *et al.* (1979) studied the effects of sedentary work in two different climates: 40°C with 40% relative humidity, and 32°C with

80% relative humidity. They concluded that the common indices such as WBGT did not discriminate between the two climatic conditions while indices based on heat balance analysis or calculations of sweat rate could predict more accurately the physiological and psychological strain.

### The Swedish Wet Bulb Globe Temperature Index (SWBGT)

This was developed (Eriksson and Olander, 1974) as a modification of the American WBGT for practical application in industry. Different formulae are recommended for air movement less than or higher than $1·5$ m s$^{-1}$. On the whole, the same comments made for the WBGT are also applicable for the SWBGT index.

### The Wet Bulb Temperature (WBT)

The WBT index uses the Botsball thermometer, introduced by Botsford (1971). This is an instrument which combines air temperature, humidity, wind and thermal radiation into a single reading, expressing the thermal stress of the environment. The instrument is quite simple and easily adapted to industrial use. It consists of a small 60 mm diameter hollow black globe, covered with a double layer of black cloth, which is continuously moistened by water seeping from a reservoir tube attached to the globe. The stem of a dial thermometer passes through a plastic tube along the centre line of the water reservoir tube and into the globe, thus sensing its temperature.

When placed in a hot area, the globe is heated by the surrounding air and by radiant heat from any hot surfaces in the surroundings. It is cooled by the evaporation of water from the globe surface, depending on air humidity and wind or air movement. The wet globe reaches an equilibrium temperature when these heating and cooling effects come into balance, which usually takes about 5 minutes. The 'Botsball' temperature is read directly on the dial thermometer.

The Botsball thermometer is light in weight and may be mounted on a simple stand or hung in a suitable, representative location at the place of work. Vibrations transmitted through the support may cause an excessive amount of water to escape from the reservoir. This problem is avoided by hanging the globe on elastic bands. Errors in the thermometer readings are less than $0·5°C$, but the thermometer should be calibrated regularly to maintain this accuracy. If the Botsball thermometer is to be used in a strong magnetic field, an error of up to $1·7°C$ may be expected. A shielded Botsball thermometer is to be preferred in such locations.

This Botsball temperature has been found to correlate quite well with other comparable indices. In our experiences the WGT readings are in reasonably good agreement with the WBGT under conditions of dry, radiant heat, and since the former is simpler and quicker, we consider it preferable for practical field applications.

## Health hazards of heat exposure

Heat exposure, both occasional and regular, may present health hazards of different kinds.

Best known are the ill effects of acute exposure, i.e., fainting heat exhaustion, heat cramp and heat stroke. How frequently these disorders occur is not known, due to the inconsistency of diagnostic criteria and reporting routines. Dinman and Horvath (1984) estimate occupational heat-induced morbidity to range between 0·12 and 1·4 cases per 1 000 man-years. In a survey of 19 US aluminium plants during a 10-year period, they reported that no cases of heat stroke occurred during that period, involving 244 770 man-years at risk. In this connection it is of interest to note that according to Lind (1970), no cases of heat stroke occurred in European coal mines between 1955 and 1965 among 36 010 miners, while if the same number of South African gold miners had worked under the same environmental conditions, one fatal and four or five non-fatal cases of heat stroke would have been expected per year. For a further discussion of heat stroke, the reader is referred to Knochel *et al.* (1961) and Khogali and Hales (1983).

The different ill-effects which have been claimed to occur in individuals occupationally exposed to heat stress over prolonged periods are less well established (Dukes-Dobos, 1981). In 1973, Kloetzel *et al.* reported their finding of a relationship between hypertension and prolonged exposure to heat. One of the authors had noticed an unusually high prevalence of hypertension in workers exposed to high heat levels, particularly furnace workers, in a small steel plant in Brazil. Later on they pursued this observation in more detail in another steel plant in Brazil. In a preliminary survey they examined blood pressure readings obtained during the past years in a group of selected workers, on the basis of records from the files. Their figures showed relatively more cases of blood pressures over 140/90 in heat exposed workers than in clerical staff employees, apparently supporting the initial impression. The subsequent study, the results of which formed the basis for their report, involved a total of 330 individuals, of whom 90 were exposed to no heat. It should be noted that the median age of the 'no-heat' exposed individuals was 25·5, as against 36·8 years in the 'extreme-heat' exposed ones. It should also be noted that the blood pressure readings were made only once on each subject, at the end of 30 minutes rest, in connection with routine medical examinations on the work force. The salt intake was not recorded, but the authors claim that the salt dispensers were avoided by the workers on account of the rumours that salt tablets may cause impotence! They found a statistically significant higher incidence of blood pressure over 140/90, and over 160/95 in the heat exposed group compared with the group not exposed to heat. They also observed an increasing incidence of hypertension with the number of years the individuals had been exposed. The authors suggested further studies in an attempt to identify the causes of hypertension in individuals exposed to high levels of heat. As far as one can see, no such studies have as yet been published.

Lund-Larsen and Dahlberg (1981) were unable to confirm the observation of Kloetzel *et al.* (1973) in their study of 551 men at an aluminium plant in Norway (Årdal–Sundal Verk) in 1975–76. Nor did we in a study of the workers at the Fiskaa Verk ferroalloy plant in 1978 (Magnus *et al.*, 1980). There was no statistically significant difference between the blood pressure of 50 heat exposed workers and 50 workers who were not exposed to heat. This finding has been confirmed in a more extensive longitudinal study, still in progress, by a team headed by Erikssen *et al.* (1986, unpublished results).

In the study of Magnus *et al.* (1980) a significantly increased plasma renin activity was observed in heat exposed workers, but this was also the case in other workers at the same plant who were engaged in hard physical work in a cool environment. The elevated plasma renin activity observed in this study was therefore interpreted as being an expression of reduced kidney blood flow due to intense heat or due to strenuous physical work, both of which are known to cause reduced kidney blood flow (Åstrand and Rodahl, 1986). A drop in kidney blood flow is observed during heat exposure at rest, and more so if the heat exposed individual is subject to physical activity (Radigan and Robinson, 1949). At maximal work rates in the heat, a 70% drop in renal blood flow has been observed (Rowell, 1974). Kosunen *et al.* (1976) observed a twofold increase in renin concentration after 20 minutes in a Finnish sauna bath at a temperature of 85–90°C; the increase was already significant after 10 minutes. Finberg and Berlyne (1977) have shown that the increase in renin is less in heat acclimatized than in non-acclimatized individuals. An increase in renin concentration was observed in physically active subjects at normal room temperature, at work rates corresponding to 70% $\dot{V}O_2$ max (Kotchen *et al.* 1971). It thus appears that, although it is generally accepted that elevated renin production over long periods of time may lead to hypertension, the elevated renin levels observed in our heat exposed subjects are not associated with hypertension.

In the course of the mentioned study, headed by Erikssen (1986), a transient elevation of blood pressure occurred in some of the individuals included in the study during a period when a reduction in the labour force was being discussed due to a serious decline in the ferroalloy market. The feared reduction did not materialize, however, and the blood pressure readings returned to normal. A similar observation was made by Mundal *et al.* (1983), who recorded significantly elevated blood pressures in air traffic controllers involved in a major labour relation conflict in 1981 (136/90). This was interpreted as an autonomous nervous system response to stress, for when the conflict was resolved the blood pressure returned to normal levels (116/75). A similar observation had previously been made by Rognum and Erikssen (1980) among employees of a company on the verge of bankruptcy. When the crisis was over, the blood pressures were restored to the levels prevailing before the onset of the crisis. At the examination in 1977, the year prior to the crisis, the mean systolic blood pressure was 138. The following year, at the time of the crisis, it was 149. In 1979 it had returned to 138. The differences are statistically highly significant ($P < 0.001$).

It is thus evident that occupational conflicts cause sustained non-permanent elevation of blood pressure. The interesting question is how long such a transitory elevation of blood pressure has to persist before it becomes permanent hypertension, if that is the case.

Resmond *et al.* (1978) made an extensive study of the mortality of workers in different heat stress categories in 59 000 steelworkers. They observed that workers in jobs involving higher environmental heat showed less risk of death from cardiovascular disease, the risk decreasing with increasing length of exposure. Work load did not appear to be an independent contributory factor. Workers with less than 6 months exposure showed higher risks of death from cardiovascular diseases. These higher risks were considered to be indicative of a relationship between inability to work in jobs involving heat stress and health. Finally they observed an increased risk of mortality from digestive disease for workers employed in jobs in high heat categories. Pavlenko (1971), on the other hand, found that the incidence of circulatory diseases was significantly higher among workers in hot shops than among those working in cold shops in the metallurgical industry.

In a study of 1000 workers at 20 different foundries, Hernberg *et al.* (1976) found that the systolic and diastolic blood pressures of carbon monoxide-exposed workers were slightly higher than those of other workers, when age and smoking habits were taken into consideration. However, exposure to heat could not be separated from exposure to CO in this study.

Gusic *et al.* (1969) have reported clinical and histological changes in the mucosa of the nose in workers in an electric light bulb factory who were exposed to heat.

An association between heat exposure and cataract has been claimed since the early 18th century. Wallace *et al.* (1971) examined two groups of steel workers who differed widely in their heat exposure. A higher prevalence of the common form of cataract was found in the heat-exposed group than in the non-exposed group, and two cases of cataract were found that would be generally accepted as occupational in origin. David and Popescu (1972) detected a significant degree of accommodation insufficiency in heat exposed men working in the forge section of an iron and steel mill. Jebava *et al.* (1980) reported a high incidence of cataract in glassblowers, three quarters of whom neglected to wear protective glasses. A decreasing trend in the occurrence of glass-blowers' cataract was observed due to improved glass technology in Czechoslovakia.

Cutaneous implications of excessive heat in the workplace have been described by Olumide *et al.* (1983) and by Dibeneditto and Worobec (1985).

Heat exposure has been claimed to reduce fertility due to impairment of semen quality and sperm abnormalities. For references, see Baird (1985). According to Macleod and Hotchkiss (1941), a body temperature rise to more than 40°C for a short period of time (1–2 hours) induces a reduction in the sperm count which may persist for some 2–3 months. Elevated body temperatures for prolonged periods of time, i.e., several days, as in the case of typhoid fever or viral diseases, may in some cases cause a transient suppression of spermatogenesis (Macleod, 1967). While the adverse effect of testicular hyperthermia is well known, the question remains as to whether or not the changes described in the case of hyperthermia are in fact caused by the elevated temperature as such or by the infectious agents and their toxins underlying the fever of the patient.

## 6. *Practical application of existing knowledge*

The safeguarding of workers employed in hot industries requires an understanding of the physiological reactions to heat stress, and more so when industrial heat exposure is superimposed on a tropical climate. It is also essential to distinguish between the dry heat often prevailing in Nordic industrial environments and the humid heat encountered in many other parts of the world, especially in tropical or semitropical areas. To cope with heat stress may therefore to a considerable extent be a matter of education, knowing how to behave, how to space work and rest, and how to replace sweat loss by fluid intake. In this respect the US Department of Labor and the US Department of Health and Human Services' pamphlet *Hot Environments* (1980) is a good example. Fernandez (1980) has described a device called the Thermoguide, i.e., a card containing information concerning the hazards of excessive heat exposure, what to do, and what to expect. Zal (1984) has prepared a recommended programme in an easily understandable language for employees exposed to extremes of heat.

In laboratory experiments, Kamon (1979) studied different exercise cycles in six heat-acclimatized male subjects exercising in hot ambient conditions. Based on the levelling off of heart rate and rectal temperature, and setting the limit of rectal temperature to 38°C, a schedule of exercise and rest periods of 20 minutes was adequate for a hot-dry ambient condition, with a dry bulb temperature of 50°C, wet bulb 25°C, irrespective of whether the resting took place under the same ambient conditions as the exercising, or at a neutral ambient condition (dry bulb temperature 23°C, wet bulb temperature 16°C). The work rate was 40% $VO_2$ max. Essentially similar observations were made in real life in a Norwegian magnesium plant at a somewhat lower work load with a work schedule of 20 minutes work interspaced with a 10 minute cooling–off period at prevailing outside ambient temperatures (Figures 2.14 and 2.15).

Considerable efforts have been devoted to the development of cooled garments suitable for use in hot workplaces in industry. A number of studies have been made to evaluate the efficiency of such garments (Schvartz *et al.*, 1974; Mairiaux *et al.*, 1977; Hausman and Petit, 1979; Pasternack, 1978; Engel *et al.*, 1984). This includes the value of reflecting layers in clothing (Kerslake, 1969). Goldman (1974) has reviewed the problems involved in clothing design for comfort and work performance in extreme thermal environments. Thomas (1974) showed that simple heart rate measurements were useful in the evaluation of the efficiency of compressed air for heat exposed maintenance workers. Martin and Callaway (1974) have shown that the wearing of a protective face mask imposes a significant additional heat stress, manifested by increased sweat rate, increased rectal temperature, increased skin temperature and increased heart rate. Shvartz *et al.* (1974) studied the effect of cooling 10 different body regions by circulating water, and concluded that for an efficient design of a whole body cooling suit, the head and upper extremities should receive 15% of the total amount of water tubing used, the torso 38% and the lower extremities 32%.

It appears well established that cooling the skin of the head affects the thermoregulatory response in general, including a decreased sweating over the entire body. Evidently this is caused by a change in the hypothalamic temperature (Baker, 1982). Morales and Konz (1968)

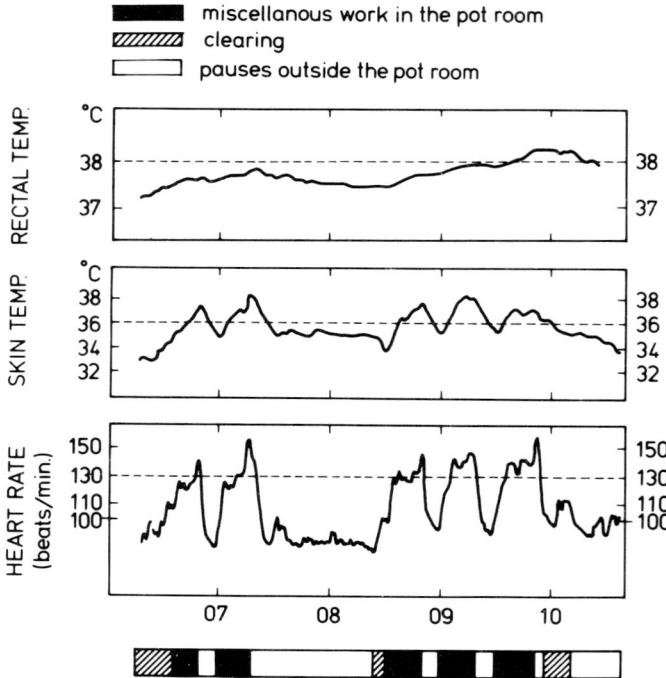

*Figure 2.14. The original work schedule of a worker removing slag in a magnesium plant.*

showed that in subjects exposed to heat, the use of a water-cooled hood caused the head temperature as well as core and skin temperatures to be lower. It caused reduced sweating and permitted longer exposure to the stress environment. Furthermore, Konz and Gupta (1969) showed that localized cooling of the head during heat exposure in the form of a cooled hood caused less decline in mental performance, the sweat rate was reduced, and the increase in heart rate was less.

The effects of cooling the head were essentially confirmed by Shvartz (1970, 1976), Nunneley *et al.* (1970) and Greenleaf *et al.* (1980). Brown and Williams (1982) showed in climatic chamber experiments that head cooling prevented an increase in both auditory canal and oesophageal temperatures during heat exposure. Riggs *et al.* (1981) observed that facial cooling in exercising subjects resulted in a significant lowering of the heart rate, whereas no difference was detected in blood pressure or rectal temperature.

Evidently, brain function appears to be especially vulnerable to heat (Baker, 1982). It is therefore of particular interest to note that a rather unique, selected cooling of the brain is possible due to the special

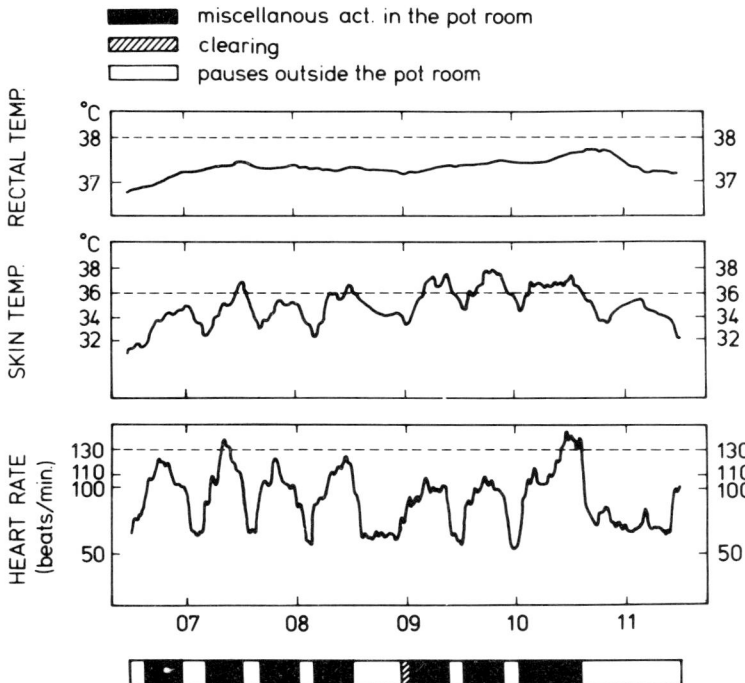

*Figure 2.15. The modified work schedule (20 minutes work, 10 minutes cooling off break) of a worker removing slag in a magnesium plant.*

vascular arrangement in the head (Cabanac, 1986). During strenuous physical work or exercise, the heat production increases in proportion to the energy output, causing an increase in body temperature, most marked in the muscle. However, the temperature inside the head, i.e., in the brain, does not rise as high as in the rest of the body (Figure 2.16). Evidently this is the result of a special arrangement in the vascular system in the head that results in a selective cooling of the brain. As pointed out by Cabanac (1986), innumerable anastomoses connect the rich subcutaneous venous plexus of the cephalic head with the intracranial sinuses. The most obvious ones are vena ophthalmica draining blood from the forehead and upper face, vena emissaria mastoidea draining blood from the lower temporal vein and external ear, and vena emissaria parietalis draining blood from the scalp on the back and crown of the head. The direction of the flow in these veins has been examined with the aid of a Doppler flow probe. During hyperthermia, when the skin is fully vasodilated and cooled by sweating, cooled blood flows inwards, from skin to brain. During hypothermia

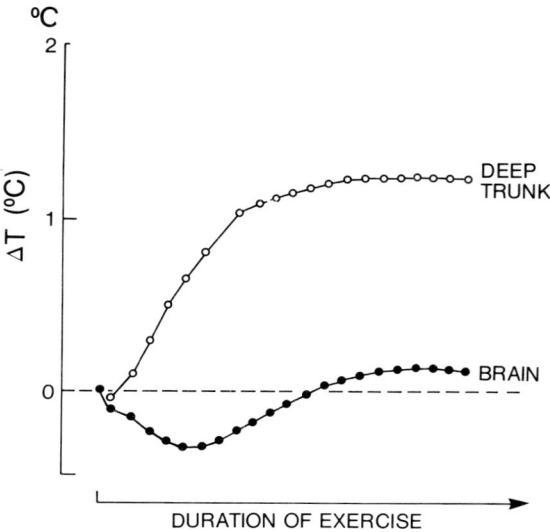

*Figure 2.16.   Time course of changes in deep trunk temperature (○) and deep intracranial temperatures (●) during exercise in human subjects (modified from Cabanac, 1986).*

the flow is in the opposite direction, or there may be no flow at all. When the hyperthermia is caused by prolonged exercise such as running, the face is exposed to increased convection, which augments cooling by the evaporation of the sweat. Bald individuals may have the additional advantage of heat loss due to sweating from the hairless scalp, since sweating from a bald scalp can be as great or greater than that of the forehead (for references see Cabanac, 1986).

The solution of the problems associated with industrial heat exposure may be based on a combination of practical measures such as: shielding the heat source; reducing each period of heat exposure to about 20 minutes, interspersed with brief 10 minute cooling-off breaks; eliminating strenuous physical work close to the heat source; introducing mechanical aids to take the place of manual labour wherever possible; and maintaining the industrial process properly at all times so as to avoid complications that will necessitate drastic measures causing excessive exposure to heat stress. A key point is to avoid prolonged intense heat exposure, which will lead to profuse sweating. This avoidance of sweating will reduce the fluid loss, and thereby correspondingly reduce the need for fluid intake. At any rate, the fluid loss should be replaced as it is being lost, and the aim should be for the worker to leave his place of work fully hydrated, able to enjoy his leisure time.

Existing plant health personnel should be supplemented with qualified work physiologists with the necessary medical background to make specific on-site measurements and observations in the industry. The technical managers and administrative directors should, in the course of their training, be given sufficient knowledge of the general principles of occupational health and physiology to make them want to listen to the work physiologists, to take their recommendations seriously, and to use their results to improve the working conditions for better productivity, well-being and motivation, in the interest of both labour and management.

# References

Abeysekera, J.D.A., 1981, Thermal environment and subjective discomfort of glass-factory workers in Sri Lanka. *J. Human Ergol.*, **10**, 185.

ACGIH (American Conference of Governmental Industrial Hygienists), 1976, *Threshold Limit Values for Chemical Substances and Physical Agents in the Workroom Environment with Intended Changes for 1976*, Cincinnati, OH.

Åstrand, I., Axelson,O., Eriksson, U. and Olander, L., 1975, Heat stress in occupational work. *Ambio*, **4**, 37–42.

Åstrand, P.O. and Rodahl, K., 1986, *Textbook of Work Physiology*, 3rd ed. (New York: McGraw-Hill).

Babaev, A. and Khodzhibaev, Yu., 1972, Peculiar working conditions of drivers of heavy dump trucks under conditions of hot climate. *Gig. Sanit.*, **37**(3), 25.

Baird, D.D., 1985, Occupational exposure to heat or noise and reduced fertility. *J. Amer. Med. Assn*, **253**(18), 1643.

Bakaev, A. and Bobokhodzhaev, Sh.A., 1977, Occupational hygiene of tower crane operators in areas with torrid climate. *Gig. Tr. Prof. Zabol.*, **2**, 8.

Baker, M.A., 1982, Brain cooling in endotherms in heat and exercise. *Ann. Rev. Physiol.*, **44**, 85.

Bass, E.E., 1963, Thermoregulatory and circulatory adjustments during acclimatization to heat in man. In *Temperature: Its Measurement and Control in Science and Industry*, edited by J.D. Hardy (New York: Reinhold), vol. 3, part 3, p. 299.

Beshir, M.Y., Ramsey, J.D. and Burford, C.L., 1982, Threshold values for the Botsball: A field study of occupational heat. *Ergonomics*, **25**, 247.

Bosco, J.S., Terjung, R.L. and Greenleaf, J.E., 1968, Effects of progressive hypohydration on maximal isometric muscular strength. *J. Sports Med.*, **8**, 81.

Botsford, J.H., 1971, A wet globe thermometer for environmental heat measurement. *Amer. Ind. Hygiene Assn J.*, **32**, 1.

Boulos, B.M. and Boulos, N., 1980, Physiological and biochemical changes among glass workers exposed to different grades of heat loads. *Fed. Proc.*, **39**, 308.

Brief, R.S. and Confer, R.G., 1971, Comparison of heat stress indices. *Amer. Ind. Hygiene Assn J.*, **32**, 11.

Brown, G.A. and Williams, G.M., 1982, The effect of head cooling on deep body temperature and thermal comfort in man. *Aviat. Space Environ. Med.*, June, 583.

Buskirk, E.R., Iampietro, P.F. and Bass, D.E., 1958, Work performance after dehydration: effects of physical conditioning and heat acclimatization. *J. Appl. Physiol.*, **12**, 189.

Cabanac, M., 1986, Keeping a cool head. *News in Physiol. Sçi.*, **1**, 41.

Christensen, H., 1952, Fysiologisk Arbetsvardering inom Jarnverksindustrien. *Uddeholmaren*, 1.

Ciriello, V.M., and Snook, S.H., 1977, The prediction of WBGT from the Botsball. *Amer. Ind. Hygiene Assn J.*, **38**, 264.

Colquhoun, W.P., 1970, Circadian rhythms, mental efficiency and shift work. *Ergonomics*, **13**, 558.

Costill, D.L. and Fink, W.J., 1974, Plasma volume changes following exercise and thermal dehydration. *J. Appl. Physiol.*, **37**, 521.

David, M. and Popescu, M.P., 1972, Changes in ocular diopters under conditions of work at high temperatures. *Fiziologia Normala si Patologica, Romanio*, **18**(5), 439.

Davies, C.T.M., 1981a, Thermal responses to exercise in children. *Ergonomics*, **24**, 55.

Davies, C.T.M., 1981b, Thermoregulation during exercise in relation to sex and age. *Eur. J. Appl. Physiol.*, **42**, 71.

Detsik, Yu.I., Katsis, M.V., Monastyrskaya, M.T. and Birka, I.I., 1976, Functional state of the sympathicadrenal system in workers of hot shops of an automobile-building enterprise. *Vrach. Delo*, **5**, 116.

Dibeneditto, J.P. and Worobec, S.M., 1985, Exposure to hot environments can cause dermatological problems. *Occup. Health and Safety*, **54**, 35.

Dinman, B.D. and Horvath, S.M., 1984, Heat disorders in industry. *J. Occup. Med.*, **26**, 489.

Dinman, B.D., Stephenson, R.R., Horvath, S.M. and Colwell, M.O., 1974, Work in hot environments. I. Field studies of work load, thermal stress, and physiologic response. *J. Occup. Med.*, **16**, 785.

Dolph, E.F. and Members of the Rochester Desert Unit, 1947, *Physiology of Man in the Desert* (New York: Interscience).

Dorofeyeva, V.I., 1973, Rational correction of disorders of the water-salt balance in workers of hot shops of a metallurgic plant. *Vrach. Delo*, **10**, 141.

Drinkwater, B.L., Denton, J.E., Kupprat, I.C., Talag, T.S. and Horvath, S.M., 1976, Aerobic power as a factor in women's response to work in hot environment. *J. Appl. Physiol.*, **41**, 815

Drinkwater, B.L., Bedi, J.F., Loucks, A.B., Roche, S. and Horvath, S.M., 1982, Sweating sensitivity and capacity of women in relation to age. *J. Appl. Physiol.*, **53**, 671.

Dukes-Dobos, F.N., 1981, Hazards of heat exposure. *Scand. J. Work Environ. Health*, **7**, 73.

Egeland, T., Lossius, P., Enger, N., Gundersen, N., Bolling, A. and Jebens, E., 1981, *Varmebelastningsundersokelse ved Fiskaa Verk, Kristiansand* (Oslo: Arbeidsfysiologisk Institutt).

Engel, P., Hense, W., Armonies, G. and Munscher, M., 1984, Psychological and physiological performance during long lasting work in heat with and without wearing cooling vests. In *Thermal Physiology*, edited by J.R.S. Hales (New York: Raven Press).

Eriksson, U. and Olander, L., 1974, Forslag till beatamning av Svenskt Varmeindex in Varmebelasting i Yrkesarbete, ed. I. Astrand, *Arbete ich Halsa*, **4**.

Fernandex, R.H.P., 1980, Health care of people at work. *J. Soc. Occup. Med.*, **30**, 40.

Finberg, J.P.M. and Berlyne, G.M., 1977, Modification of renin and aldosterone response to heat by acclimatization in man. *J. Appl. Physiol. Respirat. Environ. Exercise Physiol.*, **42**, 554–558.

Folprechtova, A., Jirak, Z. and Lutonska, V., 1973, The working heat load in forgemen at hammer forging and drop forging. *Pracov. Ledk.*, **25**, 386.

Fuller, F.H. and Smith, P.E., 1981, Evaluation of heat stress in a hot workshop by physiological measurements. *Amer. Ind. Hygiene Assn J.*, **42**, 32.

Gaebein, H., 1983, Climatic loads with selected underground workings in the potash mining industry of the GDR. *Zeitschr. gesamte Hygiene u. Grenzgebiete*, **29**, 156.

Gagge, A.P. and Nishi, Y., 1976, Physical indices of the thermal environment. *ASHRAE J.*, 47.

Gagge, A.P., Gonzalez, R.R. and Nishi, Y., 1974, Physical factors governing man's thermal comfort. Discomfort and heat tolerance. *Build. Internat.*, **1**, 305–331.

Gisolfi, C.V. and Copping, J.R., 1974, Thermal effects of prolonged treadmill exercise in the heat. *Med. Sci. Sports*, **6**, 108.

Goldman, R.F., 1974, Clothing design for comfort and work performance in extreme thermal environments. *New York Acad. Sci. Trans.*, **36**, 531.

Greenleaf, J.E. and Castle, B.I., 1971, Exercise temperature regulation in man during hypohydration and hyperhydration. *J. Appl. Physiol.*, **30**, 847.

Greenleaf, J.E., Van Beaumount, W., Brock, P.J., Montgomery, L.D., Morse, J.T., Shvartz, E. and Kravik, S., 1980, Fluid–electrolyte shifts and thermoregulation: rest and work in heat with head cooling. *Aviat. Space Environ. Med.*, August, 747.

Gresenz, P., Kobryn, V. and Weber, W., 1973, Arbeitsphysiologische Langzeitmessungen an Berufsgruppen der Eisen-und Stahlindustrie. *Zeitschr. gesamte Hygiene u. Grenzgebiete*, **19**, 511.

Gusic, B., Krajina, Z., Poljak, Z., Konic-Carnelutti, V. and Babic, I., 1969, The influence of heat on the upper respiratory tracts. *Acta Otolaryngologica*, **67**, 150.

Hardy, J.D., 1967, Central and peripheral factors in physiological temperature regulation. In *Les Concepts de Claude Bernard sur le Milieu Interieur* (Paris: Masson), p. 247.

Harrison, M.H., Edwards, R.J. and Leitch, D.R., 1975, Effects of exercise and thermal stress on plasma volume. _J. Appl. Physiol._, **39**, 925.

Hausman, A. and Petit, J.M., 1979. Physical work in a hot environment in coal mines. _Ann. Mines Belg._, **11**, 1109.

Hernberg, S., Karava, R., Koskela, R. and Luoma, K., 1976, Angina pectoris, ECG findings and blood pressure of foundry workers in relation to carbon monoxide exposure. _Scand. J. Work Environ. Health_, **2**, suppl. 1, 54–63.

Horstman, D.H. and Christensen, E., 1982, Acclimatization to dry heat: active men versus active women. _J. Appl. Physiol._, **52**, 825.

Hunting, W., Nemecek, J. and Grandjean, E., 1974, Die physische Belastung von Arbeitern an der Gesenkschmiede—eine Fallstudie. _Sozial- und Präventivmed._, **19**, 275.

Jansen, T., Olsen, B., Hatlevoild, M., Enger, N., Guthe, T., Johannessen, H. and Rodahl, K., 1982, _En Undersokelse av Varmeeksponeringen ved Lista Aluminiumverk_ (Oslo: Arbeidsfysiologisk Institutt).

Jebava, R., Hrochova, J. and Rencova, E., 1980, Glassblower's cataract. _Ceskoslovenska Oftalmologie_, **36**, 115.

Jebens, E. and Bolling, A., 1980, En arbeidsfysiologisk undersokelse av varmebelastningen om bord pa M/S Tarago i fart pa den Persiske Gulf, sept. 1980. _Arbeidsstress til Sjos_, edited by K. Rodahl. Rapport 80/2 System for Sikkert Skip, NTNF.

Kamon, E., 1979, Scheduling cycles of work for hot ambient conditions. _Ergonomics_, **22**, 427.

Kamon, E. and Ryan, C., 1981, Effective heat strain index using pocket computer. _Amer. Ind. Hygiene Assn. J._, **42**, 611.

Kerslake, D. McK., 1969, The value of reflecting layers in clothing. _Proc. R. Soc. Med._, **62**, 283.

Kerslake, D. McK., 1972, _The Stress of Hot Environments. Monograph Physiol. Soc._, **29** (Cambridge: Cambridge University Press).

Khogali, M. and Hales, J.R.S. (eds), 1983, _Heat Stroke and Temperature Regulation_ (New York: Academic Press).

Klimmer, F., Kylian, H., Ilmarinen, J., Ilmarinen, R., Meyer, R., Piekarski, C. and Rutenfranz, J., 1984, Belastung und Beanspruchung bei Tätigkeiten in der Eisen- und Stahlindustrie. _Arbeitsmed. Sozialmed. Präventivmed._, **3**, 49.

Kloetzel, K., Etelvino de Andrade, A., Falleiros, J. and Cota Pacheco, J., 1973, Relationship between hypertension and prolonged exposure to heat. _J. Occup. Med._, **15**, 878–880.

Knochel, J.P., Beisel, W.R., Herndon, E.G., Gerard, E.S. and Barry, K.G., 1961, The renal, cardiovascular, hematologic and serum electrolyte abnormalities of heat stroke. _Amer. J. Med._, 299.

Konz, S. and Gupta, V.K., 1969, Water cooled hood affects creative productivity. _ASHRAE J._, July, 40.

Kosunen, K.J., Pakarinen, A.J., Kuoppasalme, K. and Adlerereutz, H., 1976, Plasma renin activity, angiotensin II and aldosterone during intense heat stress. _J. Appl. Physiol._, **41**, 323–327.

Kotchen, T.A., Hartley, L.H., Rice, T.W., Hougey, E.H., Jones, L.G. and Mason, J.W., 1971, Renin, norepinephrine, and epinephrine responses to graded exercise. *J. Appl. Physiol.*, **31**, 178–184.

Krosnoschekov, N.N. and Chervyakov, G.M., 1973, Industrial microclimate and its effect on the physiological functions and morbidity rate of workers at paper mills. *Gig. Sanit.*, **38**, 44.

Kuno, Y., 1956, *Human Perspiration* (Springfield, IL: Charles C. Thomas).

Ladell, W.S.S., 1955, The effects of water and salt intake upon the performance of men working in hot and humid environments. *J. Physiol.*, **127**, 11.

Leithead, C.A. and Lind, A.R., 1964, *Heat Stress and Heat Disorders* (London: Cassell).

Libert, J.P., Candas, V. and Vogt, J.J., 1983, Modifications of sweating responses to thermal transients following heat acclimation. *Eur. J. Appl. Physiol.*, **50**, 235.

Lind, A.R., 1970, The lack of heatstroke in European miners working in hot climates. *Amer. Ind. Hygiene Assn J.*, **31**, 460.

Ljungberg, A.S., Enander, A. and Holmer, I., 1979, Evaluation of heat stress during sedentary work. *Scand. J. Work Environ. Health*, **5**, 23.

Locati, G., Vicini, A. and Secondi, E., 1974, Correlation between various microclimatic indices. Results obtained from the study of 200 work places. *Med. Lavoro*, **65**, 451.

Lund-Larsen, P.G. and Dahlberg, B.E., 1981, Blodtrykk hos ansatte i spesielt varme og i vanlige avdelinger ved aluminiumsmelteverket i Ardal. *Tidsskrift Norske Laegeforening*, **33**, 1892.

Macleod, J., 1967, Male infertility. In *Advances in Gynecology*, vol. 1 (Baltimore, MD: Williams and Wilkins).

Macleod, J. and Hotchkiss, R.S., 1941, The effect of hyperpyrexia upon spermatozoa counts in men. *Endocrinology*, 28.

Maehlum, S., Bolling, A., Huser, P.O., Jebens, E. and Tenfjord, O., 1980, En arbeiidsfysiol. undersokelse av varmebelastn. om bord i M/S Tarn i fart pa den Persiske Gulf, Sept. 1978. In *Arbeidsstress til Sjos*, edited by K. Rodahl. Rapp. 80/2 System for Sikkert Skip, NTNF.

Magnus, P., Bolling, A., Eide, I., Enger, N., Gundersen, N., Guthe, T., Jebens, E., Knutsen, K.E. and Rodahl, K., 1980, *Varmestress i Ferrolegeringsindustrien. En Orienterende Undersokelse ved Fiskaa Verk, Kristiansand* (Oslo: Arbeidsfysiologisk Institutt).

Mairiaux, P., Nullens, W., Fesler, R., Brasseur, L. and Detry, J.M., 1977, Evaluation des effects d'un vetement refroidissant sur l'adaptation aux efforts prolonges realises a haute temperature par des ouvriers mineurs. *Rev. Inst. Hygiene Mines*, **32**, 99.

Mairiaux, P.H., Sagot, J.C. and Candas, V., 1983, Oral temperature as an index of core temperature during heat transients. *Eur. J. Appl. Physiol.*, **50**, 331–341.

Malchaire, J.B., 1976, Evaluation of natural wet bulb and wet globe temperatures. *Amer. Occup. Hyg.*, **19**, 251.

Martin, H. de V. and Callaway, S., 1974, An evaluation of the heat stress of protective face mask. *Ergonomics*, **17**, 221.

Martinson, M.J., 1977, Heat stress in Witwatersrand gold mines. *J. Occup. Accid.*, **1**, 171.

Minard, D. and Copman, L., 1963, Elevation of body temperature in disease. In *Temperature: Its Measurement and Control in Science and Industry*, edited by J.D. Hardy (New York: Reinhold), vol. 3, part 3, p. 253.

Minard, D. and Goldsmith, R., 1971, Physiological evaluation of industrial heat stress. *Amer. Ind. Hygiene Assn. J.*, **32**, 17.

Mitchell, D., 1973, Prediction of heat stress from heat transfer. *Arch. Sci. Physiol.*, **27**, 285.

Morales, I.V. and Konz, S., 1968, The physiological effect of a water cooled hood in a heat stress environment. *ASHRAE Trans.*, **74**, 49.

Mundal, R., Ramstad, H., Erikssen, J. and Rodahl, K., 1983, Health and stress of air traffic controllers. *Proceedings of the 31st International Congress of Aviation and Space Medicine*, p. 202.

Mutchler, J.E. and Vecchio, J.L., 1977, Empirical relationships among heat stress indices in 14 hot industries. *Amer. Ind. Hygiene Assn J.*, **38**, 253.

Mutchler, J.E., Malzahn, D.D., Vecchio, J.L. and Soule, R.D., 1976, An improved method for monitoring heat stress levels in the work place. *Amer. Ind. Hygiene Assn J.*, **37**, 151.

Nadel, E.R., Pandolf, K.B., Roberts, M.F. and Stolwijk, J.A.J., 1974, Mechanisms of thermal acclimation to exercise and heat. *J. Appl. Physiol.*, **37**, 515.

Nielsen, B., 1969, Thermoregulation in reset and exercise. *Acta Physiol. Scand.*, Suppl., 323.

Nunneley, S.A. and Flick, C.F., 1981, Heat stress in the A-10 cockpit: flights over desert. *Aviat. Space Environ. Med.*, **52**, 513.

Nunneley, S.A. and Stribley, F., 1979, Heat and acute dehydration effects on acceleration response in man. *J. Appl. Physiol.*, **47**, 197.

Nunneley, S.A., Webb, P. and Troutman, S.J., 1970, Head cooling during work and heat stress. *Ergonomics*, **13**, 527.

Olumide, Y.M., Oleru, G.U. and Enu, C.C., 1983, Cutaneous implications of excessive heat in the work-place. *Contact Dermatitis*, 9360.

Pafnote, M. and Vaida, I., 1981, Thermal stress in industries processing metal by heat. *Revista Igiena Bacteriol. Virusol. Parazitol. Epidemiol. Pneumofiziol.*, **30**, 43.

Paltsev, Yu.P., 1982, Occupational hygiene of present-day heat electric power stations. *Gig. Tr. Prof. Zabol.*, **8**, 5.

Parikh, D.J., Pandya, C.B. and Ramanathan, N.L., 1976, Applicability of the WBGT index of heat stress to work situations in India. *Indian J. Med. Res.*, **64**, 327.

Parikh, D.J., Pandya, C.B. and Ramanathan, N.L., 1977, Thermal strain in foundry operations in the small scale sector. *Indian J. Med. Res.*, **65**, 900.

Pasternack, A., 1978, Die Kuhlweste—ein weiterer Schritt in der Humanisierung des Arbeitslebens. *Dragerheft*, **310**, 17.

Pavlenko, M.E., 1971, Cardiovascular morbidity among workers in hot shops. *Vrach. Delo*, **6**, 1.

Pepler, R.D., 1963, Performance and well-being in heat. In *Temperature: Its*

*Measurements and Control in Science and Industry*, edited by J.D. Hardy (New York: Reinhold), vol. 3, part 3, p. 319.

Pitts, G.C., Johnson, R.E. and Consolazio, F.C., 1944, Work in the heat as affected by intake of water, salt and glucose. *Amer. J. Physiol.*, **142**, 253.

Plunkett, J.M. and Carter, R.P., 1974, Practical problems in the use of WBGT for heat stress evaluation. *Amer. Ind. Hygiene Assn J.*, **35**, 287.

Pulket, C., Henschel, A., Burg, W.R. and Saltzman, B.E., 1980, A comparison of heat stress indices in a hot-humid environment. *Amer. Ind. Hygiene Assn J.*, **41**, 442.

Radigan, L.R. and Robinson, S., 1949, Effects of environmental heat stress and exercise on renal blood flow and filtration rate. *J. Physiol.*, **4**, 185.

Ramanathan, L.N., 1964, A new weighing system for mean surface temperature of the human body. *J. Appl. Physiol.*, **19**, 531.

Famsey, J.D. and Chai, C.P., 1983, Inherent variability in heat stress decision rules. *Ergonomics*, **26**, 495.

Ramsey, J.D., Dayal, D. and Chahramani, B., 1975, Heat stress limits for the sedentary worker. *Amer. Ind. Hygiene Assn J.*, **36**, 259.

Rapp, R. and Aubertin, G., 1985, Thermal load prediction—calculation and measurement of mean radiation temperature in workshops. *Cahiers de Notes Documentaires—Securite et Hygiene du Travail*, **121**, 429.

Reischi, U., Marschall, D.M. and Reischi, P., 1977, Radiotelemetry based study of occupational heat stress in a steel factory. *Biotelemetry*, **4**, 115.

Resetjuk, A.L., Vanin, L.G., Vasilkov, V.N., Tarasenko, V.T., Oniscenko, L.P., Repnevsky, S.M., Vedmedenko, N.I., Glikin, V.M. and Noskov, V.I., 1976, Efficiency of physiologically based work-rest schedules for miners at hard headings and coal faces in the Donety mines. *Gig. Tr. Prof. Zabol.*, **12**, 6.

Resmond, C.K., Emes, J.J., Mazumdar, S. and Magee, E., 1976, Mortality of steelworkers employed in hot jobs. *J. Environ. Pathol. Toxicol.*, **2**, 75–96.

Riggs, C.E., Johnson, D.J., Konopka, B.J. and Kilgour, R.D., 1981, Exercise heart rate response to facial cooling. *Eur. J. Appl. Physiol.*, **47**, 323.

Robinson, S., 1963, Circulatory adjustments of men in hot environments. In *Temperature: Its Measurement and Control in Science and Industry*, edited by J.D. Hardy (New York: Reinhold), vol. 3, part 3, p. 287.

Robinson, S., Turrell, E.S., Belding, H.S. and Horvath, S.M., 1943, Rapid acclimatization to work in hot climates. *Amer. J. Physiol.*, **140**, 168.

Rodahl, K., 1981, Heat stress: Norwegian experience in health protection in primary aluminium production. In *IPAI*, vol. 2, edited by J.P. Hughes (London: International Primary Aluminium Institute).

Rognum, T.O. and Erikssen, J., 1980, Blood pressure elevation caused by stress at work. *Tidskrift Norske Laegeforening*, **3**, 168.

Rowell, L.B., 1974, Human cardiovascular adjustment to exercise and thermal stress. *Physiol. Rev.*, **54**, 75.

Saltin, B., 1964, Aerobic work capacity and circulation at exercise in man. *Acta Physiol. Scand.*, **62**, Suppl., 230.

Senay, L.C., Jr, 1972, Changes in plasma volume and protein content during exposures of working men to various temperatures, before and after acclimatization to heat. Separation of the roles of cutaneous and skeletal muscle circulation. *J. Physiol.*, **224**, 61.

Shapiro, Y., Pandolf, K.B., Avellini, B.A., Pimental, N.A. and Goldman, R.F., 1981, Heat balance and transfer in men and women exercising in hot-dry and hot-wet conditions. *Ergonomics*, **24**, 375.

Shishlyannikova, G.J., Dementyevo, L.N. and Popova, N.A., 1960, Body functions and some specific functions in women engaged in finishing work at construction sites in hot climatic zones. *Gig. Tr. Prof. Zabol.*, **4**, 22.

Shvartz, E., 1970, Effect of a cooling hood on physiological responses to work in a hot environment. *J. Appl. Physiol.*, **29**, 36.

Shvartz, E., 1976, Effect of neck versus chest cooling on responses to work in heat. *J. Appl. Physiol.*, **40**, 668.

Shvartz, E., Aldjem, M., Ben-Mordechai, J. and Shapiro, Y., 1974, Objective approach to a design of a whole-body, water-cooled suit. *Aerospace Med.*, **45**, 711.

Skoldstrom, B., Elnas, S. and Holmer, I., 1980, Varmebelastning vid glas-blasning. *Arbetarskyddstyrelsen Undersikningsrapport*, **15** (Solna, Sweden).

Stebel, E., 1981, Der Einfluss des Grubenklimas auf den Menschen. *Gluckauf*, **117**, 140.

Stolbnun, B.M., Kleiner, A.M., Berseneva, A.P. and Gireva, E.V., 1970, Cardiovazcular function in heat shop operators of a metallurgical work. *Gig. Tr. Prof. Zabol.*, **14**, 12.

Strydom, N.B., 1980, Heat intolerance: its detection and elimination in the mining industry. *S. African J. Sci.*, **76**, 154.

Taylor, N.B.G., Kuehn, L.A. and Howat, M.R., 1969, A direct-reading mercury thermometer for the wet bulb globe temperature index. *Can. J. Physiol. Pharmacol.*, **47**, 277.

Thomas, N.T., 1974, Physiological and psychological effects of industrial protective clothing. *Ergonomics*, **17**, 565.

Tymanski, S., Zaborski, L., Matersyenski, J. and Maryn, J., 1975, Engine room staff and exposure to thermal radiation on ships. *Bull. Inst. Mar. Trop. Med.*, **32**, 87.

Vanin, L.G., 1972, The state of body functions and efficiency of heading-drivers in deep collieries. *Gig. Tr. Prof. Zabol.*, **16**, 12.

Vogt, J.J., Foehr, R., Kurtzinger, E., Seywert, L., Libert, J.P., Candas, V. and van Peteghem, T., 1977, Improvement of the working conditions at blast-furnaces. *Ergonomics*, **20**, 167.

Wallace, J., Sweetnam, P.M., Warner, C.G., Graham, P.A. and Cochrane, A.L., 1971, An epidemiological study of lens opacities among steel workers. *Brit. J. Ind. Med.*, **28**, 265.

Williams, C.G., Wyndham, C.H. and Morrison, J.F., 1967, Rate of loss of acclimatization in summer and winter. *J. Appl. Physiol.*, **22**, 21.

Wyndham, C., 1973, The physiology of exercise under heat stress. *Ann. Rev. Physiol.*, **35**, 193.

Wyndham, C.H., 1974, Research in the human sciences in the gold mining industry. *Amer. Ind. Hygiene Assn J.*, **35**, 113.

Wyndham, C.H., Strydom, N.B., Benade, A.J.S. and Rensburg, A.J., 1973, Limiting rates of work for acclimatization at high wet bulb temperatures. *J. Appl. Physiol.*, **35**, 454.

Wyon, D.P., Andersen, I. and Lundqvist, G.R., 1979, The effects of moderate heat stress on mental performance. *Scand. J. Work Environ. Health*, **5**, 352.

Yaglou, C.P. and Minard, D., 1957, Control of heat casualties at military training centres. *Arch. Ind. Health*, **16**, 302.

Zal, H.A., 1984, Recommended program for employees exposed to extremes of heat. *Occup. Health Nursing*, **32**, 293.

Zoller, H., May, B., Weiss, M. and Gross, W., 1981, Wie stark belastet Hitze—Schwerarbeit den Stoffwechsel? *Med. Klin.*, **76**, 378.

# 3. Thermal Comfort: A Review of Some Recent Research

**Larry G. Berglund**

## 1. Introduction

In the early 1920s Houghton and Yaglou introduced the Effective Temperature (ET) scale. This temperature index became very useful to designers and environmental engineers because it related comfort to both temperature and humidity. For their work Houghton and Yaglou (1923a,b) used two climatic chambers, one providing a fixed reference environment while the temperature and humidity of the second was altered. The test subjects passed back and forth between the two chambers and indicated if the second was warmer, cooler or equally comfortable compared to the first. From the results Houghton and Yaglou mapped out lines of equal comfort on a psychrometric chart. The temperature where these loci of equal comfort intersected the saturation or 100% RH line was designated the line's Effective Temperature.

Later Winslow *et al.* (1937, 1938a,b, 1939), working at the Pierce Laboratory, related subjective and physiological indicators of thermal comfort and pleasantness to environmental conditions. They found that, although body temperatures were important for thermal sensation and comfort in cool environments, skin moisture was a major contributor to the unpleasantness of warm conditions where perspiration occurs.

In the late 1960s Fanger (1968) developed a technique that has become the most widely used index for predicting thermal sensation. It is a technique based on the heat balance equation and his extensive experiments at Kansas State University. Fanger stated that the average thermal sensation or Predicted Mean Vote (PMV) is a function of the difference between metabolism (M) and energy losses (L) with losses calculated as if the person is comfortable:

$$\text{PMV} = \alpha\,(M/A - L/A)$$

where $\alpha = 0{\cdot}0303\,[e^{0{\cdot}36M/A} + 0{\cdot}0276]$; $A$ = the surface area of the body (m$^2$); and $L = \text{Work} + \dot{Q}_{\text{res}} + \dot{Q}_{\text{c}} + \dot{Q}_{\text{r}} + E_{\text{diff}} + E_{\text{rsw}}$ (Watts). The terms of the energy loss expression are: Work, defining the rate at which energy leaves the body to do mechanical work; $\dot{Q}_{\text{res}}$, the rate of respiratory heat loss; $\dot{Q}_{\text{c}}$ and $\dot{Q}_{\text{r}}$, the rates of convective and radiative heat loss, respectively; $E_{\text{diff}}$, the rate of heat loss from the diffusion of water vapour through the skin; and $E_{\text{rsw}}$, the rate of heat loss from sweat evaporation.

The losses are calculated from the skin temperature and sweat rate of a comfortable person at the concomitant metabolic rate. Evaluating the terms algebraically and rearranging, PMV becomes a function of metabolism, clothing insulation (clo), air speed, operative temperature and humidity. Charts and graphs facilitate PMV determinations.

## 2. *Warm discomfort prediction with PMV*\*

Gagge *et al.* (1986) observed that PMV is not very sensitive to humidity in warm environments and introduced the PMV\* index to characterize and predict warm discomfort better. The improvement is accomplished by substituting the ASHRAE Effective Temperature (ET\*) in place of the environment's operative temperature. ET\* is the temperature of an isothermal environment at 50% RH that will result in the same heat loss, skin temperature and skin wettedness as the actual environment. At neutral and cool temperatures, the slopes of ET\* lines on a psychrometric chart are similar to air temperature but become more affected by humidity (humidity ratio) as the temperature increases.

Figure 3.1 shows PMV and PMV\* plotted against air temperature for 20, 50 and 80% RH. The dotted line just above and below the solid PMV and PMV\* lines at 50% RH corresponds to PMV at 80 and 20% respectively. The PMV\* lines closely parallel the dotted discomfort lines (Disc). Warm discomfort depends on more than thermal sensation and the Disc relationship evolved from Winslow *et al.* (1939) who observed that in warm climates a sense of 'pleasantness' and 'comfort' was associated with low values of skin wettedness. Skin wettedness is defined as the ratio $\dot{E}_{\text{sk}}/\dot{E}_{\text{max}}$, where $\dot{E}_{\text{sk}}$ is the actual evaporation rate from the skin and $\dot{E}_{\text{max}}$ is the evaporation rate when the skin is

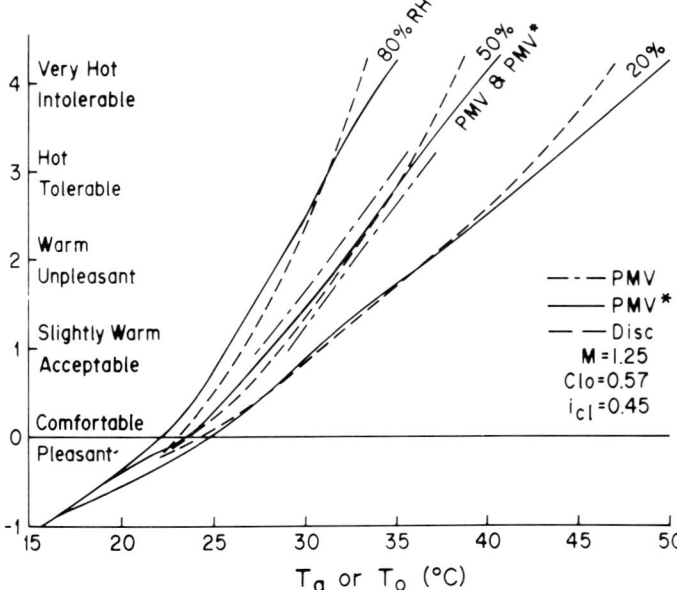

*Figure 3.1.*   *Predicted PMV, PMV\* and Disc subjective response values for slightly active (1·25 Met) persons in thin trousers and long sleeved shirt.*

completely covered with water. Warm discomfort (Disc) is linearly proportional to skin wettedness ($w$), or the fraction of the skin wet with perspiration. Indeed, Gagge shows that the PMV* expression for warm environments can also be expressed as a linear function of skin wettedness:

$$PMV^* = \alpha\beta(w - w_o)$$

where $w_o$ is the skin wettedness level for comfort at the given metabolic rate and $\beta$ is $E_{max}$ for the conditions. In neutral and cool environments where sweating does not occur PMV and PMV* values are equal.

## 3. Thermal non-uniformity

### Asymmetric radiation

Non-uniformities in the thermal environment can cause non-uniform body heat losses and skin temperatures resulting in local physiological strain on the individuals, thus decreasing the acceptability of the

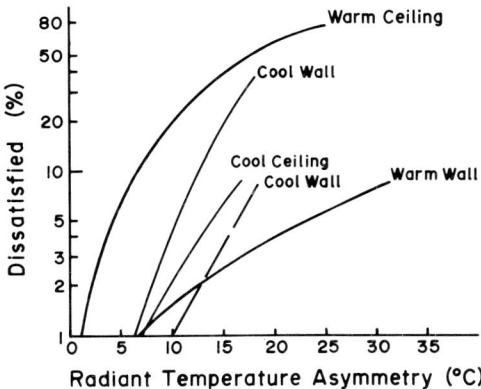

*Figure 3.2. The relationship between occupant dissatisfaction and the radiant asymmetry of various situations.*

environment to them. A common source of non-uniformity is asymmetric radiation from hot and cold surfaces. The asymmetry of a given situation is typically quantified through the term Radiant Temperature Asymmetry (RTA) which is defined as the difference in the mean radiant temperatures experienced on the opposite sides of a person or a small flat plate. Researchers at the Technical University of Denmark have been active in recent years defining the relationship between RTA and discomfort resulting from various configurations of radiant heating and cooling. Figure 3.2 shows some of their results together with one dashed curve labelled 'cool wall' (developed from data taken at the Pierce Foundation Laboratory (Berglund and Fobelets 1985, 1987). Dissatisfaction increases with asymmetry. For some reason there is a large difference between response to warm wall and warm ceiling RTA. The slope of the log % dissatisfaction response to RTA for cool walls and ceilings is steeper than for the warm walls. That is, test subjects appear more sensitive to cold radiation than to warm. Asymmetric cold radiation probably initiates local vasoconstriction which further reduces skin temperature and increases discomfort, while asymmetric warm radiation probably results in local vasodilation which would tend to stabilize skin temperature.

## Draught

Air movement is another source of thermal non-uniformity caused by variations in convective heat loss from the body surface, especially from exposed skin, that may lead to the unacceptable perception of a draught even in an overall neutral environment. Recent studies at the

*Environmental Ergonomics*

Pierce Foundation Laboratory (Berglund and Fobelets 1985, 1987) and similar studies at the Technical University of Denmark (Fanger and Christensen, 1986) have defined further the relationship between draught sensation and environmental parameters. In both studies a small percentage of subjects indicated they experienced a draught even in still air conditions. Subtracting this percentage from the percentage feeling a draft at higher air speeds results in response data of those subjects experiencing a draught due to air movement.

A summary of results from two draught studies is presented in Figure 3.3. The black dots are from the Pierce study and represent the mean response of 50 subjects under neutral conditions after 2 hours at the air temperatures indicated. The solid lines are the best multiple linear regression lines through these data ($r^2 = 0.84$). The fit is not perfect but shows that, at any air speed, the percentage of persons experiencing a draught increases with decreasing air temperature. Also shown are dotted lines of draught perception from a study at the Technical University of Denmark. The two studies shown in this figure are basically in agreement between 0·15 and 0·25 m s$^{-1}$. However, agreement becomes progressively poorer at 0·3 m s$^{-1}$ and above. Interestingly, two 19°C data points from the Pierce study agree well with the Danish figures. The Danish study solicited draught reports only from the ankles and the back of the neck while the Pierce study did not focus on specific body locations.

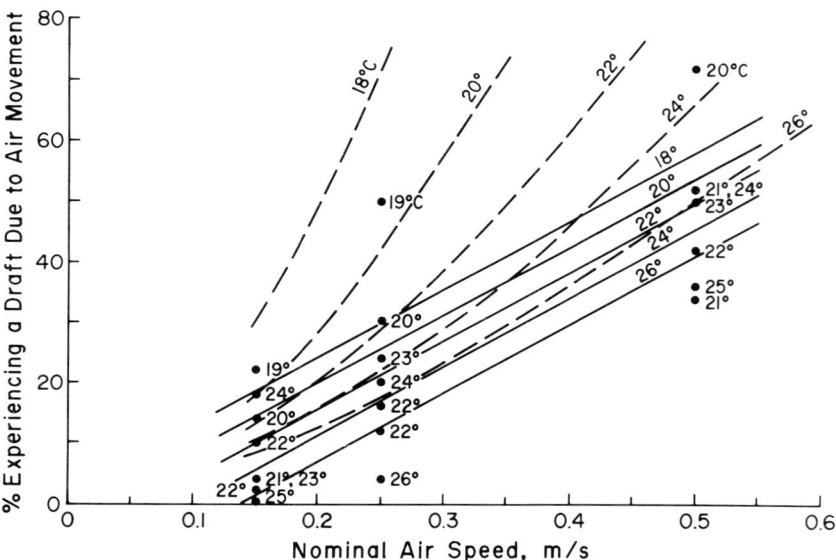

*Figure 3.3. The percentage of people experiencing the discomfort of a draught at various air speeds and temperatures.*

## Asymmetric clothing insulation

Non-uniform clothing insulation causes non-uniform body heat losses much like a non-uniform thermal environment does. Gwosdow *et al.* (1986) studied human responses to neutral and cool environments with clothing insulation distributed both uniformly and asymmetrically on the body. For the asymmetric case the fabric covering one arm and one leg from the uniform insulation ensemble was removed and placed on the other arm and leg. The clothing insulation in both cases measured 0·6 clo.

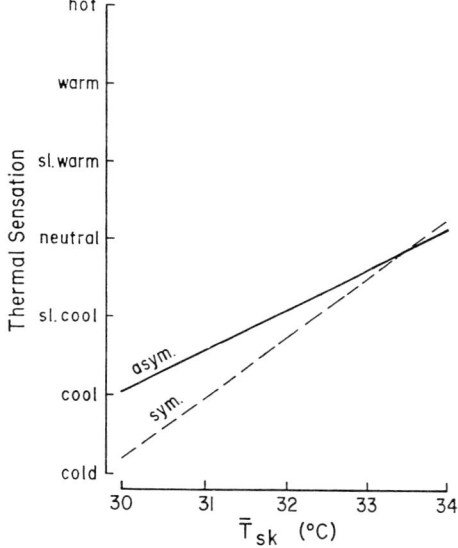

*Figure 3.4. Regression lines of the subjective sensitivity of sedentary men to mean skin temperature while wearing uniformly and asymmetrically distributed clothing of 0·6 clo.*

When wearing clothing insulation asymmetrically subjects were subjectively less sensitive to changes in skin temperature both regionally and overall. The statistically different regression lines for whole body thermal sensation shown as a linear function of mean skin temperature are drawn in Figure 3.4 for uniform and asymmetric clothing insulation. The slope of the regression line for the subjects wearing asymmetrical clothing is less steep. Therefore on entering a cool situation in such attire, the subjects would be slower to sense skin temperature changes and slower to initiate behavioural changes to prevent further body cooling than when uniformly clothed.

## 4. Evaporative heat loss and skin wettedness

In the evaporative heat loss process, water is brought to the surface of the skin by sweat glands where it evaporates and passes through the clothing to the environment. If the ambient humidity is low and the vapour resistance of the clothing is small, evaporation takes place at the top of the sweat gland ducts and the skin remains relatively dry. With increasing humidity and vapour resistance, the water 'puddle' at the top of the gland spreads out on the skin until evaporation from its surface equals the sweat delivery rate. Small sensors of many types (resistance type dew point, optical dew point and relative humidity sensors) are now being used to monitor the actual vapour pressure ($P_{sk}$) next to the skin under the clothing. When the skin is completely covered with water the vapour pressure next to the skin equals the saturation pressure of water at skin temperature ($P_{ssk}$). Then the evaporation rate ($E_{max}$) from the surface of the skin is maximum and expressable as:

$$E_{max} = h_e \, (P_{ssk} - P_a)$$

where the evaporation coefficient ($h_e$) from the skin to the ambient includes clothing and $P_a$ is the ambient vapour pressure. When not completely covered with sweat, the vapour pressure next to the skin is less than saturation and the evaporation rate ($E_{sk}$) per unit area of skin is:

$$E_{sk} = h_e \, (P_{sk} - P_a)$$

By combining the $E_{max}$ and $E_{sk}$ equations with the definition of skin wettedness ($w = E_{sk}/E_{max}$), a convenient expression (the ratio of the actual to the maximum vapour pressure differences) for evaluating local skin wettedness from measurements is formed:

$$w = (P_{sk} - P_a)/(P_{ssk} - P_a)$$

The average skin wettedness under the clothed body may be calculated from the area weighted average of the local skin wettedness values determined from measurements of $P_{sk}$ and skin temperature over the surface of the body.

Average skin wettedness may also be determined from the subject's rate of weight loss ($W_{sk}$) corrected for respiration, ambient humidity, skin temperature and the total vapour resistance of the clothing ensemble:

$$w = W_{sk}/(h_e \, (P_{ssk} - P_a))$$

The weighing technique for determining average skin wettedness is best applied to steady-state situations where the rate of weight loss is relatively constant. The multiple humidity sensor technique is not just restricted to steady-state situations, as illustrated in the following example, nor is it necessary to know the mass transfer coefficient ($h_e$).

## 5. Skin moisture during exercise and rest

The new multiple skin humidity sensor technique was used by Hoeppe *et al.* (1985) to evaluate skin wettedness in relation to subjective responses of comfort, thermal sensation and skin dampness during exercise and rest. In environments that were approximately neutral for sedentary individuals (22°C and 26°C at 45% RH) male subjects first exercised for 35 min and then rested for 60 min, while all the time being seated on a cycle ergometer. The exercise consisted of an initial 5 min warm up phase with increasing work rate followed by 30 min of exercise with a constant work rate of 85 W at 50 rpm. During exercise the subjects' total heat production determined from oxygen consumption measurement was about 500 W or 5 mets. The subjects wore training or jogging suits of approximately the same clo value but with different moisture properties. The suits were made of 100% cotton, 100% polyester, a semipermeable laminate and polyurethane coated nylon. Oesophageal, skin and outer clothing surface temperatures, dew point temperatures under the clothing next to the skin, oxygen consumption and rate of weight loss were measured continuously. Subjective responses of thermal sensation, comfort and perceptions of dampness were obtained every five minutes.

Figure 3.5 displays the results of the mean skin wettedness measurements under the training suits for 26°C and 45% RH environment with 1·4 m s$^{-1}$ air speed. The skin wettedness increases under the cotton and polyester suits were about the same during exercise and both decreased similarly during rest. Under the semipermeable garment skin wettedness was higher during exercise but during the following rest period skin wettedness decreased slowly, eventually reaching the level under the woven fabrics. The wettedness under the impermeable rain suit was still higher, reaching the 90% level during exercise, and remained high during rest, probably because the vapour pathways for drying the space between the skin and clothing were limited to the small openings at the collar and cuffs. Figure 3.6 shows how well the subject's feeling of dampness corresponded with measured skin wettedness. The subjects appear to have been very consistent at sensing changes in skin moisture.

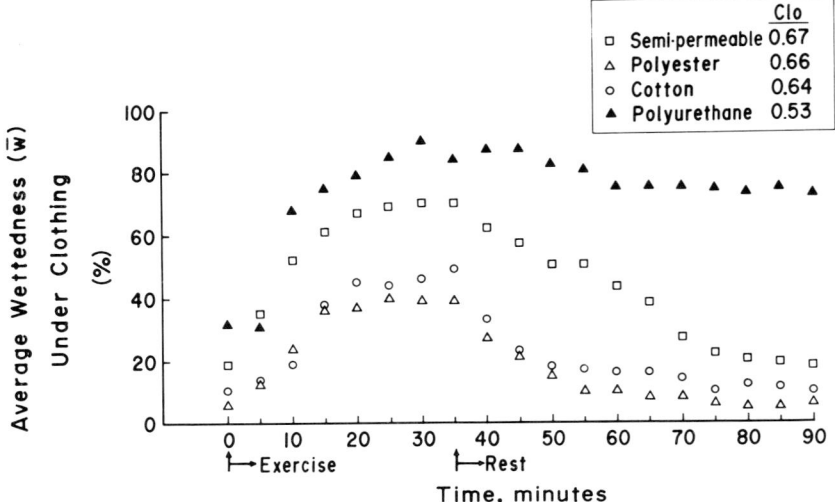

*Figure 3.5.* Measured skin wettedness during and after exercise in a 26°C environment with a 13°C dew point and 1·4 ms$^{-1}$ air movement.

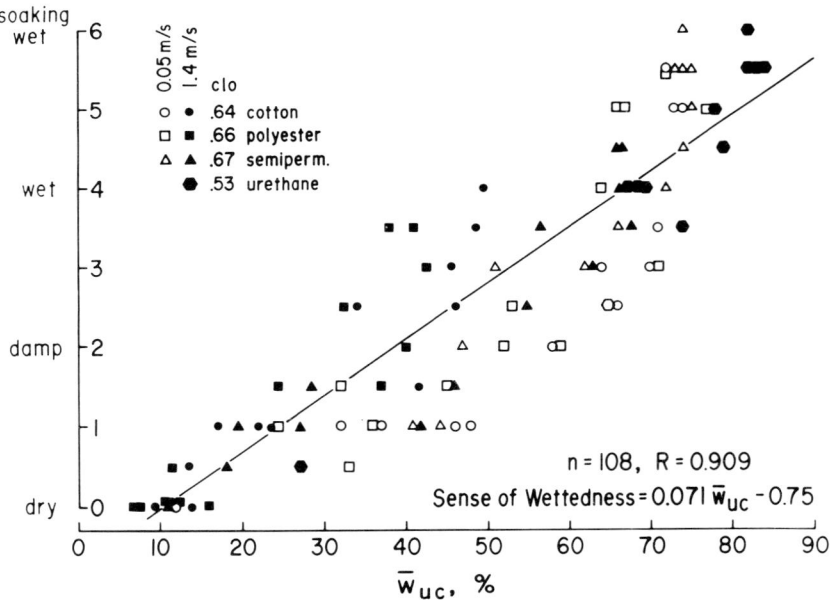

*Figure 3.6.* Perceived skin moisture during an exercise and rest experiment related to the measured average skin wettedness under clothing.

The regression equation relating perceived skin wettedness to measured skin wettedness ($r = 0.91$) is:

$$\text{Sense of wettedness} = 7.1\ w_{uc} - 0.75$$

where $w_{uc}$ is average skin wettedness under clothing expressed as a decimal fraction.

Discomfort (Disc) also increased with the measured skin wettedness under clothing as shown in Figure 3.7. These comfort judgements both during and after exercise are related to the measured skin wettedness by the regression equation ($r = 0.81$):

$$\text{Disc} = 3.6\ w_{uc} - 0.25$$

which indicates that, in warm environments, discomfort is strongly related to skin moisture. Notice that at a skin wettedness of 25%, Disc is 1.4 or a little more uncomfortable than 'slightly uncomfortable'. Thus discomfort and feelings of dampness both increased linearly with the increased measured skin wettedness.

The thermal sensations for these tests were found to be a function of the deviation of the measured mean skin temperature ($T_{sk}$) from the predicted mean skin temperature for comfort ($T_{skc}$) represented by:

$$\text{Thermal sensation} = 0.4 + 0.41\ (T_{sk} - T_{skc})$$

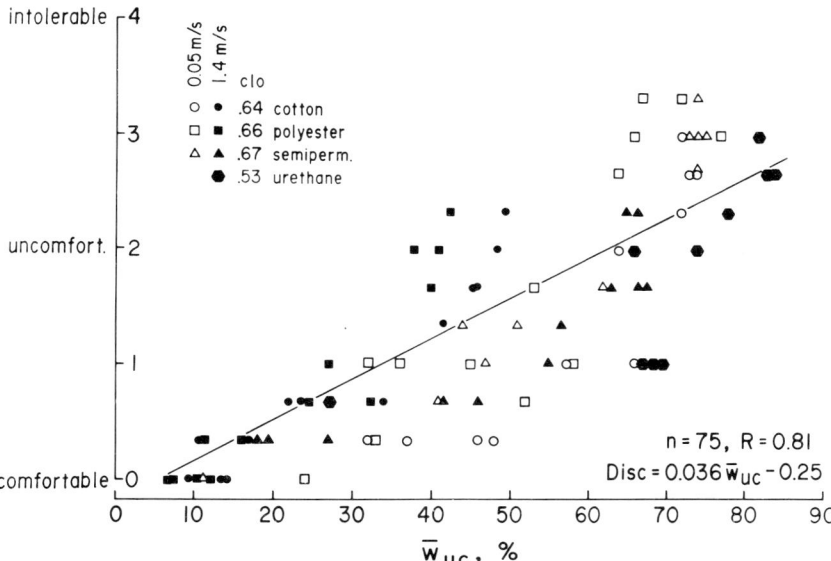

*Figure 3.7.* *Discomfort responses during an exercise and rest experiment as a function of the measured average skin wettedness under clothing.*

where $T_{skc}$ (°C) = 35·7 − 0·0276 $(H/A)$ is from Fanger (1972) and $H/A$ is the measured internal heat production per unit surface area (W m$^{-2}$). Thermal sensation is scaled here on the traditional ASHRAE 5-point scale (0 neutral, −1 slightly cool, +1 slightly warm, etc.).

The vapour resistances $(R_c)$ of the training suits was the principal cause of the skin wettedness level differences between ensembles. Their values and thermal resistances (clo) calculated from measured data during steady state (Berglund *et al.*, 1986) are given in Table 3.1.

*Table 3.1    Vapour resistance of different types of training suits.*

| | Cotton | Polyester | Semipermeable | Urethane on nylon |
|---|---|---|---|---|
| **Exercise at 5 mets** | | | | |
| $R_c$ (torr m$^{-2}$ W$^{-1}$) | 0·06 | 0·05 | 0·10 | 0·15 |
| **Rest** | | | | |
| $R_c$ (torr m$^{-2}$ W$^{-1}$) | 0·11 | 0·10 | 0·45 | 0·46 |
| clo | 0·64 | 0·66 | 0·67 | 0·57 |
| icl | 0·42 | 0·48 | 0·11 | 0·08 |

The insulation value of the clothing in clo units was evaluated only for the resting phase from oxygen consumption, rate of weight loss, skin and outer clothing temperature measurements. It is interesting that the clothing's vapour resistance values determined during exercise were less than those during rest. This could be due to the bellowing of clothing during exercise which may increase mixing of the space between skin and clothing and induce a flow of fresh ventilation air through apertures at the neck, wrists, etc. Also displayed in Table 3.1 is the calculated vapour permeability index (icl), defined as the ratio of the dry thermal resistance to the vapour resistance of the clothing:

$$icl = 0·155 \ clo/(2·2 \ R_c)$$

The dimensionless permeability values determined at rest compare favourably with those reported by Oohori (1984). The vapour permeability, icl, of typical woven fabrics in a stationary condition is about 0·45.

## 6. *Warm response differences between some men and women*

In a somewhat similar study, Cunningham *et al.* (1985) studied the physiological and subjective responses of two sedentary men and women to a sequence of increasingly severe thermal environments as

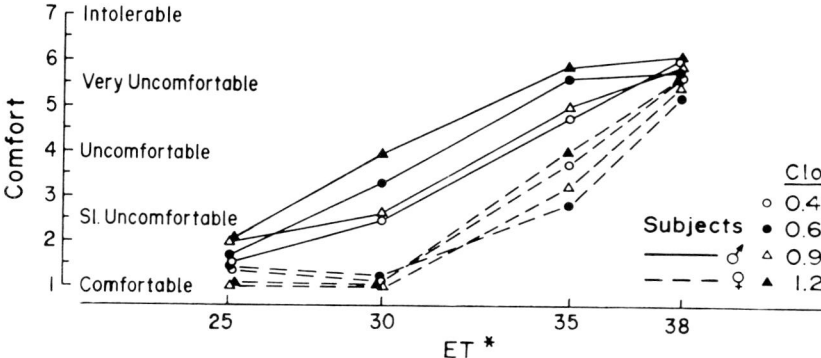

*Figure 3.8.   Discomfort responses of sedentary men and women experienced during increasingly severe climates.*

measured by the ASHRAE effective temperature scale ET*. The ET* levels were 25, 30, 35 and 38°C. The test sequence was repeated with subjects wearing four ensembles of 0·4, 0·6, 0·9 and 1·2 clo. Local dew points and skin temperatures were measured to determine the average skin wettedness under clothing and the mean skin temperature. A higher level of thermal sensation and discomfort in the various environments was experienced by the men than the women (Figure 3.8). However, the physiological responses to the heat were different, the men exhibiting a much greater level of skin wettedness. Interestingly, the sensations of the men and women relative to skin wettedness (Figure 3.9) and skin temperature were essentially the same. The regression equation of discomfort to skin wettedness for the men and women is:

$$\text{Disc} = 5{\cdot}06 \, w_{\text{uc}} + 0{\cdot}09$$

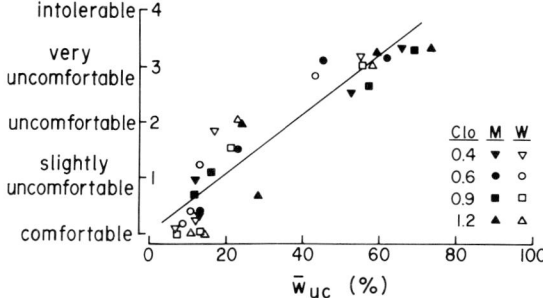

*Figure 3.9.   Discomfort responses of men and women from increasingly severe climates related to their measured skin wettedness under clothing.*

The correlation coefficient of this relationship is 0·91. The group as a whole would have reached the 'intolerable' discomfort level at a skin wettedness of about 76%. At the 25% skin wettedness level, Disc is 1·35 or a little more than slightly uncomfortable. This is similar to the results of Hoeppe (1985). The 'sense of skin wettedness' in relation to measured skin wettedness was also the same for men and women, the regression equation being:

$$PSW = 6·99 \, w_{uc} + 0·21$$

In this study the discomfort differences between men and women were due to physiological rather than sensory causes. Though maximum oxygen consumption was not measured on these subjects, the sweating and skin wettedness response differences may have been due to the level of fitness of the groups since the men of this study were athletic while the women had more sedentary backgrounds.

## 7. Experimental relationships between discomfort and skin moisture

Various comfort studies in which skin wettedness was measured have been reviewed and summarized by Berglund and Cunningham (1986). In these, skin moisture was measured by different techniques and methods by different investigators, under different conditions. The regression lines of discomfort as a function of mean skin wettedness

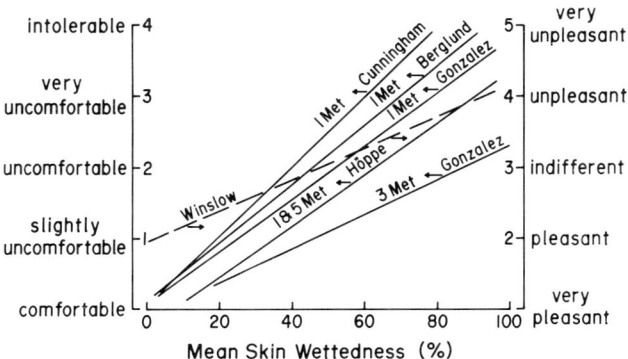

*Figure 3.10.    Comparison of discomfort–skin wettedness regression results from various studies.*

from several of these studies are presented in Figure 3.10, where it may be seen that the results are in good agreement. The regression line developed by Winslow *et al.* was the first relating comfort to skin moisture in warm environments. These results are somewhat difficult to compare numerically to the others because they employed a pleasantness scale from 1 to 5, which is dissimilar from the ASHRAE scale used in later studies. All of the regression equations of Figure 3.10 are listed in Table 3.2.

Table 3.2   *Regression equations of subjective responses*

| Source | Met | Clo | Equation | R |
|--------|-----|-----|----------|---|
| Winslow (1938) | 1 | 0·05 | $P = 2·2\,w + 1·95$ | 0·51 |
| Gonzalez (1973) | 1 | 0·05 | $Disc = 3·71\,w + 0·086$ | |
| Gonzalez (1978) | 3 | 0·05 | $Disc = 2·4\,w - 0.1$ | 0·69 |
| | | | $Disc = 0·97\,w + 0·216\,T_{sk} - 6·8$ | 0·97 |
| Berglund (1984) | 1 | 0·4–0·9 | $Disc = 4·13\,w + 0·13$ | 0·61 |
| | | | $Disc = 4·3\,w + 0·593\,T_{sk} - 20·8$ | 0·68 |
| Cunningham (1985) | 1 | 0·4–1·2 | $Disc = 5·06\,w + 0·09$ | 0·91 |
| | | | $Disc = 5·87\,w + 0·85\,T_{sk} - 28·6$ | 0·97 |
| Hoeppe (1985) | 1 & 5 | 0·64 | $Disc = 3·6\,w + 0·25$ | 0·81 |

(Skin wettedness ($w$) is a decimal fraction)

# 8. Skin moisture and clothing friction

Gwosdow *et al.* (1986) pulled various kinds of cloth across the forearms of eight subjects while they experienced a sequence of neutral, hot and dry, hot and humid, and neutral thermal environmental exposures. The force required to pull the fabric strip across the skin, the skin wettedness under cloth adjacent to the test site and the subjects' reports of pleasantness and fabric coarseness were recorded. The pleasantness level or Hedonic rating expressed while each cloth was pulled across the arm is displayed in Figure 3.11. Silk produced the most pleasant sensations but these became less so in hot, humid environments. Other fabrics produced similar responses. The coefficient of friction was determined from the contact angle and the recorded forces. These are plotted in Figure 3.12 in terms of the measured skin wettedness. There was an abrupt increase in friction at the higher skin wettedness levels. Even silk evoked a similar response. The coarseness rating of the cloth as it moved across the skin also increased (felt rougher) as skin wettedness increased. Thus, it appears that the friction between skin and clothing may contribute to the overall discomfort of warm humid situations and partially explain how skin wettedness causes unpleasantness.

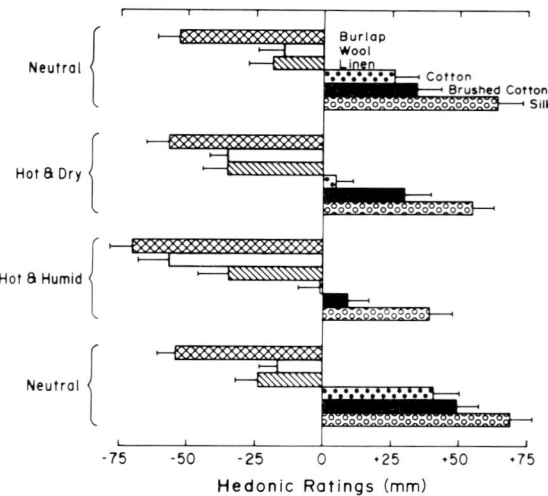

*Figure 3.11.   Hedonic pleasant (right) – unpleasant (left) rating experienced as fabrics slide across the underside of the forearm in various sequential environments.*

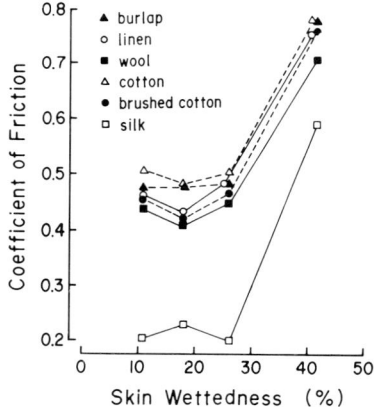

*Figure 3.12.   Coefficient of friction in terms of the measured skin wettedness.*

## 9. Conclusion

The quantification of the effects of thermal nonuniformity on human subjective response has progressed especially with regard to asymmetric radiation and draught. The development of the PMV* index for the prediction of warm discomfort improves the characterization of comfort at conditions above neutrality.

Skin wettedness has received considerable attention. In addition to the traditional weighing technique, it is now possible to monitor skin wettedness from miniature sensors on the skin. The sensors permit the assessment of skin moisture during non-steady state situations and with fewer analytical assumptions. The two measurement procedures have led to similar relationships between discomfort and skin wettedness. Measurements reveal that skin wettedness increases the sliding friction of fabrics over the skin. This friction probably contributes to the whole body discomfort–skin wettedness relationship.

# References

Berglund, L.G. and Cunningham, D.J., 1986, Parameters of human discomfort in warm environments. *ASHRAE Trans.*, **92**(2B), 732–746.

Berglund, L.G. and Fobelets, A., 1985, A study to determine subjective human response to low level air currents and asymmetric radiation of the lower boundary of human comfort. *Tech. Rep. ASHRAE Research Project 353* (Atlanta, GA: ASHRAE).

Berglund, L.G. and Fobelets, A., 1987, Subjective human response to low level air currents and asymmetric radiation. *ASHRAE Trans.*, **93**(1), 497–523.

Berglund, L.G., Hoeppe, P., Fobelets, A. and Gwosdow, A., 1986, In situ measurements of the vapor resistance of clothing. In *Proceedings of the Twelfth Annual Northeast Bioengineering Conference* (IEEE Catalog No. 86CH2329–1), pp. 67–70.

Cunningham, D.J., Berglund, L.G. and Fobelets, A., 1985, Skin wettedness under clothing and its relationship to thermal comfort in men and women. In *Clima 2000*, V.V.S. Congress, Copenhagen, edited by P.O. Fanger, vol. 4, pp. 91–96.

Fanger, P.O., 1968, Calculation of thermal comfort: Introduction of a basic comfort equation. *ASHRAE Trans.*, **73**(2) III 4.1–III 4.20.

Fanger, P.O., 1972, *Thermal Comfort* (New York: McGraw-Hill).

Fanger, P.O., 1986, Radiation and discomfort. *ASHRAE J.*, **28**, 33–34.

Fanger, P.O. and Christensen, N.K., 1986, Perception of draught in ventilated spaces. *Ergonomics*, **29**, 215–235.

Gagge, A.P., Fobelets, A.P. and Berglund, L.G., 1986, A standard predictive index (PMV*) of human response to the thermal environment. *ASHRAE Trans.*, **92**(2B), 709.

Gwosdow, A.R. and Berglund, L.G., 1986, Clothing distribution influences thermal responses of men in cool environments. *Therm. Biol.*, **12**, 2.

Gwosdow, A.R., Stevens, J.C., Berglund, L.G. and Stolwijk, J.A.J., 1986, Skin friction and fabric sensations in neutral and warm environments. *Text. Res. J.*, **56**, 574.

Hoeppe, P., Oohori, T., Berglund, L.G., Fobelets, A. and Gwosdow, A., 1985, Vapour resistance of clothing and its effect on human response during and after exercise. In *Clima 2000*, V.V.S. Congress, Copenhagen, edited by P.O. Fanger, vol. 4, pp. 97–102.

Houghton, F.C. and Yaglou, C.P., 1923a, Determining lines of equal comfort, *ASHVE Trans.*, **29**, 163.

Houghton, F.C. and Yaglou, C.P., 1923b, Determination of the comfort zone. *ASHVE Trans.*, **29**, 363.

Oohori, T., Berglund, L.G. and Gagge, A.P., 1984, Comparison of current 2-parameter indices of vapour permeation of clothing as factors governing thermal equilibrium and human comfort. *ASHRAE Trans.*, **90**(2A), 85–101.

Winslow, C.E.A., Herrington, L.P. and Gagge, A.P., 1937, Relations between atmospheric conditions, physiological reactions and sensations of pleasantness. *Amer. J. Hyg.*, **26**, 103–115.

Winslow, C.E.A., Herrington, L.P. and Gagge, A.P., 1938a, The relative influence of radiation and convection upon the temperature regulation of the clothed body. *Amer. J. Physiol.*, **124**, 51–56.

Winslow, C.E.A., Herrington, L.P. and Gagge, A.P., 1938b, The reactions of the clothed human body to variations in atmospheric humidity. *Amer. J. Physiol.*, **124**, 692–703.

Winslow, C.E.A., Herrington, L.P. and Gagge, A.P., 1939, Physiological reactions and sensations of pleasantness under varying atmospheric conditions. *ASHVE Trans.*, **44**, 179–194.

# 4. Design and Evaluation of Clothing for Protection from Heat Stress: An Overview

**Sarah A. Nunneley**

## 1. Introduction

The human body compensates well for moderate climatic heat stress, but artificial environments often block or overwhelm physiological defence mechanisms. Examples from industry include combinations of high air temperature and extreme radiant load in smelters, foundries and glassworks; elevated wet bulb temperatures which cause problems in very deep mines, ship engine compartments and textile drying rooms. Workers cannot tolerate such environments indefinitely without some relief from thermal stress.

Another source of heat stress is clothing worn for protection from nonthermal hazards. Examples are the sealed, pressurized suits or other highly specialized protective ensembles which are required to preserve life in hostile environments such as toxic, radioactive, or hypoxic atmospheres, at altitude and for extravehicular activity in space. In these cases the clothing tends to trap metabolic heat, and thermal balance is possible only in the coolest environments.

Thermoprotective clothing is defined as a wearable system that ameliorates unacceptable heat stress. Since such systems carry significant ergonomic and economic penalties, a 'brute force' approach is rarely feasible. It is therefore necessary to consider the many factors which determine the nature of the heat stress and to tailor design and testing to the specific problem at hand. Steps in the process include setting appropriate thermal goals, analysing the heat stress problem, selecting protective measures and testing candidate systems.

## 2. Setting thermal goals

Although we would like to keep workers continuously comfortable, that is not always possible or even necessary. Therefore, an important step in system design is to determine the degree of thermal stress which is acceptable for a given situation. Four potential goals can be distinguished.

*Comfort.* A comfortable microclimate is desirable where the worker is expected to perform critical mental tasks. In some cases it may be desirable not only to prevent sweating but also to minimize cutaneous vasodilation with its attendant shift in cardiac output.

*Long-term health maintenance.* The goal here is prevention of cumulative fatigue and morbidity which would adversely affect productivity. This goal is primarily of concern where a prolonged, stressful exposure must be repeated on a daily basis. An upper limit of 38°C core temperature is often used in industry for this purpose.

*Core temperature tolerance.* Excessive heat storage raises core temperature to levels likely to produce physical collapse. An upper limit of 39·5°C rectal temperature is often used for acute heat exposures in the laboratory.

*Skin temperature control.* Radiant heating can rapidly raise skin temperature to the pain threshold, about 45°C.

## 3. The heat stress triad

Heat stress problems generally require analysis in terms of three possible contributing factors: work rate, environment and clothing. Unacceptable heat stress may be produced by one of these factors or by two or three of them in combination, as illustrated in Figure 4.1.

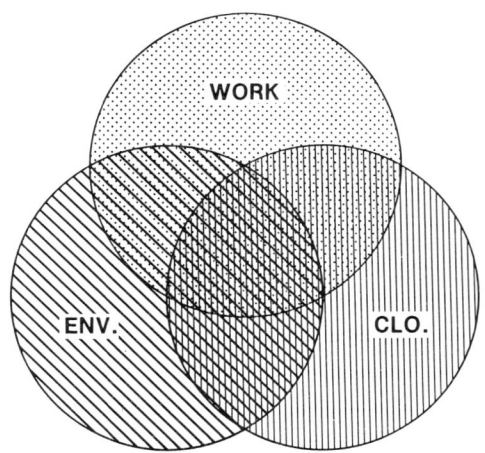

*Figure 4.1.    The heat stress triad, consisting of work, environment and clothing.*

The rise in core temperature which normally accompanies sustained work is not in itself a threat, but problems develop when environmental conditions and/or clothing prevent dissipation of excess metabolic heat and thus interfere with achievement of a tolerable steady-state condition (Leithead and Lind, 1964).

Whenever possible, design of thermoprotective systems should include options for lowering metabolic heat production. Possibilities include providing mechanical aids, dividing the work among more people, or scheduling regular rest breaks. A marginal situation may also be improved by implementing controls on the physical fitness and heat acclimatization of workers, so that the existing stress represents a small percentage of individual capacity.

Mechanisms for heat exchange between body and environment include conduction, convection, radiation and evaporation. Planning for intervention in heat stress requires a clear understanding of the contribution of each of these types of heat transfer. Excellent reviews are available in this area (Goldman, 1978; Kerslake, 1972; Leithead and Lind, 1964).

Clothing interferes with heat transfer between the skin and the environment, creating a complicated series of thermal exchanges (Figure 4.2). Characterization of clothing requires determination of its thermal insulation value, its resistance to transmission of water vapour, and its wind permeability. The exchange of air or 'pumping' associated with body movements is also a major factor.

In the case of clothing which is impermeable to water vapour, thermal equilibrium is possible only in environments cool enough to remove the necessary heat through the suit by a combination of conduction and convection. Heat transfer from the skin to the suit may

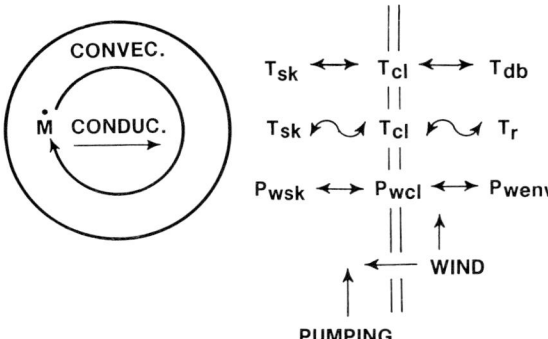

*Figure 4.2.   A simplified diagram of heat exchange between body and environment with an intervening layer of clothing. Major symbols: $M$ = metabolic heat production, $T$ = temperature, $P$ = vapour pressure. Subscripts: sk = skin, w = water, db = dry bulb, r = mean radiant, env = environmental.*

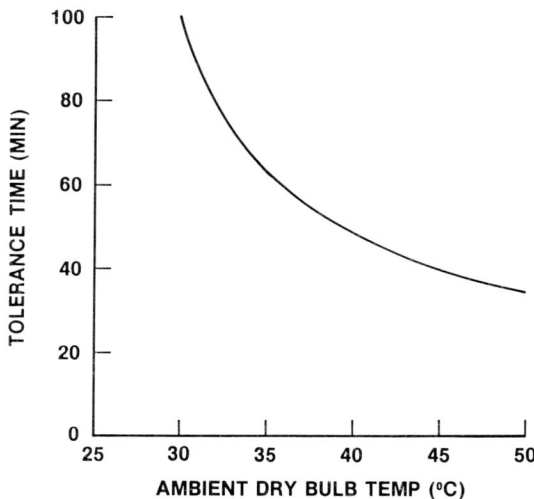

*Figure 4.3.    Tolerance time limits for men wearing impermeable suits and working at $V_{O_2} = 1\ l\ min^{-1}$ at various environmental temperatures. After Shvartz and Benor (1971).*

involve evaporation of sweat from the skin followed by condensation on the cool inner surface of the suit (Crockford, 1968). Limits on human tolerance time for moderate work in impermeable suits are indicated in Figure 4.3.

## 4. *Possible solutions*

Thermoprotective clothing can be divided into two major categories, passive and active, the latter having moving parts and requiring attachment to an energy source.

### Passive systems

Conventional clothing offers some protection from external heat loads. For instance, the desert dweller's burnoose reduces solar heat load without blocking air flow. Extreme radiant heating may be counteracted by reflective materials lined with heavy insulation; a good example is the firefighter's 'bunker', a garment made of aluminized asbestos (now Kevlar) and used for work near high-temperature fires.

Phase change of water or another substance can be used for passive cooling. An example is the ice vest, a garment which contains pockets of frozen water. The ice vest cools the microclimate through melting the ice and warming the resulting water (Van Rensburg *et al.*, 1972). The very low starting temperature of the heat sink means that the vest must be worn over an insulation garment to prevent skin chilling. It is also advantageous to provide external insulation so that less heat is absorbed directly from the environment. Vests containing 4·5 kg ice are used in the South African gold mines to prolong tolerance for hard work in an environment with a wet bulb temperature of 33−36°C (Strydom, 1974).

Another application of phase change is the evaporation of water from a wettable layer worn outside an impermeable suit, a technique which requires the rather paradoxical combination of large quantities of water and an environment dry enough to accept water vapour.

## Active systems

These systems use an external heat sink to cool a fluid (usually air or water) which is then pumped through the clothing system to provide microclimate conditioning (Figure 4.4). The heat picked up by the cooling loop may come from metabolism and from the environment (Fonseca, 1976; Shvartz, 1972). External heat sinks can be based on many different mechanisms, some of which are listed in Table 4.1.

Design of active cooling systems should· include a failure mode analysis and, specifically, consideration of consequences if cooling is

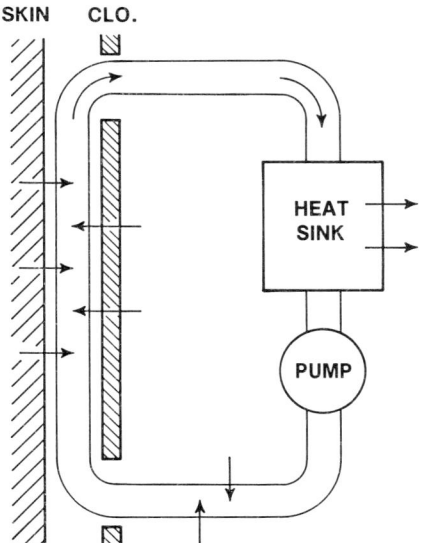

SKIN    CLO.

HEAT
SINK

PUMP

*Figure 4.4.    Diagram of an active cooling system. The arrows indicate movement of heat from the body to the fluid stream and thence to the heat sink, which dumps it into the environment.*

*Table 4.1    Potential heat sinks for the air-ventilated suit (AVS) and for the liquid-conditioned garment (LCG).*

| Heat sink | AVS | LCG |
|---|---|---|
| Ambient air | X | |
| Compressed air (± vortex) | X | |
| Evaporation (water, other) | X | |
| Super-cold liquid (air, other) | X | |
| Vapour cycle refrigration | X | X |
| Thermoelectric device | X | X |
| Ice (water, $CO_2$) | | X |

interrupted. Thus, the cooling garment should be included in measurements of insulation and permeability for the entire clothing ensemble.

### Air-ventilated suit (AVS)

The AVS is supplied with a flow of gas which is distributed over the body by a system of ducts or by a spacer garment; in either case, the air paths must be structured to maintain patency. For permeable suits, air may exit the clothing through the fabric and at openings such as neck, wrist and ankles. Impermeable suits generally have one-way valves to dump air to the environment.

Air cools the body by convection and/or evaporation. Since these cooling mechanisms are under physiological control, adequate air conditioning of the microclimate should allow the body to fine tune heat exchange in the normal manner.

Factors which determine AVS performance include the temperature and humidity of the air supply, mass flow and the effective surface area available for heat exchange. Convective cooling is a relatively weak mechanism because the specific heat of air is low. Theoretically, cooling could be improved by supplying extremely cold air, but in practice inlet temperature must remain above freezing to prevent discomfort from local chill near air vents and also to avoid condensation and freezing of water in piping and valves. Evaporation is a strong cooling mechanism, removing $0.58$ kcal $g^{-1}$ of water evaporated, but the person must first become hot enough to sweat, and therefore experiences continuous discomfort and eventual dehydration. In any case, air flow through the suit is limited by problems with noise, wind and a tendency for the suit to develop positive pressure and inflate to awkward bulkiness.

The AVS may be combined with either permeable or impermeable clothing and may cover the entire body or only part of it, as in a ventilated jacket or hood. Air cooling has been used with some success for fliers wearing anti-exposure suits and recently for army tank crews operating in hot weather, both of which involve limited physical work. The AVS is also used in industry for hot trades.

Ventilation was inadequate as a cooling mechanism for suited astronauts performing extra-vehicular work, but a special factor was the low barometric pressure in the suit which significantly reduces convective thermal exchange at the same time that it enhances evaporative capacity.

Where an extremely hot environment allows wear of permeable clothing, Crockford *et al.* (1974) showed that there are advantages to using radial air flow through the material to provide 'dynamic insulation' which removes impinging heat before it reaches the body (Figure 4.5). This work indicates that this is a more efficient use of the conditioned air than is the conventional axial flow pattern.

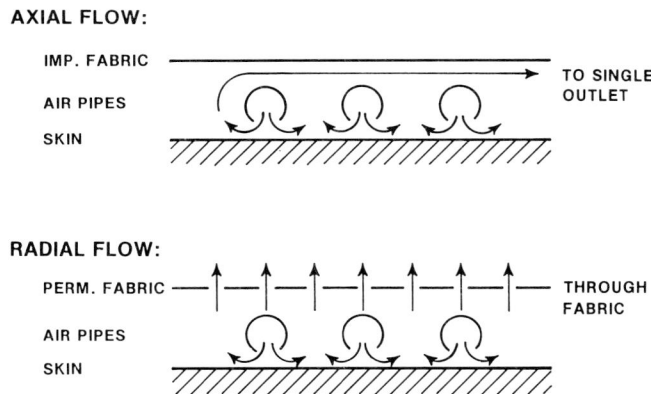

*Figure 4.5.   Diagram of ventilating air flow in axial and radial patterns. The latter can be used only with air-permeable materials.*

Selection of an air source is an important aspect of AVS design. Portable ventilation systems exist, but they weigh 6–8 kg, provide only about 30 l min$^{-1}$ of air at ambient conditions, and the work of carrying the equipment negates a significant portion of the cooling provided. A portable, cooled system based on liquid air has also been demonstrated (Gleeson and Pisani, 1967). Nevertheless, tethering is the usual means of supplying air despite the inevitable limitations on movement. Piping for chilled air must be insulated and kept short to minimize environmental heat pick-up. Longer lines can be used to connect compressed air to a cooling device co-located with the subject. The most common device for this application is the vortex tube, which uses the energy of expansion to divide the air flow into separate streams of cold and hot air (Brown, 1965; Van Patten and Gaudio, 1969).

*Liquid-conditioned garment (LCG)*

The LCG is worn next to the skin or over thin underwear and contains small-diameter tubing or heat-sealed vinyl patches through which liquid circulates to provide convective cooling. Several review articles on LCGs have appeared (Harrison and Belyavin, 1978; Midwest Research Institute, 1975; Nunneley, 1970).

The superior cooling capacity of water means that a liquid-based system requires much lower mass flow and less pumping energy than does a system using air, and therefore has a lower overall weight. Unlike the AVS, LCG cooling is relatively independent of fluid flow but is highly sensitive to inlet temperature (Harrison and Belyavin, 1978). Experiments have demonstrated that a full-coverage tubing suit connected to an unlimited heat sink can keep a subject comfortable regardless of environment (Shvartz and Benor, 1971) or work load (Webb and Annis, 1968). The design question then becomes one of tailoring the garment and the heat sink to the particular application.

Disadvantages of the LCG include the need for a reasonably close fit on the subject and possible ill-effects should the liquid loop spring a leak. Humid environments will produce condensation on the LCG, thus contributing to dampness of the clothing and stealing some of the cooling capacity of the system. The patch type of LCG forms a vapour barrier which could cause problems if cooling were cut off.

Again, heat sinks may be either portable or fixed. Current US space suits incorporate a back-mounted heat sink which sublimates ice to vacuum. Earthbound portable systems use melting ice as the external heat sink, but the weight of the system, the need for frequent replacement of ice, and the logistic consequences of this all conspire to limit applicability. Miniature vapour-cycle refrigerators powered by batteries or gasoline engines are currently under development and would improve the logistics of the system.

Tethering the LCG to a heat sink is best adapted to seated operators or those doing physically circumscribed work. An alternative application is intermittent cooling, under study in our laboratory, in which the worker wears the LCG continuously but is attached to the heat sink only during rest breaks. Attempts to use this technique with air cooling have met with little success because core temperature simply remains elevated during the break, but the stronger cooling offered by an LCG offers some promise for enhanced thermal recovery.

The great cooling power of LCGs means that it may not be necessary to cool the entire body, particularly if work rate is low. In that case the area covered by the garment and the size of the external heat sink must be adapted to the particular heat stress condition.

The powerful cooling offered by the LCG brings with it the potential problem of overcooling, with associated discomfort and/or paradoxical heat storage. While subjects can learn to control inlet temperature in an appropriate manner, automatic systems have been demonstrated in which inlet temperature is controlled with reference to some combination of garment temperature, skin temperature or heart rate (Kuznetz, 1980).

## 5. Some special considerations

### Regional cooling

Various body regions differ in their capacity for delivering heat to a cooled garment. Relevant characteristics include peak conductance, vasoconstriction threshold, preferred temperature and subjective comfort weighting (Crawshaw *et al.*, 1975). The face, hands and feet show a strong vasoconstrictive response to cooling, combined with subjective awareness of discomfort; these areas are therefore not only inconvenient but physiologically unsuitable for systemic cooling. In contrast, the head and neck do not vasoconstrict until very cold and are well suited to cooling (Nunneley *et al.*, 1971). Various body areas can thus be ranked according to the efficiency of cooling those sites (Shvartz, 1972).

Skin temperature normally varies over the body, and comfort is associated with a temperature gradient of several degrees from the cooler extremities to the warmer torso and head. This pattern can be accommodated by delivering a cooling medium (air or water) to the extremities and collecting it centrally, a pattern which should be reversed in garments used for heating.

### Effects of physical fitness

A high level of aerobic fitness may provide improved tolerance for work-heat stress and may also enhance the effectiveness of artificial cooling. For the fit individual, a given task uses a lower fraction of work capacity and therefore produces a lower equilibrium rectal temperature ($T_{re}$), a greater heart rate reserve and less cumulative fatigue. The activities which induce and maintain a high level of fitness also involve partial acclimation to heat stress, with accompanying changes in sweat production, lowering of sweat electrolyte content, and increased plasma volume. All of these may assist thermoregulation in ventilated systems.

We find that LCGs also appear to be more effective in very fit individuals. Although their leanness could be a factor, skinfold thickness should not greatly alter heat dissipation. Nor should improved sweating play a major role in heavily clothed persons. We speculate that in fit individuals work-induced hyperthermia produces a more active cardiovascular response including enhanced blood flow to the skin, thus improving heat transport from core to LCG.

## Paradoxical effects

Early work on artificial cooling produced a strong debate on whether head cooling might 'fool the hypothalamus' into inappropriate suppression of sweating (McCaffrey, 1975). The fact that head cooling lowers tympanic temperature was cited early on as evidence for a direct effect on brain temperature, but it is now known that the ear canal receives part of its blood supply from the face and is therefore affected by external temperatures (McCaffrey, 1975; Nunneley *et al.*, 1971). Other experiments indicate that the body somehow regulates sweating according to net heat load, even in the presence of strong regional temperature differences (Nunneley *et al.*, 1971; Williams and Chambers, 1971; Williams and Shitzer, 1974). Head cooling does strongly influence comfort and might therefore impair a subject's judgement regarding the severity of heat stress and the extent of physiological reserves.

## Heat stress and mask intolerance

Some workers must wear respirators during heat stress, for example, firefighters and mine rescue personnel. Several studies show that work in heat causes a high incidence of dyspnoea and mask intolerance even among highly trained personnel. This phenomenon may reflect hyperventilation, which normally develops as core temperature rises above 38°C, or may involve some other mechanism(s) (Petersen and Vejby-Christensen, 1973).

## 6. Evaluation of thermal protection

Thermoprotective systems can be evaluated in a number of ways, including the guarded hot plate, heated manikins, computer models, chamber simulations and field trials. Each has its own limitations and advantages, and it is necessary to select the appropriate step or series of steps to produce the desired information at the lowest possible cost in time, money and risk to human subjects. Often there is complex, iterative interaction among one or more of these techniques. In some cases, alternate systems can be compared, as in an RAF study of air and water cooling techniques for pilots (Allan *et al.*, 1971).

Final selection of thermoprotective systems requires consideration of several aspects of their performance.

1. Effectiveness. Whether the proposed solution produces the desired thermal result, reliably reaching the goal selected earlier.
2. Efficiency. Assurance that the final system is the best thermal answer in physical and physiological terms.
3. Compatibility. Whether the system will fit into the workplace without adversely affecting productivity.
4. Practicality. Assurance that the solution will work, including considerations of logistics, ruggedness and repairability.
5. Cost. Consideration of economic consequences including both initial investment and upkeep.

In conclusion, the design and evaluation of thermoprotective systems clearly involve a mixture of skills and expertise ranging from physics and computer programming through ergonomics and physiology to engineering and cost analysis. Technological improvement is still possible in many areas.

# References

Allan, J.R., Allnutt, M.F., Beeny, M., Hanson, R. de G., Morrison, J., Needham, R.W.J., Robertson, D.G. and Short, B., 1971, A laboratory comparison of three methods of personal condition. *RAF Institute of Aviation Medicine, Report No. FPRC/1307.*

Brown, J.R., 1965, Impermeable clothing and heat stress. *Med. Serv. J. Canada*, **21**, 518–532.

Crawshaw, L.I., Nadel, E.R., Stolwijk, J.A.J. and Stamford, B.A., 1975, Effect of local cooling on sweating rate and cold sensation. *Pflugers Arch.*, **354**, 19–27.

Crockford, G.W., 1968, Industrial pressurized suits. *Ann. Occup. Hyg.*, **11**, 357–365.

Crockford, G.W. and Awad El Karim, M.A., 1974, Assessment of a dynamically-insulated heat-protective clothing assembly. *Ann. Occup. Hyg.*, **17**, 111–121.

Fonseca, G.F., 1976, Effectiveness of four water-cooled undergarments and a water-cooled cap in reducing heat stress. *Aviat. Space Environ. Med.*, **47**, 1159–1164.

Gleeson, J.P. and Pisani, J.F., 1967, A cooling system for impermeable clothing. *Brit. J. Industr. Med.*, **24**, 213–219.

Goldman, R.F., 1978, Prediction of human heat tolerance. In *Environmental Stress*, edited by S.J. Folinsbee (New York: Academic Press).

Harrison, M.H. and Belyavin, A.J. 1978, Operational characteristics of liquid-conditioned suits. *Aviat. Space Environ. Med.*, **49**, 994–1003.

Kerslake, D.McK., 1972, *Stress of Hot Environments* (Cambridge: Cambridge University Press).

Kuznetz, L.H., 1980, Automatic control of human thermal comfort by a liquid-cooled garment. *Biomech. Eng.*, **102**, 155−161.

Leithead, C.S. and Lind, A.R., 1964, *Heat Stress and Heat Disorders* (Philadelphia: Lea & Febiger).

McCaffrey, V., Davis, F.A., Geis, G.S., Chung, M. and Wurster, D., 1975, Effect of isolated head heating and cooling on sweating in man. *Aerospace Med.*, **46**, 1353−1357.

Midwest Research Institute, 1975, *Liquid Cooled Garments*. NASA CR-2509, 51 pp.

Nunneley, S.A., 1970, Water cooled garments: a review. *Space Live Sci.*, **2**, 335−360.

Nunneley, S.A., Troutmas, S.J., Jr and Webb, P., 1971, Head cooling in work and heat stress. *Aerospace Med.*, **42**, 64−68.

Petersen, E.S. and Vejby-Christensen, H., 1973, Effect of body temperature on steady state ventilation and metabolism in exercise. *Acta Physiol. Scand.*, **89**, 342−351.

Shvartz, E., 1972, Efficiency and effectiveness of different water cooled suits—a review. *Aviat. Space Environ. Med.*, **43**, 488−491.

Shvartz, E. and Benor, D., 1971, Total body cooling in warm environments. *J. Appl. Physiol.*, **31**, 24−27.

Strydom, N.B., Mitchell, D., Van Rensburg, A.J. and Van Gran, M., 1974, Design, construction, and use of a practical ice-jacket for miners. *J. S. Afr. Inst. Min. Metall.*, **74**, 5 pp.

Van Patten, E. and Gaudio, R., Jr, 1969, Vortex tube as a thermal protective device. *Aerospace Med.*, **40**, 289−292.

Van Rensburg, A.J., Mitchell, D., Van Der Walt, W.H. and Strydom, N.B., 1972, Physiological reactions of men using microclimate cooling in hot humid environments. *Brit. J. Industr. Med.*, **29**, 387−393.

Williams, A. and Chambers, A.B., 1971, Effect of neck warming and cooling on thermal comfort. *Second Conference on Portable Life Support Systems.* NASA SP-302.

Williams, A.B. and Shitzer, A., 1974, Modular liquid-cooled helmet liner for thermal comfort. *Aerospace Med.*, **45**, 1030−1036.

Webb, P. and Annis, J.F., 1968, Cooling required to suppress sweating during work. *J. Appl. Physiol.*, **25**, 489−493.

# 5. Standards for Human Exposure to Heat

**Ralph F. Goldman**

## 1. Introduction

One may divide standards into five types:

1. Absolute physical standards such as the triple point of ice, water and vapour form the first category.

2. Instrumentation standards, fairly well defined although established by a consensus process; the precision and accuracy of the parameters measured by an instrument can be specified once agreed upon by a group of experts.

3. Standards can be established from statistical sampling techniques, specifying the number of measurements made, the location of the measurements, the frequency of the measurements, etc.; again, these consensus standards are fairly easily derived, and subject to ready validation.

4. Consensus standards derived from experimental data or field experience.

5. Consensus standards based only on expert opinion.

Heat stress standards fall in these last two categories and are problematic except for limited exposure conditions, e.g., the saturated, low air motion conditions found in deep mines (Wyndham and Heyns, 1971), or select populations wearing specified clothing, e.g., military troops (US Army, 1980), particularly in specified settings such as a crew compartment. Otherwise, there are too many variables contributing significantly to the heat stress problem to develop a universal standard.

The search for a standard which can be used to minimize or eliminate heat stress has gone on for over 80 years (Lee, 1980). Various combinations of factors have been used as a 'heat stress index'; i.e., a single number expression which integrates the various factors contributing to heat stress. In theory, such an index can be used to limit the duration of exposure, reduce the level of work allowed or require that the worker meet certain standards of heat acclimation and/or fitness. Table 5.1 provides a chronological listing of a number of heat stress indices.

There have been a great number of reviews of heat stress indices (Belding, 1970; Bergland, 1983; Brief and Confer, 1971; Gagge and Nishi, 1976; Goldman, 1973; Gonzalez *et al.*, 1978; Goromosov, 1958; Jensen and Heins, 1976; Lee, 1980; Metz, 1976; Mutchler and Vecchio, 1977; Ramsey, 1975a, 1975b, 1978; OSHA, 1974; Witherspoon and Goldman, 1974), many generated by proposals to adopt a heat standard for the USA (Dukes-Dobos and Henschel, 1973; OSHA, 1974; Horvath and Jensen, 1976; Anon, 1985).

Generally, such indices have been derived along four approaches:

1. Physical indices based on one or more of the physical factors of the environment (temperature, humidity, air motion and radiant load).

2. Subjective indices based on assessments of thermal sensation ('Effective' and 'Corrected Effective' Temperature, Comfort Vote).

3. 'Rational' indices based on the human heat balance equations (HSI; Index of Relative Strain; Index of Thermal Strain; Humid Operative Temperature).

4. Physiological indices based on physiological strain (P4SR; % Sweat Wetted Area; Index of Physiological Strain; Predicted Rectal Temperature and Heart Rate).

The major focus of the first three approaches has been on the capacities or demands for heat transfer of the environment, in a few cases adjusted for the heat production demanded of the worker, but with few adjustments for differences in clothing. The importance of the capacities of the work force, as modified by acclimation, hydration, physical fitness, clothing and the like, has been introduced principally in the fourth approach and has been a focus of professional work since 1970 (cf. Table 5.1).

The experience gained during this time suggests that heat stress is generally poorly understood and that the most promising new approach for resolution of heat stress problems is through prediction modelling. Such modelling can build on the concept that heat stress results from an imbalance between the demands imposed on the worker by the task and the environment, and the worker's capacity to eliminate the heat load as modified by clothing. In the author's concept, as the ratio of demand to capacity exceeds 20% (whether the demand is for sweat production, heart rate increase, sweat evaporation or maximum oxygen uptake), the worker is moved from a 'comfortable'

Table 5.1. Systems for rating heat stress (adapted from Belding, 1970).

| Date | Index | Author(s) | Source | Description |
|---|---|---|---|---|
| 1905 | Wet-bulb temperature | Haldane | J. Hygiene, 5, 494 (1905) | A better indicator of physiological effect than dry bulb temperature in hot, wet confined spaces. |
| 1916 | Katathermometer | Hill et al. | Phil. Trans., 207, 183 (1916) | Rate of cooling of a previously warmed thermometer covered with a wetted wick is related to radiant, convective and maximal evaporative heat exchange. |
| 1923 | Effective temperature (ET) | Houghton and Yaglou | J. Amer. Soc. Heat. Vent. Engrs, 29, 515 (1923) | Combinations of DB, WB and velocity which yield equal sensations of comfort are assigned an ET equal to that of saturated, low V air which yields the same sensation. |
| 1937 | Operative temperature (OT) | Winslow et al. | Amer. J. Physiol., 120, 1 (1937) | Uses heat transfer coefficients to reduce the effect of prevailing $T_{rad}$, $T_a$ and $V$ to equivalent temperature if $T_{rad} = T_a$ with minimum air movement. |
| 1945 | Thermal acceptance ratio | Ionides et al. | OQMG, Environ. Protection Report, Sept 17 1945 | Index defined as ratio $= H_a/M$; $M$ is rate of body heat production, $H_a$ is rate of acceptance of heat by environment for unclothed subjects with $T_s = 97°F$. |
| 1945 | Index of physiological effect (I) | Robinson et al. | Amer. J. Physiol., 143, 21 | Combinations of $T_a$ and $T_{wb}$ at 3 levels of activity, which impose equivalent average demand in terms of elevation of $H_R$, $T_s$, $T_{rc}$ and sweat rate. |
| 1946 | Corrected effective temperature (CET) | Bedford | Med. Res. Council Memo 17, HMSO, London (1946) | Modifies ET scale, with 6″ black globe temperature ($T_g$), for thermal radiation. |
| 1947 | Probable 4-hour sweat rate (P4SR) | McArdle et al. | Med. Res. Council R.N.P. Rep., 47, 391 (1947) | Uses sweating to indicate physiological strain; predicts 4 h rate for combinations of $M$, $T_g$, $T_{wb}$ and $V$; clothing adjustment and computation method suggested later. |
| 1948 | Resultant temperature (RT) | Missenard et al. | Chaleur et Industrie, Jul-Aug (1948) | Same basis as ET except exposures were longer and equilibrium obtained. Intended for rest only. Better estimation of effect of humidity than ET. |
| 1950 | Craig index | Craig | USA Chem. Corps Med. Div. Res. Rpt, 5 (1950) | Modifies Robinson 1948 index of physiological effects so that $I = T + S + H/100$, where $H$ = heart rate, $T = \Delta T_{re}$ in °C h⁻¹ and $S$ = sweat rate in kg h⁻¹. |

*Table 5.1 Continued.*

| Date | Index | Author(s) | Source | Description |
|---|---|---|---|---|
| 1955 | Heat stress index (HSI) | Belding and Hatch | *Heat. Pip. Air Cond.*, **27**, 129 (1955) | Ratio between heat load ($M + R + C = E_{req}$) and evaporative capacity of environment ($E_{max}$; first use); uses coefficients of Nelson *et al.*, *Amer. J. Physiol.*, **151**, 626 (1947). |
| 1957 | Wet-bulb globe temp (WBGT) | Yaglou and Minard | *AMA Arch. Ind. Health*, **16**, 302 (1957) | $WBGT = 0.7\,T_{wb} + 0.2\,T_g + 0.1\,T_{db}$, where $T_{wb}$ is non-aspirated, 'natural' wet bulb. |
| 1957 | Oxford index (WD) | Lind and Hellon | *J. Appl. Physiol.*, **11**, 35 (1957) | $WD = 0.85\,T_{wb} + 0.15\,T_{db}$. Formula for deriving estimate of ET from $T_a$ and $T_{wb}$. |
| 1957 | Discomfort index (DI) | Thom | *Air Cond. Heat. Vent.*, **54**, 73 (1957) | Discomfort index, $DI = 0.4\,(T_{db} + T_{wb}) + 15$. |
| 1958 | Thermal strain index | Lee and Henschel | *Arid Zone Rsch*, **10**, (UNESCO, Paris, 1958) | Empirical equation involving R, C, M, clothing and V. Similar to HSI. |
| 1962 | Index of thermal stress | Givoni | UNESCO Symp. on Arid Cond. (India, 1962) | Predicts sweat rate on basis of heat load, utilizing concept of efficiency of sweating or its reciprocal. |
| 1965 | Heat strain predictive systems | Lustinec | Thesis, Czech. Univ. (1962) | Predicts HSI under effects of variable clothing. |
| 1966 | New nomographs for HSI | McKars and Brief | *Heat. Pip. & Air Cond.*, **38**, 113 (1966) | Based on refined coefficients where $E_{req} = M + 17.5\,(t_w - 95) + 0.756\,V^{0.6}\,(t_a - 95)$: $E_{max} = 2.8\,V^{0.6}\,(42 - VP_a)$ in Btu, °F, fpm and mmHg. |
| 1967 | Effective radiant field | Gagge *et al.* | *ASHRAE Trans.*, **73**, 1 (1967) | Combines operative temperature and radiant heat. See Chapter 18, ASHRAE 1982 Applications. |
| 1970 | Comfort vote & % dissatisfied | Fanger | *Thermal Comfort* (Danish Tech. Press, 1970) | Uses modified heat balance to predict comfort vote (PMV) and % predictably dissatisfied (PPD). |
| 1970 | Prescriptive zone | Lind | *J. Appl. Physiol.*, **28**, 57 (1970) | Uses inflection point of $T_{re}$ increase. |
| 1971 | Humid operative temperature | Gagge *et al.* | *ASHRAE Trans.*, **77**, 247 (1971) | Utilizes humid operative temperature ($T_{oh}$) to derive new $T_{eff}$, based on an interaction of constant skin wettedness loci with $T_{db}$ for the 50% RH curve. |
| 1971 | Wet globe temperature (WGT) | Botsford | *AIHA J.*, **32**, 1 (1971) | A copper sphere covered by a wetted black cloth provides one number for heat stress. Comparable to WBGT. |

| Year | Index | Author | Reference | Description |
|---|---|---|---|---|
| 1972 | Predicted body temperature ($T_{re}$) | Givoni and Goldman | J. Appl. Physiol., **38**, 812 (1972) | Utilizes function of $E_{req} - E_{max}$ to predict $T_{re}$ response for given environment, clothing and work. |
| 1972 | Skin wettedness | Kerslake | Stress of Hot Environments (Cambridge Univ. Press, 1972) | Expression for skin humidity based on Gagge's concept of skin wettedness, Amer. J. Physiol., **120**, 277 (1937). |
| 1973 | Standard effective temperature (SET) | Gagge et al. | Thermal Comfort Mod. Heat Stress. Proc. Build. Res. Estab. Conf. (1973) | Translates the reference for effective temperature from the original 100% RH base to a subjectively more familiar 50% RH. |
| 1973 | Predicted heart rate | Givoni and Goldman | J. Appl. Physiol., **34**, 201 (1973) | Predicts HR response for given environment, clothing and work using predicted $T_{re}$; acclimatization modification added J. Appl. Physiol., **35**, 875 (1973). |
| 1976 | Heart rate | Dayal and Ramsey | Proc. 6th Intl Ergon. Congress (Maryland, 1976) | Suggests using heart rate as an index of heat stress. |
| 1978 | Skin wettedness | Gonzales et al. | J. Appl. Physiol., **44**, 889 (1978) | Measurement of skin wettedness by dew point sensors as index of physiological strain. |
| 1978 | Fighter index of thermal stress (FITS) | Stribley and Nunneley | Aviat. Space Env. Med. (1978) | FITS $= 0.828 T_{wb} + 0.355 T_{DE} + 504$ predicts plane cockpit conditions from ground readings. |
| 1982 | Predicted sweat loss | Shapiro et al. | Euro. J. Appl. Physiol., **48**, 83 (1982) | Uses $E_{req}$ and $E_{max}$ to predict sweat rate. |
| 1985 | Required sweating | ISO 7933 | ISO (AFNOR) (Paris, 1985) | Calculated sweat production for temperature regulation as an index of physiological strain; limits, based on heat storage, adjust exposure time. |

condition to one of discomfort. As the demand/capacity ratio increases to 40−60%, the work is subject to performance decrements. Above 60%, work will usually be discontinued if that option exists; it can only be continued for a limited period. Above about an 80% ratio there is a substantial risk of heat illness.

Rather than provide another review of the available heat stress indices, this chapter will attempt to define heat stress, and the associated heat illnesses, in the simplest of terms and suggest why the concept of a simple, universal heat stress index (Belding, 1970; Lee, 1980) is unworkable. It will emphasize the role of clothing, particularly protective clothing, in producing heat stress, discuss how acclimation, hydration and work practices alter heat stress, and evaluate the role of environmental and subject monitoring. Finally, it proposes that either simple 'decision trees' or more complex, computer-based predictive models are better approaches than monitoring the environment and can provide better *a priori* protection than monitoring the worker.

## 2. Heat stress: an outline

### What is heat stress?

Heat stress simply means that the body cannot get rid of all the heat it needs to. The worker may be working too hard and producing too much heat to get rid of it all, or the temperatures can be so hot and the humidity so high that the body cannot get rid of any heat but may also be gaining heat from the outside, or the clothing worn to protect against the cold (or toxic chemicals in the environment) may prevent sweat evaporation from providing the cooling required by the body's heat production.

Heat stress does not begin at any specific temperature. It has occurred at temperatures below −5°C in troops carrying heavy loads while marching through deep snow; the men were producing well over 500 W of heat and, even though it was cold, the Arctic clothing they were wearing provided so much insulation that they could not lose all the heat they were producing.

### Who gets it?

Anyone can get heat stress. People who are overweight and not in good physical condition, people who are hungover or who haven't been taking enough water, and people who are not used to working in the heat have a greater chance of getting heat stress than anyone else, but anyone can get heat stress, even well-conditioned workers. Smaller individuals performing heavier physical work, older individuals, or people who are working very hard, sweating heavily and not drinking enough water or eating food with enough salt, are more likely to suffer heat stress. People with recent immunizations, infections or febrile illness are at greater risk.

## How does the body get rid of heat?

First, the amount of blood flowing to the skin from the muscles is increased tremendously; with a warmer skin, there is a better chance of getting rid of heat to the cold air. Second, the body begins to produce large amounts of sweat. However, sweating does not provide cooling unless the sweat can evaporate, preferably from the skin. The protective clothing many workers wear today limits sweat evaporation to a considerable degree (Goldman, 1984).

## When will heat stress occur?

Obviously, not many workers will suffer heat illness in the winter although it can occur during hard work at any season, as well as at rest in hot weather. It is most likely to occur when wearing full protective clothing, because anything which prevents air from getting to the skin also limits the body's ability to evaporate sweat. The risk is highest during the first days of heavy physical work under hot conditions when the body is less physically fit and not acclimatized to heat.

## Where will it occur?

It is most likely to occur under full sun, between 11 and 3 o'clock; when there is little or no wind to help remove heat from the surface of the clothing; when the work is physically hard and the workers are therefore producing a lot of heat; where there are furnaces or other large, high-temperature sources of radiant heat; or where the humidity builds up to high levels.

## Why does it occur?

While reading, the body produces about 100 W; more if one is bigger, less if smaller. With more activity, the body produces more heat. One can produce 500 W for 3–4 hours while working reasonably hard, and 700 W or more for periods of an hour or two during heavy work. With such a high heat production, even wearing regular clothing can be uncomfortably hot when the temperature is about 21°C or higher, especially if the humidity is also high. The added insulation of wearing a complete protective clothing ensemble is like adding 2°C to the outside air temperature (Goldman, 1984).

## 3. Heat illnesses

If workers cannot get rid of the heat they produce, their body temperature must rise. As they get hotter, the body has only two mechanisms to use in an attempt to get rid of the extra heat. The heart beats faster, pumping more blood from the muscles to the skin, to help get rid of heat; that will not work if the skin cannot lose the heat, through the clothing to the environment. The skin can also increase the amount of sweat it releases, putting out as much as 2–3 l of sweat an hour, at least for the hour or so before the workmen suffer heat exhaustion or 'dehydration exhaustion', which is simply the result of loss of more water from the body than is replaced by drinking. Sweat also contains a lot of salts, and some workers may lose too much salt from their body. Such problems result in a variety of heat illnesses (Minard, 1973) which, although they can be differentially diagnosed and categorized, tend to represent a continuum of effects rather than separate entities. Figure 5.1 provides a diagram of heat stress and its effects.

### Heat exhaustion

As the heart rate increases to over 160–180 beats per minute, there is no longer enough time between beats to completely refill the heart; with each beat, it pumps a little less blood to the brain and the worker begins to feel dizzy, nauseous and weak. If one sits down to take a break, or slows down enough, so much blood is out in the skin, not being pumped back into the central circulation by body activity, that there isn't enough blood coming back to the heart to fill it. The result is that there is not enough blood to be pumped to the brain. The worker gets 'heat exhaustion collapse' and simply faints or blacks out. Blood flow to the brain is restored when one is lying on the ground—and splashing some cold water on the body, coupled with massaging the legs and arms, may help get some blood out of the skin and back into the central circulating blood volume. The worker might even be able to go back to work, but the heat stored in the body hasn't been removed; if the worker is allowed to go back to work, there is an increased risk of a second collapse from heat exhaustion, or of heatstroke which could be fatal.

### Dehydration exhaustion

Those who lose much more water than they drink will be among the first to get into trouble. There are only about 5 l of blood to circulate

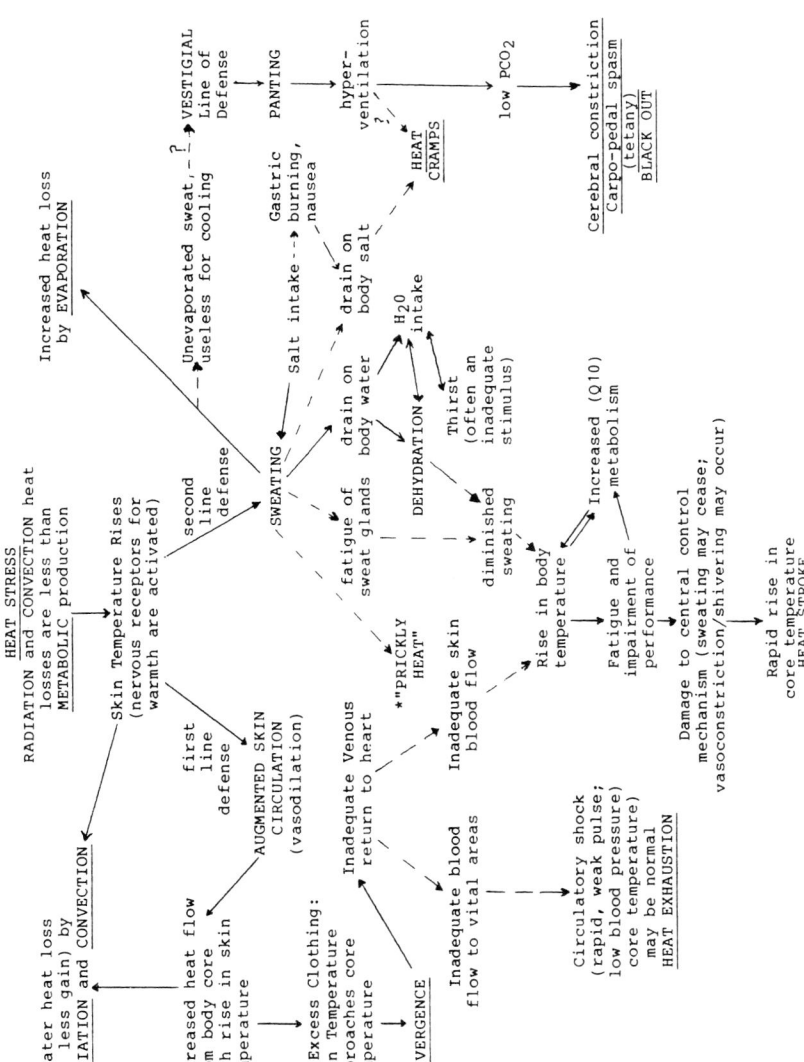

*Figure 5.1.  A diagrammatic representation of heat stress (adapted from Belding, 1970).*

around the adult male body; most of it is water and most of the sweat produced comes, initially, from this circulating blood reservoir. If the worker doesn't replace it by drinking enough water, 'dehydration exhaustion' and collapse are more likely.

While the tremendously tired feeling that accompanies heat exhaustion and dehydration, and the collapse that can occur, is frightening it is not terribly dangerous in itself. Some men have suffered heat exhaustion collapse during studies in the field for 5 or 6 days in a row with no residual damage (Joy and Goldman, 1968). However, they were allowed to recover completely (cold shower, full rehydration and overnight rest) before being re-exposed to work in the heat the next day.

## Heat stroke

The most serious danger of the body's inability to get rid of enough heat is not heat exhaustion collapse. Some individuals, frequently those in the best physical condition as well as those in the worst physical condition, may reach deep body temperatures above 41°C. That represents a serious risk since the body is unable to deal with such high temperatures and may suffer permanent damage to its ability to sweat, its ability to increase the blood flow to the skin, and its ability to regulate its temperature. In essence, the body's 'thermostats' which control shivering, sweating and skin blood flow may be damaged and, once damaged, there is no way to replace them. If the body temperature gets high enough and stays high long enough, 'heat stroke' can be fatal, from the damage to the brain which body temperatures above 41°C can produce (Bynum *et al.*, 1978).

## Heat cramps

Individuals who are not eating enough, or who are not adding some extra salt to their food during the first few days of hard work in the heat, may lose enough body salt that their muscles begin to be sore, and cramp up on them. They suffer from 'heat cramps' and are not able to continue. Such cramps are nothing like a single 'charley horse' that can be worked out. Instead, muscle cramps may occur in a number of sites at once and the worker can be totally disabled. Salt pills are not the answer at that point, or at any time (US Army, 1980); drinking more, lightly salted, water is the way to prevent heat cramps and, in most cases, to cure them if they occur.

## Salt depletion dehydration

There is a serious, but fortunately relatively infrequent, problem of prolonged, day-after-day exposure to heat stress for some individuals who fail to take in enough salt to replace their bodies' losses. The body may reach a new, but abnormal balance, 'salt depletion dehydration', which is very difficult to treat. Taking extra salt may not work since the body does not recognize it is short of salt and may simply dump the extra salt provided; it may also dump any extra water given because it has set up this new balance of salt and water. Ensure that extra salt is taken by heavy use of a salt shaker during the first few days of work in the heat, thereby avoiding this problem by maintaining body salt levels, i.e., prevention is the simplest solution.

**Overbreathing**

Some individuals exposed to heat stress (particularly hot humid environments) will attempt to pant (like a dog) to get rid of extra heat. However, it doesn't work for humans; instead, they may simply blow off so much of the carbon dioxide normally in the blood that they start feeling dizzy, have a tingling of the lips, and may feel faint. Some people can black out, i.e., faint, just from this alone. The small muscles of the hand may start to cramp so badly that the thumb is pulled into the palm, the fingers are stiff and the hand can't be used at all. Slow, deep breathing, or breathing in and out of a paper bag may be necessary to recover from this overbreathing or 'hyperventilation'.

## 4. The role of dehydration in heat stress

The humidity builds up in the 'microclimate' inside such clothing quite rapidly during heavy work in temperatures above about 15°C. The body produces more and more sweat in an attempt to get cooling from sweat evaporation. This further increases the humidity inside the clothing, while at the same time drawing water from the 5 l or so of circulating blood that the body has available. Once the body has lost about 1 l of water that is not drunk back, one begins to feel tired; it is difficult to perform well once 3 l have been lost or to continue when 4 l or more have been lost. This can occur within 1–2 hours during work outdoors in full protective gear when the air temperature is above 27°C with high humidity; even sooner doing continuous heavy exercise while not drinking any water.

To the extent that workers fail to replace all the water they are losing as sweat (Shapiro *et al.*, 1982), there will be increasing problems. Dehydration alone will become the major problem under temperature conditions where work time can be longer than 2−3 hours. Loss of water is usually expressed as a percent of loss of body weight; a 100 kg worker who loses 1 kg of water (i.e., drinks 1 l less water than he

*Table 5.2.    A diagrammatic representation of heat stress (adapted from Belding, 1970).*

| Weight loss (kg) | Level of dehydration (% of 70 kg body wt) | Degree of distress |
| --- | --- | --- |
| 2 | 3% | Some physiological upset |
| 3·5 | 5% | Risk of heat/dehydration exhaustion |
| 5 | 7% | Dangerous hallucinations, etc. |
| 7 | 10% | Very dangerous high risk of heat stroke; total incapacitation likely |

produces as sweat) is considered to be 1% dehydrated. The effects of various levels of dehydration are shown in Table 5.2.

Dehydration can occur within a single work shift, be cumulative over several days of work, or can result from a single night's alcohol consumption. Variability in state of worker hydration is a major stumbling block in prescribing an environmentally-based index as a heat stress standard.

## 5. The role of heat production in heat stress

The ability of an individual to do physical work is restricted at the upper end by a physiological limit called the maximum oxygen uptake, i.e., the ability of the body to deliver oxygen to the working muscles. The maximum oxygen uptake, in litres per minute, is a function of fitness, modified by age and sex; use of 1 litre of oxygen per minute is equivalent to 5 kilocalories per minute of working metabolic demand. The average capacity ranges from just under 10 kcal min$^{-1}$ (600 kcal h$^{-1}$ or 700 W) for less fit women to just over 20 kcal min$^{-1}$ (1200 kcal h$^{-1}$ or 1350 W) for young men of good physical fitness. However, maximum work capacity is only sustainable for a brief period of time, usually of the order of 6 minutes or less.

The most fit and highly trained athletes may be able to work at 70% of the VO$_2$ max for 4–5 hours, but most individuals are able to perform sustained work at 50% of their individual maximum work capacities or less. Our studies have shown that when the work is limited to 3 or 4 hours, even if the workers are well motivated by incentive pay, team spirit or outstanding leadership, the voluntary hard work level sustained by individuals tends to average only 45% of their maximum work capacity (Hughes and Goldman, 1970). Extensive tables exist for the energy cost of almost all human activities. These energy expenditure rates suggest that the hourly average level of heat production during 'hard' work will seldom exceed 500 W; indeed, in the absence of heavy manual labour it will seldom exceed 300 W (258 kcal h$^{-1}$). Heat production is a function of body weight at rest ($1.5$ W kg$^{-1}$), and of the sum of body plus load weight moved and the velocity during work ($2$ W kg$^{-1}$ × $V^2$); adjustments for grade, terrain and heavier loads allow prediction of working heat production (Givoni and Goldman, 1971) or, if hard work is involved, of probable sustained work rate (Hughes and Goldman, 1970).

While there may be a slight rise in the energy expended for a given task as the worker becomes hotter, this will usually be offset by the individual reducing his activity level unless pushed by machismo, or an uninformed supervisor. However, as heat exhaustion collapse becomes imminent, there can be a dramatic rise in metabolic heat production (Goldman, 1965) as the worker becomes unco-ordinated, stumbles, fights to remain conscious, etc. Workers at this advanced stage of work problems may be sufficiently 'obtunded', i.e., suffering a lack of blood carrying oxygen to the brain, that they are working on reflex. They may not be aware of any difficulty, be highly aggressive and may resist being removed from work. Obviously, such individuals are at severe risk and part of supervisor training should include being alert for these symptoms/signs of incipient heat illness.

Differences in the heat production demands of various tasks, and in the heat production between individuals, or even in the same individual as a result of immunizations or infection, again suggest that environmentally-based heat stress standards will fail to prevent heat illness.

## 6. The role of clothing in heat stress

The heat stress associated with any clothing item can be directly related to three factors. The first is the overall thickness of the materials and air spaces between the skin and the environment. This includes: the insulation of the still air trapped within the clothing materials; the insulation of the still air layers trapped between the clothing materials and the skin, and between each layer of clothing; and the still air insulation provided by the surface air layer that adheres to the outside of the garment. This outside still air layer provides roughly $0.8$ clo units of insulation to an unclothed man. This unit of insulation, the clo, is directly analogous to the $R$ value conventionally used with building insulation materials, simply being some 14% different in magnitude. The major advantage of using the clo as the unit of insulation is that one can then express the heat loss that will occur for an average adult male, who has $1.8$ m$^2$ of surface area, using the simple relationship that such an individual will lose 10 kcal h$^{-1}$ of heat by radiation and convection for every degree (°C) difference between his average skin temperature and the air temperature with 1 clo unit of insulation; half that (i.e.,

5 kcal h$^{-1}$) will be lost with 2 clo units of insulation, etc. (Goldman, 1978; Goldman, 1984).

The extent to which air can penetrate the clothing is a key feature in the extent to which the insulation of a clothing ensemble is altered by wind or wearer motion. Air and/or worker movement can reduce the insulation of the external air layer (from 0·8 clo in essentially still air to as low as 0·2 clo in a 5·4 m s$^{-1}$ breeze), and also can break down the outer layer of insulation thus shunting heat through the insulation if: (*a*) the weave is open enough; or (*b*) the closures or design of the garment allow sufficient air pumping by wearer movement to shunt hot air near the skin to the outside without going through the materials at all. This change in insulation with wind or wearer motion can be characterized by a pumping coefficient (*p*). This coefficient simply describes the rate of loss of insulation with increasing effective air motion, where effective air motion is defined as a composite of the air motion generated by wind plus the air motion generated by wearer activity (Givoni and Goldman, 1972).

Ideally, the sweat produced by the wearer should evaporate and transfer to the ambient environment by unrestricted passage through the clothing. However, some sweat usually evaporates into the intra-clothing space and the humidity inside the garment may build up. With less permeable clothing this microclimate humidity build–up may reach the point where further sweat evaporation becomes limited. The ability of sweat to be evaporated through clothing can be characterized by a permeability index ($i_m$). In essence, a permeability index of '1' implies that an individual has the same capacity for evaporative heat loss as a psychrometric wet bulb thermometer (a thermometer with a fully wetted surface, ventilated by a fan or slung at the end of a chain to produce maximum evaporative cooling). If one were to wrap such a thermometer in a vapour barrier membrane, one would expect no evaporative cooling and the permeability index ($i_m$) would be 0.

A 100% sweat-soaked, nude man will have a permeability index of about 0·5, i.e., the still air layer next to the sweating skin will pose a limit to the rate at which moisture vapour can evaporate of about 50% of the maximum evaporative heat transfer allowed by a fully ventilated wet bulb thermometer. The permeability index found for most conventional clothing, e.g., a long-sleeved shirt and trousers, a military uniform or a surgical scrub suit, will be about 0·45.

The permeability index represents a form of diffusion coefficient for evaporative transfer. The actual net evaporative cooling obtained by the wearer will also be a function of the path length, i.e., the distance between the skin and the ambient environment. Since insulation is linearly related to the length (thickness) of the diffusion path, one can calculate the net effective evaporative cooling available to the wearer of an ensemble (or the limit imposed by a material) using the ratio $i_m$/clo. Typical values of this 'permeability index ratio' for a sweating, nude man would be about 0·6, in contrast to a man wearing a long-sleeved shirt and trousers who would have a value of about 0·32 $i_m$/clo, i.e., a nude man in a low air motion environment can get about 60% of the maximum evaporative cooling, while a clothed man can get only about 32%. The rate at which the permeability index changes with air motion appears to be a function of the same pumping coefficient that modifies clothing insulation, with insulation being decreased in relationship to the pumping coefficient [i.e., $A$ (clo)$^{-p}$] and moisture permeability being increased at the same rate [i.e., $B$ $(i_m)^{+p}$]. Thus, the effect of the pumping coefficient on the actual evaporative cooling that can be obtained is doubled, i.e., $(i_m$/clo$)^{+2p}$ (Goldman, 1984).

For values of insulation and permeability of clothing ensembles to be comprehensible, one must give the insulation and moisture permeability at a specific air velocity. Thus, the value of insulation measured at a given wind speed can be extrapolated, using the pumping coefficient, to the equivalent insulation at any effective air motion. Pumping coefficients range from 0·3 for the nude man, to 0·21 for a fatigue uniform, to 0·09 for a completely impermeable vapour barrier coverall with mask, hood and gloves. The typical 1·4 clo units of insulation measured at 0·75 m s$^{-1}$ of air motion for army fatigues provides a base $(A)$ value of 0·86 clo at 1 m s$^{-1}$. Correspondingly, the $i_m$ of about 0·45 measured at 0·75 m s$^{-1}$ becomes a base value $(B)$ for $i_m$ of 0·66 clo expressed in the basic 1 m s$^{-1}$ form relative to a pumping coefficient of 0·20.

Table 5.3 includes values (measured at 0·75 m s$^{-1}$ air flow) for a range of industrial clothing, hazardous waste protective garments, surgical gowns, military protective clothing, and types of cotton coveralls, hoods and gloves, used in single or double layers, with or without masks. A value is also given with a vapour barrier suit over all, as these items are used in industry. The clothing ensemble values range from a low insulation value of 1·3 for a surgical scrub suit by itself, to a maximum of 2·3 for a complete US military NBC overgarment worn with a fatigue uniform worn underneath, with gas mask, hood and gloves.

The permeability indices for these types of ensemble range from the 0·5 for the nude man to a low of 0·06 for an individual wearing a single cotton coverall with a complete vapour barrier suit plus a mask with double hoods and two pairs of gloves. In essence, a simple Tyvek® coverall and hood worn over a long-sleeved shirt and trousers allows an $i_m$/clo of only 0·165; i.e., a man wearing such an ensemble can get only about 16% of the maximum evaporative cooling allowed by the environment, while a sweating, nude man can get 60%. Note also that

*Table 5.3. Key clothing parameters for typical garments (measurements at 0·75 m s⁻¹).*

|  | clo | $i_m$ | p |
|---|---|---|---|
| Nude man, sweating | 0·860 | 0·510 | 0·593 |
| Scrub suit | 1·280 | 0·450 | 0·352 |
| Surgical gown #1+scrub | 1·690 | 0·276 | 0·213 |
| Surgical gown #2+scrub | 1·680 | 0·340 | 0·118 |
| Tyvek+utility+hood | 1·670 | 0·276 | 0·165 |
| Safeguard 1·5+utility+hood | 1·730 | 0·286 | 0·165 |
| Safeguard 2·25+utility+hood | 1·750 | 0·259 | 0·148 |
| Tyvek+nude+hood | 1·350 | 0·241 | 0·179 |
| Safeguard 1·5+nude+hood | 1·440 | 0·283 | 0·197 |
| Safeguard 2·25+nude+hood | 1·510 | 0·289 | 0·191 |
| US Army fatigues | 1·400 | 0·442 | 0·250 |
| NBC OG+fatigues | 2·170 | 0·280 | 0·226 |
| NBC OG+nude+mask and hood | 2·040 | 0·265 | 0·130 |
| NBC OG+fatigue+mask and hood | 2·320 | 0·194 | 0·084 |
| Cotton coveralls with mask, hood and gloves | 1·470 | 0·270 | 0·195 |
| 2 cotton coveralls with mask, 2 hoods and 2 pairs of gloves | 1·910 | 0·260 | 0·114 |
| Cotton coverall+VB suit with mask, 2 hoods and 2 pairs of gloves | 1·780 | 0·090 | 0·091 |

most single-layer clothing systems allow between 30 and 35% of the maximum evaporative cooling allowed by the environment, while most double-layer ensembles allow only between 10 and 20% of the maximum evaporative cooling. In essence, even if we consciously avoid introducing impermeable materials, the amount of insulation inherent in a two-layer clothing system reduces the maximum evaporative cooling allowed an individual to roughly 25% of that allowed a nude man, while a single-layer ensemble reduces the maximum available cooling to about half of that allowed for a nude man.

Obviously, heat stress is dramatically increased as clothing is increased. Any heat stress index which cannot handle the increasingly complex clothing required to protect workers cannot characterize the stress appropriately.

## 7. Prediction modelling of heat stress

The ability to lose heat at a given temperature, humidity and air motion is directly predictable from the physical insulation, moisture permeability and pumping coefficients discussed in the prior section. The maximum heat loss can now be compared for workers wearing typical protective clothing. A 'standard', 25-year old, 173 cm, 70 kg male worker of average fitness was assumed and a heat production of 300 W specified. It was also assumed that the worker would be fully heat-acclimatized, with typical temperatures of 35°C for his skin and 37°C for deep body, with a standing heart rate of 72 bpm at rest (Givoni and Goldman, 1973). The environment selected for comparing the effects of clothing on the worker was 30°C (85°F), 50% RH, 0·67 m s$^{-1}$ wind speed condition; this represents the low end of the heat stress zone for such work.

Table 5.4 presents the heat loss capacity for an unclothed, sweating individual, and for a sweating individual wearing just T-shirt and

Table 5.4. *Maximum heat loss for workers wearing typical protective clothing. Part I. Unclothed vs T-shirt and shorts.*

| | | |
|---|---|---|
| Height = 68·00 in | Heat production = 300 Watts | |
| Weight = 154·00 lbs | = 258 kCal h$^{-1}$ | |
| Age 25 years male | This is 31% of maximum work capacity | |
| Average fitness | Perceived exertion (RPE) = 10·2 | |
| Max. work capacity = 971 Watts | Fairly light | |
| 41·0 ml/kg min$^{-1}$ | | |

| | | |
|---|---|---|
| Skin temperature = 95°F | Days in heat 7 | |
| Initial rectal temperature = 98·6°F | Initial heart rate = 72 BPM | |
| Dry bulb temp = 85°F | Relative humidity = 50% | Wind speed = 1·50 mph |

| | Nude, sweating skin | T-shirt and shorts |
|---|---|---|
| Cu man (clo) | 0·53 | 1·04 |
| Cu man $i_m$ | 1·02 | 0·94 |
| Pumping coefficients | 0·30 | 0·28 |
| ———— Work (300 Watts) ———— | | |
| Final rectal temperature | 99·9°F | 100·1°F |
| Change in temperature due to: | | |
| Metabolic load | 2·2°F | 2·2°F |
| Radiation and convection | −0·4°F | −0·2°F |
| Evaporation | 0·0°F | 0·0°F |
| $E_{req}$ | 53 Watts | 155 Watts |
| $E_{max}$ | 2396 Watts | 1078 Watts |
| % sweat wetted area | 2% | 14% |
| Estimated sweat loss | 71 g h$^{-1}$ | 207 g h$^{-1}$ |
| Discomfort vote | −0·2 | 0·4 |
| Clo eff | 0·33 | 0·67 |
| $i_m$ eff | 1·14 | 1·04 |
| ———— Rest (105 Watts) ———— | | |
| Final rectal temperature | 98·6°F | 98·9°F |
| Change in temperature due to: | | |
| Metabolic load | 0·8°F | 0·8°F |
| Radiation and convection | −0·3°F | −0·2°F |
| Evaporation | 0·0°F | 0·1°F |
| $E_{req}$ | −54 Watts | 26 Watts |
| $E_{max}$ | 1508 Watts | 700 Watts |
| % sweat wetted area | Dry | 22% |
| Estimated sweat loss | 0 g h$^{-1}$ | 32 g h$^{-1}$ |
| Discomfort vote | −0·5 | −0·1 |
| Clo eff | 0·41 | 0·83 |
| $i_m$ eff | 0·90 | 0·84 |

shorts, using the insulation (clo), moisture permeability index ($i_m$) and pumping coefficient ($p$) as measured on a heated, sweating, copper manikin (Goldman, 1984). The predicted mean comfort vote for such individuals is well within the comfort zone and the working condition is projected to be comfortable, with the nude individual really preferring to be drier than he is when sweating; the individual wearing just a T-shirt and shorts is also not sweating at rest, but would have little requirement for sweat production and a low sweat-wetted area.

The model used (Givoni and Goldman, 1971, 1972, 1973a,b; Shapiro *et al.*, 1982) predicts the final equilibrium rectal temperature that would be reached if steady state could be achieved, and how much of the rise in rectal temperature above the initial temperature is due to metabolic heat production, how much results from non-evaporative (i.e., radiation and convection) heat transfer, and any increment in the final rectal temperature associated with difficulty in obtaining the necessary evaporative cooling. The model also projects the required sweat evaporation ($E_{req}$), compares it with the maximum sweat evaporative capacity ($E_{max}$), and presents the ratio of $E_{req}/E_{max}$ as the percent sweat-wetted area.

Sweat-wetted areas less than 20% are considered comfortable. Areas (skin relative humidity) between 20 and 40% are considered uncomfortable but in no way performance decrementing, between 40 and 60% are considered both performance decrementing and exposure limiting if the worker has an easy option to discontinue work, while few workers will continue to work at levels of skin humidity above 60% (i.e., percent sweat-wetted area greater than 0·60). The model also projects the estimated sweat losses (Shapiro *et al.*, 1982) which, in turn, can serve as a measure of replacement drinking water requirements. It will flag any unusual $E_{req}$ or $E_{max}$ values and produce a discomfort vote. Even though the original comfort vote was limited to a seven point scale (from a −3 for cold, to comfortable at 0, to hot at +3), others have extended this procedure to a +4 for very hot. The author's experience using the model indicates that conditions where subjects will stop work correspond roughly to a +5 vote; the range of votes between +5 and +10 are consistent with conditions producing heat exhaustion collapse and the range above +10 votes appear to represent increasing risk of heat stroke. Since the comfort vote procedure is based on the heat balance equation, these relations with heat stress are not surprising, although more study is required to validate them. Finally, the effective insulation and moisture permeability at the given wind speed and work rate are given; these values reflect the effects of the pumping coefficient and its expression of the effects of wind and body motion (Givoni and Goldman, 1972).

As can be seen in Table 5.4, in a 30°C (85°F), 50% RH environment individuals at rest without clothing actually are on the cool side of comfort, while wearing a T-shirt and shorts the subject is essentially comfortable at rest ($-0.1$ vote). During work at 300 W, the unclothed individual is still comfortable ($-0.2$), with only 71 ml h$^{-1}$ of water intake required. With a T-shirt and shorts, work at 300 W is still more comfortable than slightly warm, i.e., a discomfort vote of 0.4, with a 14% sweat-wetted area and a requirement ($E_{req}$) for only 155 W of evaporative cooling under conditions where potentially one could get over a kilowatt of evaporative cooling ($E_{max}$).

Table 5.5 presents the corresponding results with three types of work clothing; a fairly heavy, 8.5 oz yd$^{-2}$, long-sleeved cotton shirt and trousers (the US Army fatigue uniform); a Tyvek® coverall and hood

Table 5.5. *Maximum heat loss for workers wearing typical protective clothing.*
*Part II. Long-sleeved shirt and pants vs Tyvek® coverall and hood alone and with long-sleeved shirt and pants.*

| | Army fatigues (long sleeve shirt and pants) | Tyvek® coverall and hood without work clothes | Tyvek® coverall + work clothes |
|---|---|---|---|
| Cu man (clo) | 1·34 | 1·10 | 1·44 |
| Cu man ($i_m$) | 0·75 | 0·44 | 0·44 |
| Pumping coefficients | 0·25 | 0·31 | 0·25 |
| ----- Work (300 Watts)----- | | | |
| Final rectal temperature | 100·4°F | 100·5°F | 101·0°F |
| Change in temperature due to: | | | |
| Metabolic load | 2·2°F | 2·2°F | 2·2°F |
| Radiation and convection | −0·1°F | −0·2°F | −0·1°F |
| Evaporation | 0·2°F | 0·3°F | 0·7°F |
| $E_{req}$ | 181 Watts | 157 Watts | 185 Watts |
| $E_{max}$ | 629 Watts | 498 Watts | 345 Watts |
| % sweat wetted area | 29% | 32% | 54% |
| Estimated sweat loss | 298 g h$^{-1}$ | 297 g h$^{-1}$ | 400 g h$^{-1}$ |
| Discomfort vote | 1·1 | 1·3 | 2·4 |
| Clo eff | 0·90 | 0·68 | 0·96 |
| $i_m$ eff | 0·82 | 0·49 | 0·48 |
| ----- Rest (105 Watts) ----- | | | |
| Final rectal temperature | 99·0°F | 99·1°F | 99·4°F |
| Change in temperature due to: | | | |
| Metabolic load | 0·8°F | 0·8°F | 0·8°F |
| Radiation and convection | −0·1°F | −0·2°F | −0·1°F |
| Evaporation | 0·2°F | 0·4°F | 0·6°F |
| $E_{req}$ | 45 Watts | 29 Watts | 49 Watts |
| $E_{max}$ | 427 Watts | 311 Watts | 233 Watts |
| % sweat wetted area | 42% | 51% | 79% |
| Estimated sweat loss | 70 g h$^{-1}$ | 52 g h$^{-1}$ | 101 g h$^{-1}$ |
| Discomfort vote | 0·2 | 0·2 | 0·8 |
| Clo eff | 1·09 | 0·86 | 1·17 |
| $i_m$ eff | 0·68 | 0·39 | 0·40 |

worn without work clothing beneath it; and a Tyvek® coverall and hood worn over a long-sleeved shirt and trousers. Note that both at rest and work, the physiological effects of wearing the long-sleeved shirt and pants are quite comparable to wearing the Tyvek® coverall and hood without other clothing; these conditions are too mild to be an 'effective forcing function' for discriminating between these two garments. The somewhat slightly higher effective insulation of the heavy, long-sleeved shirt and trousers is counterbalanced by the substantially lower moisture permeability for the Tyvek® material under these relatively moderate conditions of work and temperature. If the requirement for sweat evaporation was substantially greater, as it would be at higher work rates or hotter environmental conditions, wearing the Tyvek® coverall and hood would have been substantially hotter than wearing the long-sleeved shirt and pants. An adequate forcing function to discriminate between these two ensembles would require greater dependence on obtaining cooling by sweat evaporation, i.e., as $E_{req}$ increases, the effects of any reduction in moisture permeability become much more marked.

The additional stress of the Tyvek® coverall worn over the work clothing is clearly evident in this Table. Even at this moderate temperature and work rate, the final equilibrium deep-body temperature is 0·3°C (0·5°F) higher with the addition of the long-sleeved shirt and trousers under the Tyvek® coverall and hood, while the maximum evaporative cooling is about 30% lower (500 *vs* 345 W). Thus, although the difference in required evaporative cooling is relatively small (less than 30 W), the percent sweat-wetted area increases from 32% to 54%, the discomfort vote increases from a 'slightly warm' 1·3 to a 'warm-to-hot' 2·4 value, and sweat losses increase by 100 g h$^{-1}$ at work. Even under this relatively mild temperature and work combination, workers would need to drink about one litre of water per hour to avoid dehydration.

Some other items to note in Table 5.5 are that, while the measured insulation is significantly increased by adding the work clothing under the Tyvek® coverall and hood, as stated previously the moisture permeability index ($i_m$) is unchanged. Note that the moisture diffusion of the material is limited in this case by the Tyvek® coverall and little altered by adding material with better permeability indices. However, the path length for evaporative cooling (the $i_m$/clo ratio) is increased and the evaporative resistance is greater with increasing insulation *per se*. The effects of these differences in physical measurements can be seen in the projected changes in $E_{max}$, sweat-wetted area and estimated sweat productions, as well as the change in final rectal temperature associated with the impaired evaporation in Table 5.5 for the Tyvek® coverall and hood with and without work clothing.

*Table 5.6. Clo insulation units for individual items of clothing and formula for obtaining total intrinsic insulation.*

| Men's clothing | Clo units | Women's clothing | Clo units |
|---|---|---|---|
| **Underwear** | | | |
| Sleeveless | 0·06 | Bra and panties | 0·05 |
| T–shirt | 0·09 | Half slip | 0·13 |
| Underpants | 0·05 | Full slip | 0·19 |
| **Torso** | | | |
| Shirt | | | |
|   Light, short sleeve | 0·14 | Blouse | |
|       long sleeve | 0·22 |   Light | 0·20 |
|   Heavy, short sleeve | 0·25 |   Heavy | 0·29 |
|       long sleeve | 0·29 | Dress | |
| (Plus 5% for tie or turtleneck) | |   Light | 0·22 |
| | |   Heavy | 0·70 |
| Waistcoat (vest) | | Skirt | |
|   Light | 0·15 |   Light | 0·10 |
|   Heavy | 0·29 |   Heavy | 0·22 |
| Trousers | | Slacks | |
|   Light | 0·26 |   Light | 0·26 |
|   Heavy | 0·32 |   Heavy | 0·44 |
| Sweater | | Sweater | |
|   Light– | 0·20[1] |   Light | 0·17 |
|   Heavy | 0·37[1] |   Heavy | 0·37 |
| Jacker | | Jacket | |
|   Light | 0·22 |   Light | 0·17 |
|   Heavy | 0·49 |   Heavy | 0·37 |
| **Footwear** | | | |
| Socks | | Stockings | 0·01 |
|   Angle length | 0·04 |   Any length | 0·01 |
|   Knee length | 0·10 |   Panty hose | |
| Shoes | | Shoes | 0·02 |
|   Sandals | 0·02 |   Sandals | 0·04 |
|   Oxfords | 0·04 |   Pumps | 0·08 |
|   Boots | 0·08 |   Boots | |

Total i = 0·8 Σ insulation of individual clothing items + insulation provided by the external air layer (0·8).

The insulation added by various clothing items can be estimated most readily by use of Table 5.6, the clothing insulation estimation approach developed for the American Society of Heating, Refrigerating and Air Conditioning Engineers (ASHRAE). One simply sums the individual insulation (expressed in clo units) provided by the individual items being worn, takes 80% of the total to offset the effects of overlap and compression of one item by another item, and then calculates the total insulation by adding the 0·8 clo provided by the external (surface) still air layer. As indicated above, the moisture permeability index will not be altered by the number of layers of material, but will be dramatically affected by the moisture permeability index of the least permeable material in the ensemble. The pumping coefficient will be a complex function of the design of the garment, particularly the design of apertures and closures, the stiffness of the material, and the closeness of garment fit. The nature of the outermost layer will have the greatest impact on pumping coefficient.

In summary, under warm conditions, adding any insulation compromises the ability to obtain evaporative cooling; adding any impermeable or reduced permeability insulating garments seriously compromises the ability to obtain the required evaporative cooling. Variation in clothing worn by workers at various tasks could be dealt with by specifying the clothing worn. Wyndham, favoured by saturated vapour pressure (and limited air motion), has been able to reduce heat illness dramatically for miners in South Africa (Wyndham and Heyns, 1971), while the military, with its homogenization of clothing (and physical fitness) has had similar success during training and field operations using the WBGT index (US Army, 1980). However, with introduction of complex protective ensembles, heat stress has become an increasing problem. Any heat stress standard which cannot handle clothing variables will be unable to deal with one of today's most prominent causes of heat stress.

## 8. Prevention of heat illness

There are three major elements in reducing heat illness:

1. Heat acclimatization.
2. Water intake.
3. Adjusting the work and the rest breaks.

### Heat acclimatization

As indicated above, workers in good physical condition have a better chance of avoiding heat illness. Heat acclimatization is a term used to work in the heat are considered heat acclimatized and have the best chance of avoiding heat illness. Heat acclimatization is a term used to describe a number of changes in the body that result from working in the heat over a period of days.

There is no way an individual exercising under cool conditions can attain the full acclimatizing benefits of working in the heat. While the conditioning that occurs as a result of hard physical workouts is an important part of the total heat acclimatization process, it is necessary to work for about 2 hours a day in the heat to fully acclimatize to heat. How much work and how much heat depend on the body's ability to last for 2 hours. Getting heat exhaustion collapse during conditioning is not helpful. Working more than 2–4 hours a day in the heat does not seem to help the heat acclimatization process much and working 1 hour a day in the heat is not enough. The level of work and heat should be at least that expected to be met during the work exposures; one can be partially acclimatized by being conditioned to lower levels of work and temperature than will be actually experienced. In theory, one can also be over-acclimatized, but as long as the worker is able to sustain 2–4 hours of work under the hottest conditions, there is little evidence that over-acclimatization is in any way disadvantageous.

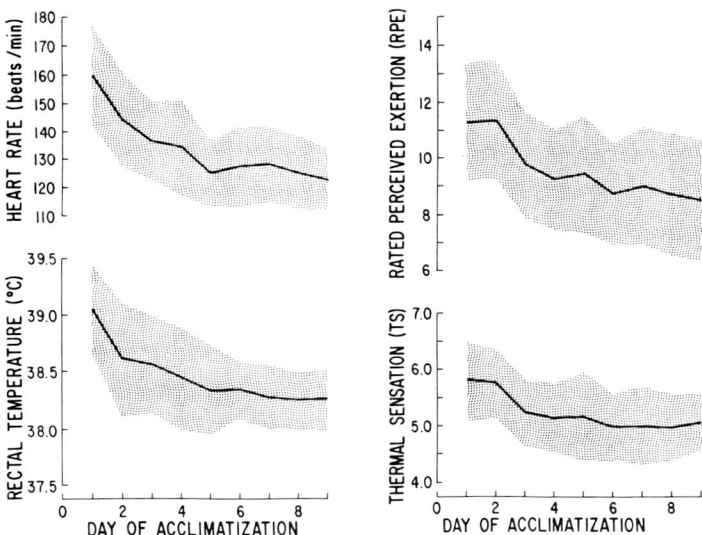

*Figure 5.2.    The pattern of heat acclimatization (Pandolf et al., 1977a).*

It takes from 5 to 7 days of working about 2 hours a day in the heat to become fully heat acclimatized. However, it is not a simple straight line relationship, since 30–40% of the acclimatizing benefits will occur after the first day, another 15–20% after the second day, another 10–15% after the third day, etc., as shown in Figure 5.2. Individuals in good physical condition from exercising under cooler conditions will start the heat acclimatization process at about the day 3 level, i.e., they will be about 50% heat acclimatized before they start working in the heat (Pandolf *et al.*, 1977).

Put simply, during heat acclimatization the body begins to sweat at a lower skin temperature, and each sweat gland produces more sweat. The sweat produced also has less salt so that there is less chance of heat cramps after the first few days of heat acclimatization.

An acclimatized individual wearing normal clothing can generally perform as well under conditions some 2°C hotter than an unacclimatized individual. However, wearing protective clothing partially blocks the effects of one of the major benefits of heat acclimatization, i.e., extra sweat production, since such clothing limits the amount of sweat that can be evaporated (Goldman, 1979). Thus, any adjustment to heat stress indices based on acclimatization to heat may be negated by clothing, another complication in the search for a universal heat stress standard.

## Water intake

Thirst is not an adequate signal for how much water it is necessary to drink; even if the work force can have all the water they want, they may still dehydrate, i.e., lose body water, by 6–8%. This is well beyond the 5% level at which performance is seriously affected (Pandolf *et al.*, 1977). The solution is to ensure adequate and frequent water intake by direction. Although difficult to ensure, the following guidelines for water intake (Table 5.7) have been suggested for workers wearing normal, permeable, work clothing. Note that these are couched in terms of the WBGT (Wet Bulb Globe Temperature) environmental index (ISO 7243, 1982; Joy and Goldman, 1968; US Army, 1980).

Simply tracking the workers' weight, fully dressed, before the day's work begins, at noon before lunch, or every few hours during the work day, to make sure they are taking enough water to maintain a relatively constant weight is one of the simplest ways to spot heat problems before they become serious, as well as to prevent them (Goldman, 1979). The requirement for sweat production is sufficiently keyed to heat stress that it has been standardized (ISO 7933, 1986) for use as a heat stress index. However, relying on the demand for sweat production as a heat stress index, without considering the worker's capacity to sustain sweat production and avoid dehydration by assuring sufficient replacement of water (sweat) losses by ingestion, does not appear to be a satisfactory way of preventing heat illness.

*Table 5.7. Suggested water intake and work/rest cycles during training for heat acclimatized workers wearing ordinary clothing (long-sleeved shirt and trousers).*

| Heat condition | WBGT(°F) | Minimum water intake (qt h$^{-1}$)[1] | Work/rest cycles (min) |
|---|---|---|---|
| Green | 82–85 | 0·5–1·0 | 50/10 |
| Yellow | 85–88 | 1·0–1·5 | 45/15 |
| Red | 88–90 | 1·5–2·0 | 30/30 |
| Black | 90 and above | 2·0 | 20/40 |

[1] 1 US quart = 0·94 l.

## 9. Maintaining physical performance

1. Drink 1 litre of water before breakfast, another at each meal, and another before any hard work.

2. Take frequent drinks, since they are more effective than all at once. Larger men need more water.

3. Replace salt loss by eating three meals per day; use extra salt, on your food or in your water, if you are not eating regularly, particularly during the first few days of work in the heat.

4. As the temperature increases, individual rest breaks must be more frequent, work rate lowered, and protective clothing may have to be reduced.

## 10. Practical modifications of work practices to reduce heat stress

Table 5.7 suggests that adjustments in work-to-rest times may be necessary to minimize heat stress during a work day. Of course, it may be useless to adjust the work/rest cycle if the work level is already low. It is then especially important to reduce the amount of work and heat load in any other way possible:

1. Provide shade at each worker's work location.

2. Rotate the workers into any available shade.

3. Rotate workers out of the protective clothing.

4. Reduce the risk by working early (dawn until 11:00 am) and late (4:00 pm to dusk) in the day to avoid the peak periods of heat.

Being in the shade instead of the sun can have the same effect as lowering the outdoor air temperature by 2°C with clothing and by as much as 7°C when little or no clothing is worn. Anything that will reduce the work lowers the heat produced by the body. Simply sitting down instead of standing can have the same effect as lowering the air temperature by 2 or 3°C (Goldman, 1984).

Heat injuries are almost totally preventable. Indeed, in the Israeli Defence Forces a heat injury is the subject of a court martial inquiry – not of the victim but of his supervising officer.

## 11. Heat stress standards

Having decried the possibility of a successful universal heat stress standard, what is left? Perhaps the simplest way to reduce heat stress is to monitor the ambient environment and use guidance from 'decision trees' (see below), or predictive modelling to deal with all four environmental variables, the three clothing variables, the multiple task variables affecting heat production, plus the numerous worker status variables. Then one can introduce work practices to moderate the level of effort, or completely eliminate work during ambient conditions associated with high risk of heat stress. However, there are significant differences between workers, with smaller individuals and those of low work capacity (low $VO_2$ max) at greater risk, while an individual worker with good resistance to heat stress may suffer temporary decreases in stress resistance as a result of upper respiratory (or more severe) infections, dehydration (usually occurring as a result of alcohol ingestion), or the loss of sweating capability associated with heat rash (US Army, 1980). Thus, environmental conditions which might be 'safe' for a physically fit, young, fully heat-acclimatized, hydrated and healthy worker, would place impaired individuals at high risk of heat illness. Therefore, personal monitoring offers a greater degree of safety when sufficient control, and rigorous supervision of the work force are allowed by work location, nature of the work and union/management agreed working relationships.

### Environmental monitoring

Environmental monitoring is now routinely done (Kleinhanz *et al.*, 1980) and has been standardized (ISO 7726, 1985; Aubertin, 1986). A wide variety of commercial, off-the-shelf systems are available for spot measurement, for scanning or for continuous recording. The effects of various combinations of the four key environmental factors (air temperature, air motion, humidity and mean radiant temperature) in producing heat stress in workers wearing normal work clothing are well understood (Juptner, 1984).

A number of environmental heat indices exist to characterize the discomfort of any environment. Most such indices only evaluate environments for resting individuals. Some indices exist which integrate the effects of different work levels with subjective comfort. Other indices allow integration of the four key environmental factors plus some adjustments for work as well as for additional conventional clothing. However, as discussed, few of these environmental indices are relevant for an individual encapsulated in totally impermeable clothing; such individuals are hardly affected by the ambient humidity or air motion. They live in the microclimate within the clothing which is generated as a result of the clothing being relatively impermeable to both wind and water vapour.

The Wet Bulb Globe Temperature (WBGT = 0·7 natural wet bulb

temperature + 0·2 Black Globe Temperature + 0·1 Dry Bulb Temperature) is the most widely used heat stress (but not heat discomfort) index. It has proven remarkably successful in reducing heat casualties for military trainees (US Army, 1980). Use of the 'natural' wet bulb temperature in the WGBT index, rather than the conventional Weather Bureau 'psychrometric' (i.e., fully ventilated) wet bulb, provides some compensation for the fact that the worker cannot be slung by the heels to gain the maximum evaporative cooling projected by the ventilated 'psychrometric' wet bulb thermometer. Also, use of the natural wet bulb incorporates an adjustment for actual air motion. Note that when wind or worker movement across the ground reaches about 3 m s$^{-1}$, the natural wet bulb is essentially the same as the psychrometric wet bulb. Inclusion of a 20% allowance in the WBGT for radiant load ($T_g$), with only 10% of the index made up by the air temperature, is appropriate for field workers exposed to direct and diffuse solar radiant heat plus the re-radiation off the heated earth.

With the recent introduction of the Wet Globe Thermometer ('Botsball'), a simple, portable device which can go to the work site and has a single dial which can be colour-coded for ease of interpretation (cf. Table 5.7), is available. The Wet Globe Temperature (WGT) is comparable to the WBGT with about a 2°C offset (Onkaram *et al.*, 1980) but can provide data where the worker is, rather than at some weather station some miles away, and the output is easily understood. The WGT has been adopted by all three of the US armed forces (US Army, 1980) and has been proven to reduce heat casualties when provided to troops in the field.

Use of an index which relies on the natural wet bulb (WBGT, WGT), and thus depends heavily on ambient vapour pressure, is however inappropriate for workers wearing protective clothing which is, to a considerable extent, impermeable to moisture vapour (Goldman, 1979; Goldman, 1984). Regardless of the ambient wet bulb, it is the microclimate wet bulb within the clothing to which the worker's skin attempts to transfer heat by sweat evaporation that is important. This microclimate has to be considered as air at close to skin temperature, with 100% relative humidity. While ambient WBGT/WGT are probably more appropriate than any of the ambient environmental indices that are based on psychrometric wet bulb (Effective Temperature (ET), Corrected Effective Temperature (CET), Predicted 4-Hour Sweat Rate (P4SR), the Temperature Humidity Index (THI) recently promulgated by the US National Weather Service, and the like), none of the indices are really relevant for an encapsulated individual.

A rule of thumb, developed by Goldman for troops wearing air-permeable, chemical protective, charcoal-in-foam overgarments, suggests adding 5°C to the ambient WBGT index as compensation for such suits (US Army, 1980; Goldman, 1984). In low winds, the difference between the heat stress associated with these garments and a truly impermeable garment is relatively slight, so the 5°C increment to the measured ambient WBGT seems reasonable to use for a worker wearing complete encapsulation, if ambient environmental monitoring is to be specified for safety assessment.

Note that, given the above limitations, almost any of the environmental indices can be selected for monitoring ambient environment. There will be little difference between them at the very warm to hot conditions of concern and they will all be equally subject to severe underprediction of heat illness given a high wind, or a high solar radiation heat load in combination with a high wind. Simply using a 15 cm black globe temperature by itself to include the air temperature and radiant heat coming from the sun, earth and any structures or engines within the 360° radiant field surrounding the worker, is probably as good as using any of the indices involving humidity measurement.

The actual critical level for various work–rest adjustments, or for terminating discretionary work until cooler conditions occur is, of course, subject to modification as a function of the level of work (i.e., heat production) of the worker. Each 25 kcal h$^{-1}$ of increased heat production roughly equates to a requirement to drop the ambient temperature by 1·5°C to compensate. Under circumstances where some workers have very high heat production, i.e., are required to do harder physical work, monitoring environmental temperature can provide only a crude guideline to the actual risk of heat stress, particularly when any clothing is worn.

Environmental monitoring would perhaps be most successful if it proves possible to measure the true environment to which the worker is exposed, i.e., the microclimate within his clothing. A measurement of the temperature of the air within the clothing is perhaps the simplest measurement to assess heat stress but measurement of the microclimate vapour pressure would be very much better if it can be obtained without too much interference with the work. Measurement of the temperature and relative humidity of the microclimate within the clothing would allow the very large data base that exists for workers wearing conventional work clothing, on tolerance, accident rates, etc., to be directly applied to workers wearing completely encapsulating protective clothing.

## Personal monitoring

Personal monitoring is an excellent way to monitor the status and track the increasing risk of heat illness incurred by individuals working in laboratory settings. It has also been used extensively in the field by the military as a way to eliminate any heat illness or, on occasion, to selectively produce a given percentage of heat exhaustion casualties without incurring heat stroke (Joy and Goldman, 1968); and to limit the risk of heat stroke during treatment of cancer patients involving induction of very high body temperatures in a hospital setting (Bynum *et al.*, 1978).

Measurement of heart rate, skin temperature, skin humidity, deep body temperature, sweat rate, or the ratio between sweat evaporation and sweat production, are all useful forecasts of heat illness if monitored on a continuing basis. They can also be used to limit heat illness if periodic measures are taken, but with less confidence because of the difficulty of detecting errors with periodic measurements. The major problem is ensuring the integrity of the signal pick-up by the sensor at the body. Telemetry of signals, although expensive, is well within the state of the art. However, the errors associated with muscle movement, with sensors slipping off hot, sweaty skin, or shifting position within body cavities, are such that a continuously displayed, 'on-line' recorded measurement, and access to the subject to be able to check sensor dislocation or malfunction, are both highly recommended. Thus, telemetry poses problems in the field and introduces unnecessary complication in the laboratory wherever a 'hard wire' connection is possible.

Table 5.8 provides suggested personal monitoring thresholds for various actions when wearing protective ensembles. The Table identifies four parameters for measurement: deep body temperature, skin temperature, skin humidity and heart rate.

*Table 5.8. Suggested personal monitoring thresholds for various actions when wearing protective ensembles.*

| | Threshold for concern | Initiate work termination | Depart work site | Leave now | Accompanied departure | Help |
|---|---|---|---|---|---|---|
| Heart rate level[1] 20–25 year old | >80 bpm | 100 | 120 | 140 | 160 | 180 |
| Older (% of max HR) | 40% | 50% | 60% | 70% | 80% | 90% |
| Skin humidity | 50% | 60% | 70%[3] | 80% | 90% | 100% |
| Deep body temperature (least reliable) | 100·5°F | 100·8°F[2] | 101·5°F | 102°F | 102·5°F | 103°F |
| Skin temperature | 95°F | >96°F | 96·5°F | >96·5°F | 97°F | >97°F |

[1]Age dependent; maximum heart rate = 220-age (years).
[2]OSHA suggested limit.
[3]Sweat begins dropping off skin.

Deep body temperature is relatively easily measured. Sites that have been used for measurement of deep body temperature include the tympanic membrane, which is not really appropriate for a field measurement since it requires packing the ear canal and using an ear defender to protect the air within the ear canal; the oesophagus, which is out of the question for field measurement; the gut, by radiosonde pill, which may on occasions be worth the problems of recovery of the unit; and the rectum, which is subject to slippage, requires considerable convincing for subjects to accept even in a research setting, generally responds slowly and, without a parallel measurement of skin temperature to indicate convergence, is unable to prevent heat exhaustion collapse when wearing encapsulating clothing. It is, however, probably the most widely used indicator of transition from heat exhaustion risk to heat stroke risk (Lind, 1963; Wyndham and Heyns, 1971).

Skin temperature is usually measured at three sites in the heat, which are conventionally the chest, forearm and calf but, since skin temperature under heat stress becomes essentially homogeneous, a single site is probably sufficient for measurement in the field to prevent heat illness. A single site can be used if one that is not directly affected by exterior radiant heat on the surface of the clothing can be found. Medial thigh temperature has been recommended as perhaps the best site for a one point determination of skin temperature in the heat. Skin temperature sensors are readily available off the shelf, as are portable skin temperature monitors. These are ready to be used immediately, but a small-scale research evaluation in the field is required to validate this approach to heat stress reduction.

The key factor in heat illness for workers wearing encapsulating clothing is the convergence of skin temperature towards deep body temperature (Pandolf and Goldman, 1978; Goldman, 1984). Seated comfortably at rest, average skin temperature is about 33°C and deep body temperature about 37°C. This 4°C difference between skin and deep body temperature allows each unit of blood flowing from the deep body where the heat is produced, to the skin from which it is eliminated, to transport about 4 kcal of heat to the skin. As skin temperature rises towards deep body temperature, the amount of heat that can be transferred per unit of blood flow is linearly reduced. When the difference between skin temperature and deep body temperature is only a few degrees, the heart rate must increase to such a level that there is inadequate filling time between beats and heat exhaustion collapse is imminent. If it does not occur beforehand, heat exhaustion collapse is almost certain to occur at or before the point where skin temperature reaches deep body temperature. It can occur with deep body temperatures as low as 38·2°C, with skin temperatures in the 37°C range (Pandolf and Goldman, 1978).

Skin humidity is a complex, laboratory measurement at the moment. Some small, experimental sensors are under evaluation, but the electronics associated with them, their requirement for calibration, problems which occur when they become saturated with sweat, and other uncertainties suggest that this will not be available for a personal monitoring device in the near future.

Heart rate, since it reflects the combination of physical work stress and heat elimination difficulty, is probably the best single measure of heat stress. However, it has been difficult to obtain simple and reliable heart rate measurement. Emergency care facilities have very elaborate and expensive EKG telemetry systems; however, it is not necessary to get the full spectrum of EKG events associated with heart function to assess heat stress. The rate at which the heart beats is all that is required. New heart rate instruments have become available recently, generally for use during physical fitness training activities. These have not been assessed for their reliability, utility for assessment of heat stress and the like. If heart rate is to be used for personal monitoring, research seems highly appropriate to select from the devices that are available, evaluate the most promising in a small-scale human subject trial in a climatic chamber, and select the most reliable for use in a field trial.

In summary, measurement of heart rate seems the best choice, given present state-of-the-art systems for personal monitoring, but research is necessary before selection and introduction of this approach.

## 12. *Decision tree projection–prediction modelling for heat risk prevention or reduction*

An alternative approach to monitoring the environment and establishing somewhat arbitrary limits based on the measured environment, or to measuring the workers' physiological response to the environment–task–clothing combination, is the use of:

1. Projections based on a few subjective responses or symptoms used with decision trees where inputs of environmental, task and/or subject status lead to a ranking of the relative difficulty or stress.
2. Modelling and prediction of the interaction between worker, task and environment.

### Decision tree projections

There is a major problem with asking workers subject to environmental heat stress how they feel. The simple question focuses their attention on their discomfort (Goldman, 1979; Goldman, 1984) and frequently

*Table 5.9. Symptom rating and clinical impression (adapted from US Army Research Institute of Environmental Medicine).*

|  | How severe? | | (Circle one) | |
|---|---|---|---|---|
|  | none | mild | moderate | severe |
| A.  Spontaneous Verbal Complaints (How are you doing?) | | | | |
| B.  Do you have? | | | | |
|    1.  Dizziness | 0 | 1 | 2 | 3 |
|    2.  Weakness/fatigue | 0 | 1 | 2 | 3 |
|    3.  Blurred vision | 0 | 1 | 2 | 3 |
|    4.  Nausea | 0 | 1 | 2 | 3 |
|    5.  Headache | 0 | 1 | 2 | 3 |
|    6.  Shortness of breath | 0 | 1 | 2 | 3 |
|    7.  Numbness/tingling | 0 | 1 | 2 | 3 |
|    8.  Muscle cramps/stomach ache | 0 | 1 | 2 | 3 |
| C.  Signs: | | | | |
|    1.  HR = | | | | |
|    2.  BP = | | | | |
|    3.  Visible sweating | 0 | 1 | 2 | 3 |
|    4.  Face flushed | 0 | 1 | 2 | 3 |
|    5.  Lack of coordination/stumbling | 0 | 1 | 2 | 3 |
|    6.  Staring glazed eyes:   −   + | | | | |
|    7.  Apathy:   −   + | | | | |
|    8.  Irritability:   −   + | | | | |
| D.  Impression: | | | | |

results in their saying, "Oh, now that you mention it, I feel so bad that I don't want to continue". The types of symptom ratings, coupled with clinical impressions that could be used as input for a decision tree to avoid heat illness, are shown in Table 5.9. This approach requires administration by a trained observer, and the ability to observe signs which may well be hidden by protective clothing, e.g., the degree of flushed face or glazed eyes. Such an evaluation format could be used to develop a decision tree to terminate work before serious heat illness occurs. In essence, this is what is currently being done on a less formalized basis, by the health professionals monitoring possible heat illness on the job.

A decision tree could be developed, based on the extent of protective clothing worn, the activity level required of the task, and the ambient environment. The output would be a rank ordering of the difficulty of performing the task without risk of heat injury. Key levels could be established that would call for such actions as:

1. Minimize physical work.
2. Work only in the shade.
3. Limit the duration of work.
4. Use auxiliary cooling or stop work.

## Prediction modelling

A nomogram for predicting heat stress, the classic Belding and Hatch Heat Stress Index (HSI) is given in Figure 5.3. There are a number of assumptions inherent in this nomogram, including that the workers are limited to wearing a typical, long-sleeved shirt and trousers. The ratio

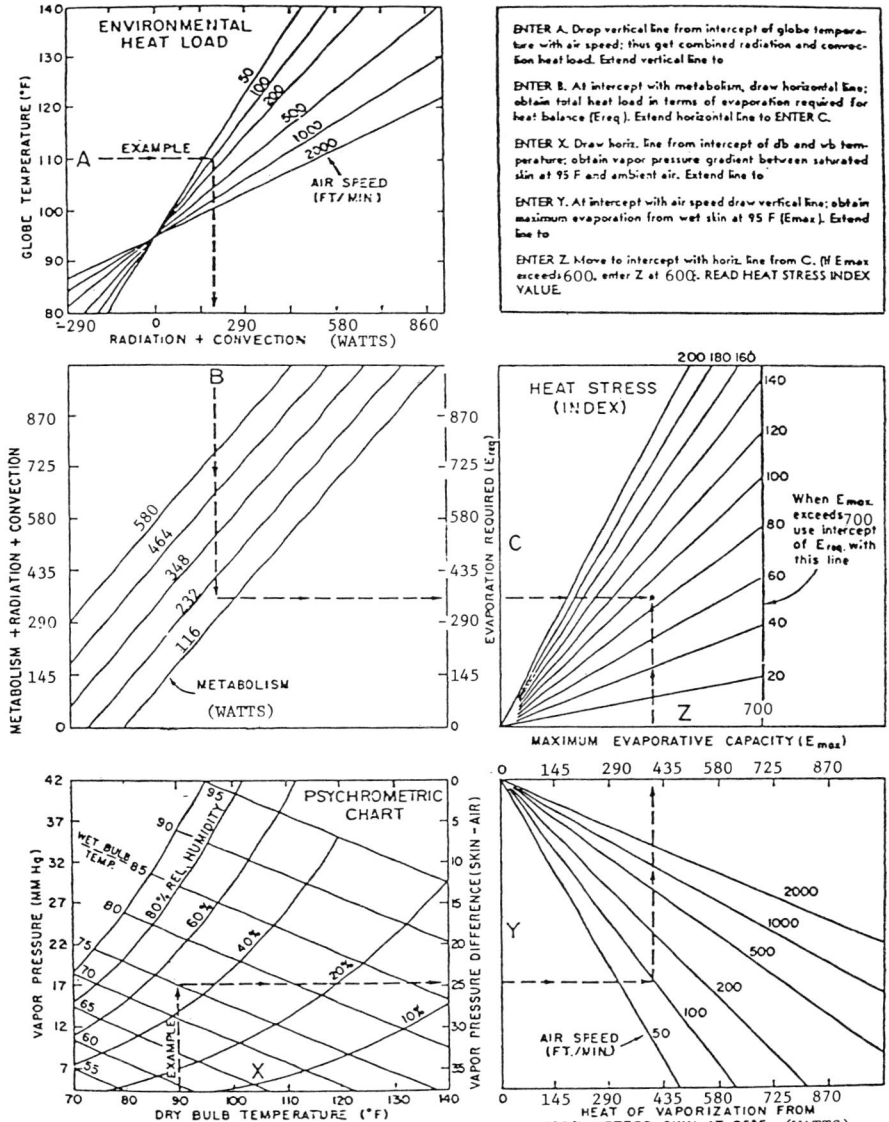

*Figure 5.3.    A modified Belding and Hatch Heat Stress index nomogram.*

*Table 5.10.   Evaluation of values in the Belding and Hatch (1955) HSI.*

| Index of heat stress (HSI) | Physiological and hygienic implications of 8 h exposures to various heat stresses |
|---|---|
| −20 −10 | Mild cold strain. This condition frequently exists in areas where men recover from exposure to heat. |
| 0 | No thermal strain. |
| +10 20 30 | Mild to moderate heat strain. Where a job involves higher intellectual functions, dexterity, or alertness, subtle to substantial decrements in performance may be expected. In performance of heavy physical work, little decrement expected unless ability of individuals to perform such work under no thermal stress is marginal. |
| 40 50 60 | Severe heat strain, involving a threat to health unless men are physically fit. Break-in period required for men not previously acclimatized. Some decrement in performance of physical work is to be expected. Medical selection of personnel desirable because these conditions are unsuitable for those with cardiovascular or respiratory impairment or with chronic dermatitis. These working conditions are also unsuitable for activities requiring sustained mental effort. |
| 70 80 90 | Very severe heat strain. Only a small percentage of the population may be expected to qualify for this work. Personnel should be selected (*a*) by medical examination, and (*b*) by trial on the job (after acclimatization). Special measures are needed to ensure adequate water and salt intake. Amelioration of working conditions by any feasible means is desired, and may be expected to decrease the health hazard while increasing efficiency on the job. Slight "indisposition" which in most jobs would be sufficient to affect performance may render workers unfit for this exposure. |
| 100 | The maximum strain tolerated daily by fit, acclimatized young men. |

between required evaporation and maximum evaporative capacity, given by carrying the nomogram through to block C, has proved to be a quite useful index of the heat stress risk. An evaluation of the values in this Belding and Hatch HSI is given in Table 5.10. A simple summation approach for obtaining the total insulation of a clothing ensemble (cf. Table 5.6), could be developed for the protective and other clothing worn by workers and used to develop a modified form of the Belding and Hatch approach.

Finally, predictive modelling of heat stress has been extremely well developed, as shown in the preceding discussion on the role of clothing in heat stress (cf. Tables 5.4 and 5.5). Models with a number of inputs have been validated by extensive laboratory studies (Givoni and Goldman, 1971, 1972, 1973a,b; Shapiro *et al.*, 1982) and by field trials for troops wearing chemical protective clothing (Goldman, 1978; Goldman, 1984). The model has been simplified for use with a simple programmable pocket calculator. A pre-programmed chip (EPROM) program used with a hand-held minicomputer could be developed to call for the necessary inputs and then provide recommendations for work modification or halt. The inputs and available outputs from such models are given in Tables 5.11 and 5.12. Such modelling works best under conditions of higher heat stress, where individual variability becomes minimal. If the goal of such modelling is not to predict the absolute responses of individuals, but rather to prevent the responses of a group of workers from reaching levels of concern, the modelling approach has been used quite successfully, particularly for the military and for workers in the nuclear power industry.

Such computer-based predictive modelling can simultaneously handle the four key environmental parameters (air temperature, humidity, motion and radiant temperature), the three key clothing parameters (insulation, moisture permeability and pumping coefficient), the key task variables (load weight, placement or lift, and frequency, speed of movement, terrain and grade), and many of the key physical and physiological variables of the worker (weight, surface area, age, physical fitness, level of acclimatization, state of hydration). However, modelling the psychological factors which may be the key to explaining why a worker pushes beyond the limits of discomfort to the extent that heat injury occurs, involves modelling motivation, need, expectation, machismo, leadership and similar terms for which even

*Table 5.11. Inputs for prediction modelling of heat stress.*

1. Anthropological make-up
     Age
     Sex
     Height
     Weight
2. Physiological status
     Heat acclimatization
     Level of hydration
     State of fitness/training
3. Clothing and equipment
     Description of clothing worn
     Nature of protective clothing
4. Scenario
     Number and type of tasks
     Duration of each task
     Loads, location, speed of movement
5. Environment
     Geographic locale
     Terrain
     Time of year and day
       *or*
     Temperature
     Humidity
     Wind speed
     Precipitation
     Solar load

*Table 5.12. Heat stress model outputs.*

A. Graphic display of time pattern response of:
  1. Heart rate
  2. Deep body temperature
B. Tabulated values as a function of time for:
  1. Heart rate
  2. Deep body temperature
  3. Mean skin temperature
C. At theoretical equilibrium (NB may be lethal):
  1. Body temperature rise from work
  2. Body temperature change by non-evaporative heat loss
  3. Body temperature rise from limited evaporation
  4. Equilibrium (final) deep body temperature
  5. Decrease if fully acclimatized
  6. Equilibrium (final) heart rate
  7. Required evaporative cooling
  8. Sweat production required
  9. Rate of dehydration, based on water intake (if any)
  10. Maximum evaporative potential
  11. % sweat wetted skin (skin humidity)
  12. Cooling power of the environment
  13. Effective clothing insulation
  14. Effective clothing moisture permeability

agreement on definitions is lacking, let alone validated paradigms suitable for modelling. This suggests that experience and common sense will continue to be essential elements in establishing heat stress standards.

## References

Anon., 1985, Threshold limit values and biological exposure indices for 1985–86. *American Conference of Government Industrial Hygienists*, Cincinnatti, OH, p. 68.

Aubertin, G., 1986, International standards for the evaluation of thermal environments. In *Ventilation 1985*, edited by H.D. Goodfellow (Amsterdam: Elsevier).

Belding, H.S., 1970, The search for a universal heat-stress index. In *Physiological and Behavioral Temperature Regulation*, edited by J.D. Hardy (Springfield, IL: C.C. Thomas).

Berglund, L.G., 1983, Characterizing the thermal environment. In *Microwaves and Thermoregulation*, edited by E.R. Adair (New York: Academic Press).

Brief, R.S. and Confer, R.G., 1971, Comparison of heat stress indices. *Amer. Indust. Hyg. Assn J.*, **32**, 11.

Bynum, G.D., Pandolf, K.B., Schwette, W.H., Goldman, R.F., Lees, D.E., Whang-Peng, J., Atkinson, E.R. and Bull, J.M., 1978, Induced hyperthermia in sedated humans and the concept of critical thermal maximum. *Amer. J. Physiol.*, **4**, R228.

Dukes-Dobos, F.N. and Henschel, A., 1973, Development of permissible heat exposure limits for occupational work. *ASHRAE J.*, **15**, 57–62.

Gagge, A.P. and Nishi, Y., 1976, Physical indices of the thermal environment. *ASHRAE J.*, **18**, 47–51.

Givoni, B. and Goldman, R.F., 1971, Predicting metabolic energy cost. *J. Appl. Physiol.*, **30**, 429–433.

Givoni, B. and Goldman, R.F., 1972, Predicting rectal temperature response to work, environment and clothing. *J. Appl. Physiol.*, **32**, 812–822.

Givoni, B. and Goldman, R.F., 1973a, Predicting heart rate response to work, environment and clothing. *J. Appl. Physiol.*, **34**, 201–204.

Givoni, B. and Goldman, R.F., 1973b, Predicting effects of heat acclimatization on heart rate and rectal temperature. *J. Appl. Physiol.*, **35**, 875–879.

Goldman, R.F., 1965, Energy expenditure of soldiers performing combat type activities. *Ergonomics*, **8**, 321–327.

Goldman, R.F., 1973, Environmental limits, their prescription and proscription. *Int. J. Environ. Sci.*, **22**, 193–204.

Goldman, R.F., 1978, Prediction of human heat tolerance. In *Environmental Stress*, edited by S.J. Folinsbee *et al.* (New York: Academic Press).

Goldman, R.F., 1979, Prediction of heat strain revisited 1979–1980. In *Proceedings of the NIOSH Workshop on the Heat Stress Standard*, Cincinnati, OH, September.

Goldman, R.F., 1984, Heat stress in industrial protective encapsulating garments. In *Protecting Personnel at Hazardous Waste Sites*, edited by S.P. Levine and W.F. Martin (Cincinnati, OH: NIOSH/Ann Arbor, MI: Butterworth–Ann Arbor Science Publishers).

Gonzalez, R.R., Berglund, L.G. and Gagge, A.P., 1978, Indices for thermoregulatory strain for moderate exercise in the heat. *J. Appl. Physiol.*, **44**, 889–899.

Goromosov, M.S., 1958, Combined indices of the effect of various meteorological factors on the organism and criticism of them. *Gigiyena i Sanitariya*, **7**, 66 (in Russian).

Horvath, S.M. and Jensen, C. (eds), 1976, *Standards for Occupational Exposures to Hot Environments*, NIOSH Publ. 76-100 (Pittsburgh, PA: NIOSH).

Hughes, A.L. and Goldman, R.F., 1970, Energy cost of "hard work". *J. Appl. Physiol.*, **29**, 570–572.

ISO 7243, 1982, *Hot Environments—Estimation of the Heat Stress on Working Man, based on the WBGT Index* (Paris: Assn Francaise de Normalisation (AFNOR)).

ISO 7726, 1985, *Thermal Environments—Instruments and Methods for Measuring Physical Quantities* (Paris: Assn Francaise de Normalisation (AFNOR)).

ISO 7933, 1986, *Hot Environments—Analytical Determination and Interpretation of Thermal Stress Based on the Calculation of Required Sweat Rate* (Paris: Assn Francaise de Normalisation (AFNOR)).

Jensen, R.C. and Heins, D.A., 1976, *Relationship between Several Prominent Heat Stress Indices*, DHEW (NIOSH) Publ. 77-109, October.

Joy, R.J.T. and Goldman, R.F., 1968, A method of relating physiology and military performance: A study of some effects of vapor barrier clothing in hot climate. *Mil. Med.*, **133**, 458–470.

Juptner, H., 1984, ISO Standards in the field of ergonomics. *Appl. Ergon.*, **15**, 211−213.

Kleinhanz, C., Piekarski C. and Haug, E., 1980, Recording and processing of various climatic parameters. *Biomed. Technik*, **25**, 12−26 (in German).

Lee, D.H.K., 1980, Seventy-five years of searching for a heat index. *Exptl Res.*, **22**, 331−356.

Lind, A.R., 1963, A physiological criterion for setting thermal environmental limits for everyday work. *J. Appl. Physiol.*, **18**, 51−56.

Metz, B.G., 1976, Ergonomics and standards. *Ergonomics*, **19**, 271−274.

Minard, D., 1973, Physiology of heat stress. In *The Industrial Environment—Its Evaluation and Control* (Cincinnati, OH: NIOSH).

Mutchler, J.E. and Vecchio, J.L., 1977, Empirical relationships among heat stress indices in 14 hot industries. *Amer. Indust. Hyg. Assn J.*, **38**, 253−263.

Onkaram, B., Stroschein, L.A. and Goldman, R.F., 1980, Three instruments for assessment of WBGT and a comparison with WGT (Botsball). *Amer. Indust. Hyg. Assn J.*, **41**, 634−641.

OSHA, 1974, *Recommendations for a Standard for Work in Hot Environments*, OSHA Standards Advisory Committee, US Dept of Labor.

Pandolf, K.B. and Goldman, R.F., 1978, Convergence of skin and rectal temperatures as a criterion for heat tolerance. *Aviat. Space Environ. Med.*, **49**, 1095−1101.

Pandolf, K.B., Burse, R.L. and Goldman, R.F., 1977a, Role of physical fitness in heat acclimatization, decay and reinduction. *Ergonomics*, **20**, 399−408.

Pandolf, K.B., Burse, R.L., Givoni, B., Soule, R.G. and Goldman, R.F., 1977b, Effects of dehydration on predicted rectal temperature and heart rate during work in the heat. *Med. Sci. Sports*, **9**, 51−52.

Ramsey, J.D., 1975a, Heat stress standard: OSHA's advisory committee recommendations. *Nat. Safety News*, June, 89−95.

Ramsey, J.D., 1975b, Threshold limits for workers in hot environments. In *Proceedings of the International Mine Ventilation Congress*, Johannesburg, September.

Ramsey, J.D., 1978, Abbreviated guidelines for heat stress exposure. *Amer. Indust. Hyg. Assn J.*, **39**, 491−495.

Shapiro, Y., Pandolf, K.B. and Goldman, R.F., 1982, Predicting sweat loss response to exercise, environment and clothing. *Eur. J. Appl. Physiol.*, **48**, 83−96.

US Army, 1980, Prevention, treatment and control of heat injury. *TB Med.*. July, 175.

Witherspoon, J.M. and Goldman, R.F., 1974, Indices of thermal stress. In *Air-Conditioning Criteria for Man's Living Environment*, ASHRAE Bulletin No. LO-73-8, pp. 5−13.

Wyndham, C.H. and Heyns, A.J.A., 1971, The assessment of heat stress in working places underground in mines. Part 8. Estimating the probability of reaching heat stroke levels of body temperature at different levels of heat stress. *South Africa Chamber of Mines Research Report*, 35/71.

# PART II
# EVALUATION OF THERMAL CHARACTERISTICS OF CLOTHING

# 6. Physiological Tests and Evaluation Models for the Optimization of the Performance of Protective Clothing

## K. H. Umbach

## 1. A system for the physiological analysis of textiles and garments

Protective clothing has to offer adequate protection against occupational hazards and should ensure a reasonable wear comfort. One of the prerequisites for good wear comfort is that a garment's thermal and moisture transport properties are adapted to the specific climatic and activity conditions in such a way that the heat and moisture exchange between body and surrounding atmosphere are balanced.

However, in order to include wear comfort as a constructional parameter in the development of textiles and garments it is necessary to be able to measure and evaluate wear comfort. To be of practical use to the fabric and garment industry this evaluation must be based on economical test procedures restricted in number and performed in the laboratory. The test results gained in the form of specific quantifiable characteristics of textiles and garments must have a universal applicability and must correlate with the experience gathered while wearing the fabrics and garments.

For this purpose a 5-level system of analysis is suggested, shown in principle in Figure 6.1.

At *Level 1* of this system the textiles are tested as fabric layers. The thermophysiological properties are determined by means of a thermoregulatory model of the human skin (skin model), simulating heat and moisture exchange from the skin. Thermal insulation and moisture transport qualities of fabrics are not only measured under stationary ('normal') wear conditions, but also under transient situations, characterized by intermittent sweat pulses resulting from increased or strenuous body activity. Thus, a complete set of specific fabric quantities is

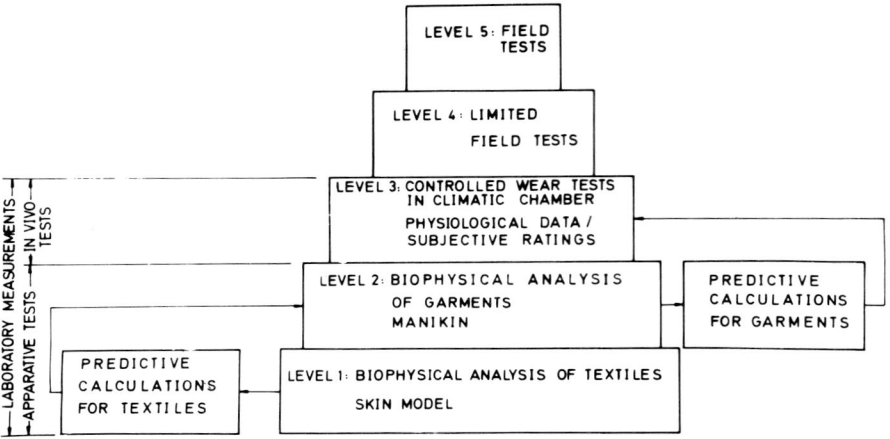

*Figure 6.1. A five-level system for the analysis of the physiological properties of textiles and garments.*

available which, inserted in predicitive formulae, yields the thermophysiological comfort of fabrics. Consequently, at this level of analysis an accurate selection among a number of textile items probably suited for a particular garment is possible. Only the best items should be considered for the following evaluation levels.

The wear properties of a clothing ensemble, consisting of several garments, are not only determined by the different fabrics included in the ensemble and their interaction with each other, but also by the interspaced air layers due to the garments' pattern. Therefore, at *Level 2* of the evaluation system a life-sized movable manikin is employed, representing a thermoregulatory model of man.

With the manikin the thermal insulation of garments can be measured and their moisture transport determined, exactly simulating their effects for the wearer in practical use. The manikin can perform body movements, and therefore ventilation effects in the microclimate between the garments and their influence on thermal comfort can be evaluated. The effect of garment patterns on man's thermoregulation can also be investigated.

Using the manikin, a set of physiological quantities specific to the total clothing system can be inserted into a predictive model, simulating the system's wear performance under all possible climatic and activity conditions. From a number of clothing systems an accurate selection of the best items can be made.

Actually the accuracy of the tests and predictive calculations of levels 1 and 2 of the system described is so high that the wear trials with subjects are in most cases expendable for the manufacturer or user. In some instances, however, it might be advisable to verify the test results of levels 1 and 2 of the system. For this purpose controlled wear trials in a climatic chamber, Level 3 of the analysing system, are performed to check the results of the predictive calculations for one particular wear situation. In case of agreement it can be induced that the model's predictions with regard to other wear situations are also valid.

In the development of clothing the wear tests of *Level 4* and *Level 5*, either with a limited or with a large number of subjects, would actually be restricted to a small number of items optimized systematically in their comfort characteristics through levels 1−3 of the system. The wear tests of Levels 4 and 5, which are conducted under the real ambient conditions of the clothing's field of application, indicate whether parameters not included in the predictive calculations have some bearing on the physiological comfort perceived by the wearer.

## 2. Thermoregulatory model of human skin (skin model)

The main component of the skin model is a perforated plate of sintered metal (stainless steel) sized $20 \times 20$ cm$^2$ (see Figure 6.2). The plate is electrically heated to man's skin temperature (35°C) and covered by the textile to be tested (see Figure 6.3). A guard ring heated to the same temperature as the metal plate acts as a temperature shield. The heat from the plate can only flow through the sample.

Distilled water fed to the plate's bottom through channels can evaporate through the porous plate, simulating sweating. In the climatic cabinet (see Figure 6.2) into which the skin model is built, air with a defined temperature, humidity and speed flows over the sample.

In analysing the thermal and moisture transport properties of fabrics we must distinguish between stationary and transient wear situations which occur in practical use. Under stationary or 'normal' wear conditions there exists a time-constant heat and moisture transfer from the wearer's body to the environment. The moisture transfer is due to insensible sweating only and, thus, is in the form of water vapour. Transient conditions are characterized by intermittent moderate or heavy sensible sweating by the wearer, resulting for instance from strenuous physical activity. Under these conditions, for good wear comfort textiles must not only possess good moisture transport properties but also a good moisture buffering capacity. The moisture either occurs as water vapour or as liquid sweat.

For each of these wear conditions, the skin model offers insight into specific properties of textiles, which are directly correlated to the physiological responses of the wearer.

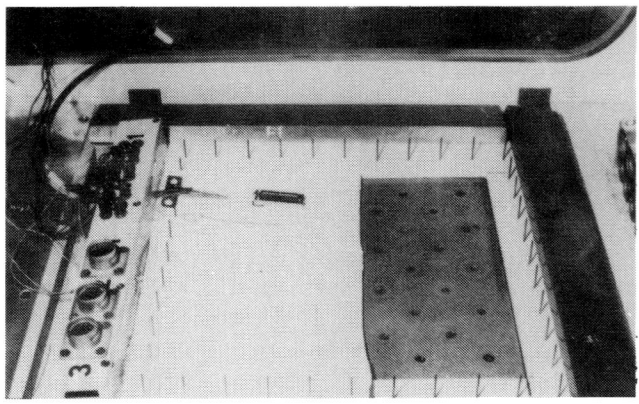

*Figure 6.2. A skin model plate with a sample arranged for the measurement of the stationary thermal and water vapour resistance of fabrics (top), and of a climatic cabinet with the skin model (bottom).*

In stationary measurements, comprising normal wear conditions, either thermal insulation ($R_{ct}$) or water vapour resistance ($R_{et}$) of the fabric layer is measured (Umbach, 1980). The former is determined from the dry heat flux from the skin model in the experimental set-up shown in Figure 6.3 (without water supply), with the air temperature in the climatic cabinet controlled at 20°C. By subtracting a 'bare-plate value', determined without a sample on the skin plate, the thermal insulation gained for the fabric is free of any erroneous increment in insulation inherent to the apparatus (Umbach, 1981a,b).

For measuring the water vapour resistance ($R_{et}$) the air temperature in the climatic cabinet is set to 35°C with a relative humidity of 40%. Water is fed to the skin plate on which a cellophane sheet (thickness

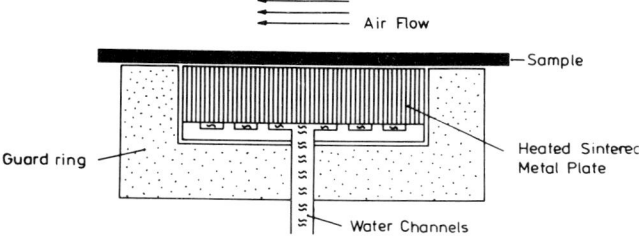

*Figure 6.3. The principle of the skin model.*

about 20 μm) is placed, shielding the sample fabric from liquid water. From the resulting evaporative or latent heat flux through the fabric its water vapour resistance is gained. Transition resistances inherent to the apparatus are also eliminated by a 'bare–plate value'. The fabric's water vapour absorbance ($F_i$) can be determined by conditioning the sample prior to the $R_{et}$ measurement. The amount of water vapour absorbed by the sample during an exposure time of 1 hour on the skin model is defined as $F_i$.

Thermal resistance ($R_{ct}$) as well as water vapour resistance ($R_{et}$) of a fabric depend directly on its thickness. In order to compare the comfort characteristics of different textile samples, it is necesssary to define a specific quantity free from the influence of the fabric's thickness. For this purpose the water vapour permeability index $i_{mt}$ is introduced. By definition (see Figure 6.4) $i_{mt}$ compares the ratio of thermal to water vapour resistance of a fabric to the ratio found with an air layer of the same thickness as the fabric. The coefficient 0·6 mbar $K^{-1}$ stems from the latter, rendering $i_{mt}$ a dimensionless number, ranging between 0 (for a fabric impermeable to water vapour) and 1. The water vapour

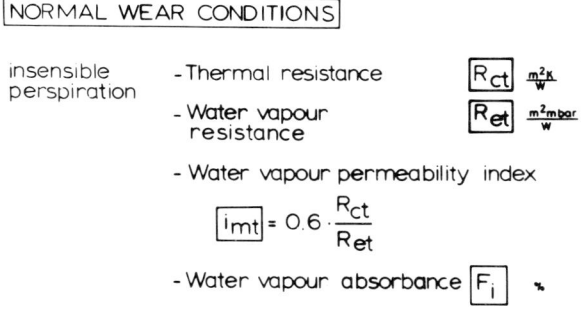

*Figure 6.4. Textile quantities relevant to thermophysiological comfort under normal wear situations.*

absorbance $(F_i)$ and $i_{mt}$ values represent criteria to judge a fabric's physiological quality under stationary wear conditions (high $i_{mt}$ and $F_i$ values are preferable).

The stationary tests with the skin model described are fixed in the German standard DIN 54 101 (1984). However, for particular fields of application it is necessary to deviate from the standardized tests. For instance, with cold weather protective garments, where the thermal insulation is effected by fleece battings, it is of interest to know how much of the insulating power is lost with external pressure on the garment. For this purpose, as shown in Figure 6.5, loads are placed on the sample on the skin model and the pressure pattern of the sample's

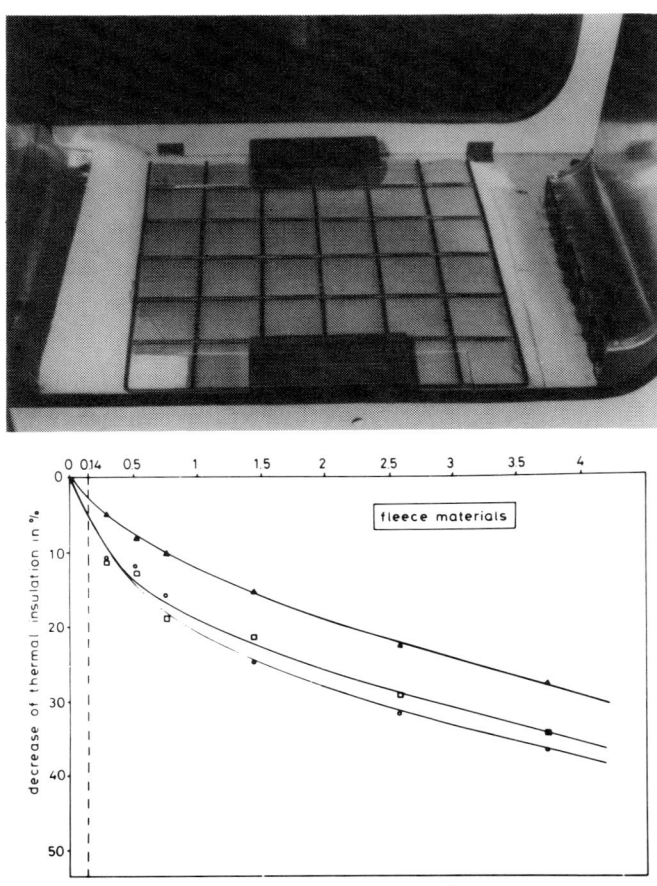

*Figure 6.5. Measurement of thermal insulation under pressure (top), to show the pressure pattern of thermal insulation of various fleece battings (bottom).*

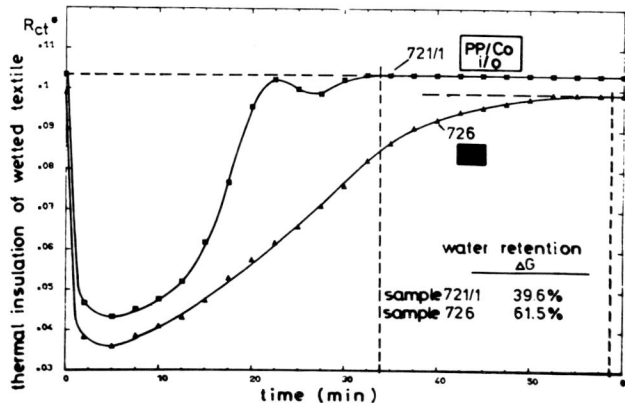

*Figure 6.6.    The time pattern of thermal insulation of wetted fabrics.*

thermal insulation is measured, enabling the selection of the comparatively higher insulating items.

For the physiological evaluation of cold-weather protective garments, the loss of the fabrics' thermal insulation when they become wet
on the skin, either by heavy sweating or by wet-weather conditions,
must be known. For this purpose the sample is wetted on the skin
model and the time-pattern of thermal insulation $R_{ct}^*$ is measured, as
shown in Figure 6.6 for the example of two different underwear
samples. From the two samples, one is physiologically more beneficial,
as it demonstrates a smaller loss of thermal insulation and a faster regain
of thermal insulation of the dry fabric, preventing more effectively a
post-exercise chill. Obviously, these tests are a criterion for the selection of physiologically optimized items.

A fabric's buffering capacity against water vapour is one of the
quality criteria under transient wear conditions for which the skin
model is used, in the arrangement shown in principle in Figure 6.7. A
sweat pulse is simulated by injecting 4 cm$^3$ of water into an absorbent
fabric layer placed on the metal plate. The fabric to be tested is held by a
frame at a distance of 1 cm from the metal plate, simulating the
microclimate next to the skin. Sensors register the time pattern of water
vapour pressure (humidity) and temperature beneath the sample. In
order to create a forced convection like the one caused by body movements, the sample is moved rhythmically by strings driven by an
electromotor.

From the registered time patterns of water vapour pressure and
temperature, shown in Figure 6.8, the moisture and temperature regulation indices $K_d$ and $\beta_T$, respectively, can be determined. These indices

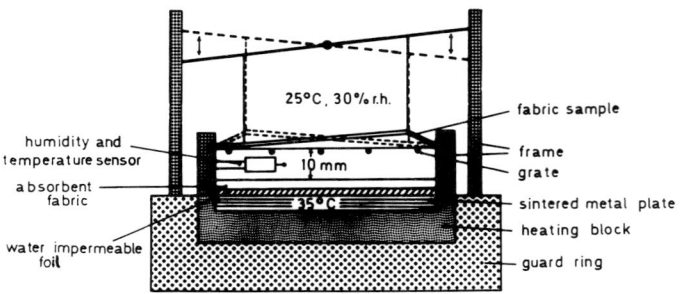

*Figure 6.7.    The measurement of a fabric's buffering capacity against water vapour.*

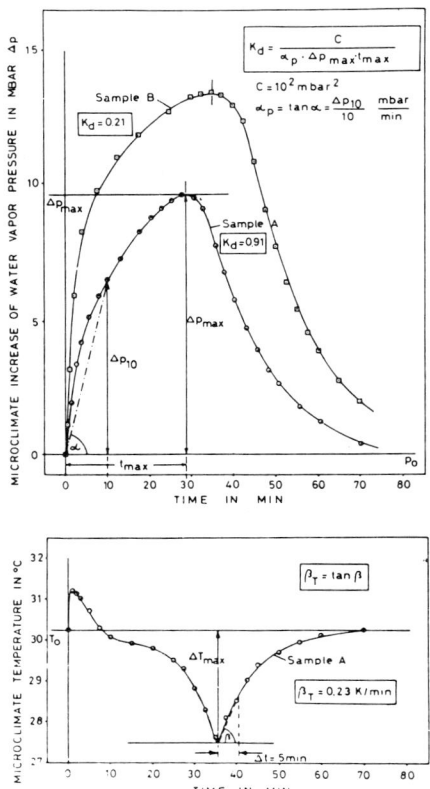

*Figure 6.8.    The definition of a fabric's moisture and temperature regulation indices, $K_d$ and $\beta_t$.*

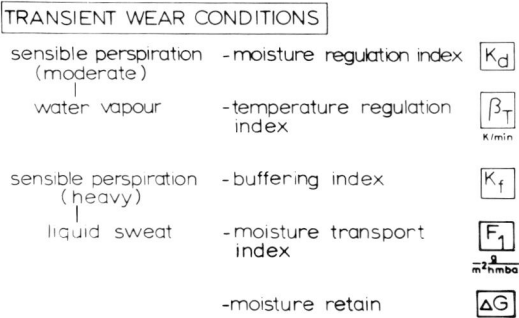

*Figure 6.9.    Textile quantities relevant to thermophysiological comfort under transient wear situations.*

describe a fabric's capacity to readjust moisture and temperature equilibrium in the microclimate next to the skin after a sweat pulse (see Figure 6.9). The higher the $K_d$ and $\beta_T$ values, the better the fabric from a physiological point of view.

For determining a fabric's buffering capacity against liquid sweat the skin model serves in the arrangement shown in Figure 6.10. The fabric can absorb and desorb liquid water, which—in the simulation of heavy sweating—is injected into an absorbent layer placed on the skin plate

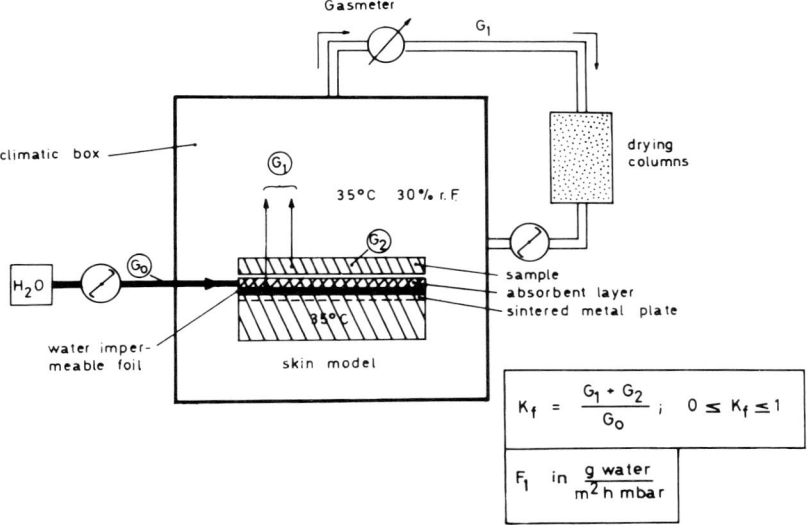

*Figure 6.10.    The measurement of a fabric's buffering capacity against liquid sweat.*

directly beneath the fabric sample. After 15 minutes we determine the portion of this original volume of water ($G_0 = 15$ cm$^3$) that has been absorbed by the fabric ($G_2$) and the amount that has been transported as water vapour into the surrounding air ($G_1$). From these data (Figure 6.10), a buffering index ($K_f$) and a moisture transport index ($F_1$) can be defined, characterizing a fabric's buffering and transport capacity of liquid water. The larger the values for $K_f$ and $F_1$, the more suitable the fabric.

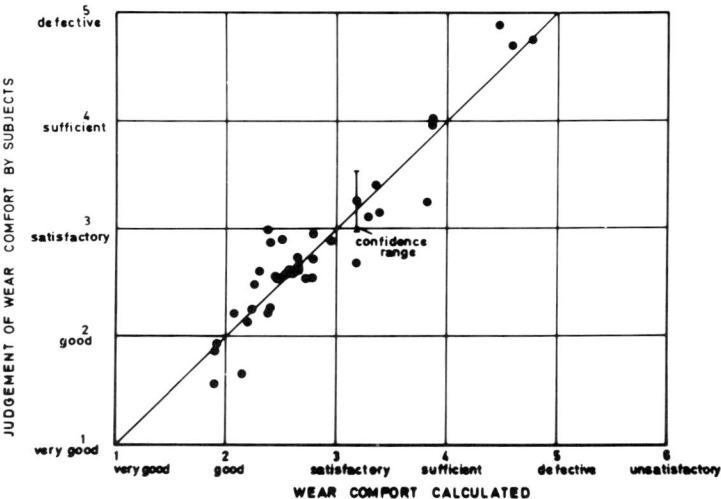

$$WC_T = \alpha_1 \cdot i_{mt} + \alpha_2 \cdot F_i + \alpha_3 \cdot K_d + \alpha_4 \cdot \beta_T + \alpha_5 \cdot K_f + \beta$$

1······6
very  unsatisfactory          $s = \pm 0.48$
good

|         | $\alpha_i$ |          | $\alpha_i$ |
|---------|-----------|----------|-----------|
| $i_{mt}$ | $-5.64$   | $\beta_T$ | $-4.512$  |
| $F_i$   | $-0.375$  | $K_f$    | $-4.532$  |
| $K_d$   | $-1.587$  | $\beta$  | $11.553$  |

| B PI | Thermophysiological wear comfort |
|------|----------------------------------|
| Hohenstein | Predictive formula for textiles |

*Figure 6.11.  Predictive formula for the thermophysiological wear comfort of underwear textiles.*

*Figure 6.12.  Correlation between the predicted and actually perceived wear comfort of fabrics.*

Because in practical use normal and transient wear conditions occur alternately, all the quantities measured with the skin model and listed in Figures 6.4 and 6.9 have to be included in the predictive model of Level 1 of the system of analysis (see Figure 6.1). This is done in the predictive formula of Figure 6.11, which was developed to express a textile's thermophysiological wear comfort under typical practical use conditions by a comfort vote ($WC_T$), ranging from 1 (very good) to 6 (completely unsatisfactory). The formula in Figure 6.11 is valid for underwear. Numerous wear trials with subjects have shown that the comfort vote ($WC_T$) predicted for a fabric is in very good agreement with the comfort sensation actually perceived by people (see Figure 6.12).

## 3. *Thermoregulatory model of man (manikin Charlie)*

At Level 2 of the system of analysis (see Figure 6.1), the thermal insulation and moisture transport properties of the whole clothing ensemble (including underwear and outerwear fabrics as well as the air layers within the microclimate between the fabrics and adhering to the clothes' outer surface) are determined by a life-sized manikin, *Charlie*, shown in Figure 6.13. The manikin's body is made from copper and electrically heated to man's skin and body temperature by wires soldered to its inner side. The manikin is divided into 16 separately controlled circuits, allowing the simulation of characteristic temperature gradients at man's body surface. Thus, the integral thermal resistance ($R_c$) of a clothing ensemble can be directly measured—via heat flux from the manikin's body to the surrounding air of a climatic chamber—and its local insulating effect at different body sections can be evaluated.

It is essential that the manikin can simulate all the body positions (i.e., standing, walking or lying) occurring in the clothing's actual use. Therefore the manikin can move its arms and legs by attached bars driven by an electromotor. Thus, the influence of convection and ventilation in the garments' microclimate, caused by body movements of the wearer, on the wear comfort of a clothing ensemble can be evaluated.

Quantitatively this influence of body movements is expressed in a set of three different thermal resistance values (see Figure 6.14). For $R_c(1)$ (in $m^2$ K $W^{-1}$) the manikin is standing still. For $R_c(2)$ it is moving with a defined walking speed, but the clothing ensemble's openings are sealed with airtight adhesive tape. Thus, only convection within the microclimate is possible, but no ventilation via the clothing's openings.

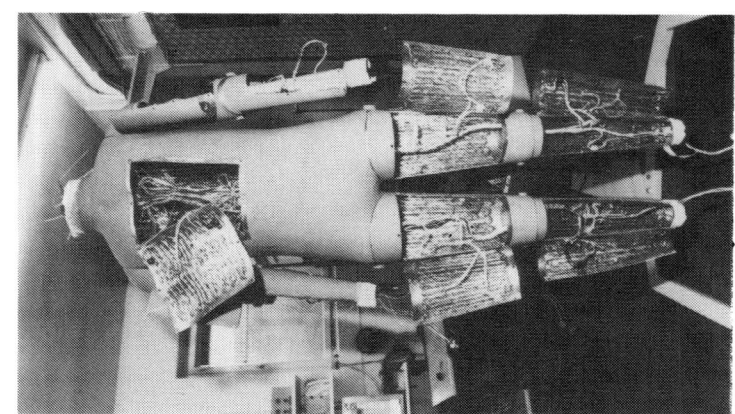

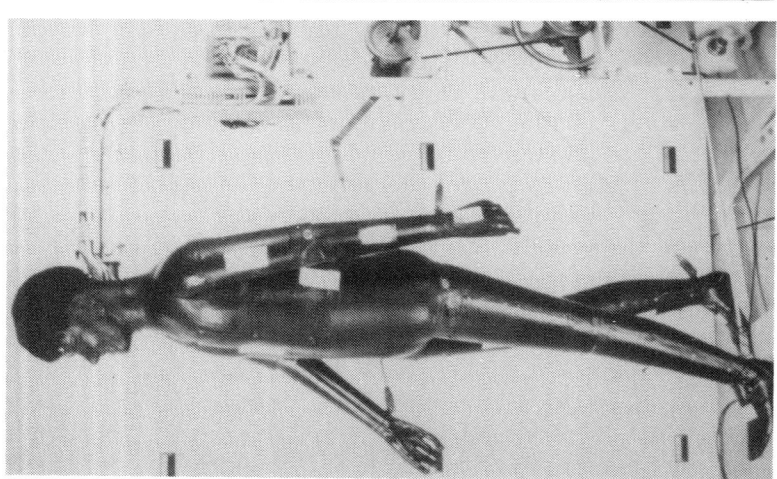

*Figure 6.13.    A movable copper manikin for the measurement of the thermophysiological properties of ready-made garment ensembles. The photograph on the right illustrates the measurement of the influence of wind speed on a garment ensemble's thermal insulation.*

After removal of the adhesive tape the influence of additional ventilation becomes effective and is contained in the $R_c(3)$ value measured. From the difference between $R_c(2)$ and $R_c(3)$ the ensemble's ventilation rate in $m^3 h^{-1}$ can be determined.

Actually the ventilation rate plays an important role in the physiological function of clothing. It enables substantial amounts of heat and moisture to be dissipated from the body without having to pass through the fabrics. The cut, pattern and fit of protective clothing should be adapted to facilitate an optimum of ventilation. This increases the wear comfort perceived in a warm climate or under strenuous physical work.

The water vapour resistance values ($R_e$, in $m^2$ mbar $W^{-1}$) corresponding to the different body positions or movements cannot be measured directly, because the manikin cannot sweat. They are determined by the application of a thermodynamic model (Mecheels and Umbach, 1977a), making use of the water vapour resistances ($R_{et}$) of the clothing ensemble's fabrics, as measured with the skin model (see Figure 6.14).

The precision of the physiological data of a clothing ensemble measured with the manikin is high. As is shown in Figure 6.15 for the example of a work suit the agreement with the $R_c$ and $R_e$ values actually found in controlled wear tests with subjects is good.

For the correct evaluation of the physiological performance of a clothing ensemble the influence of outer wind speed on thermal insulation must be known. This is particularly important for cold-weather

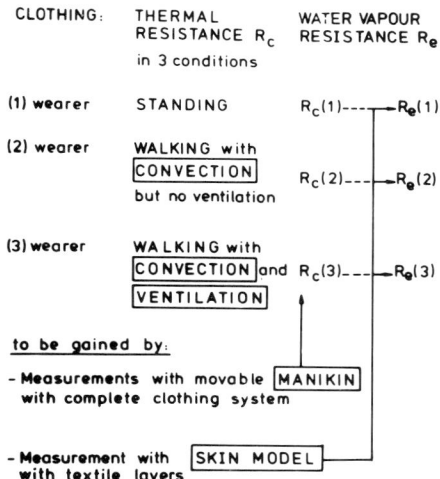

*Figure 6.14. A set of thermal and water vapour resistance values for the evaluation of the thermophysiological properties of garment ensembles.*

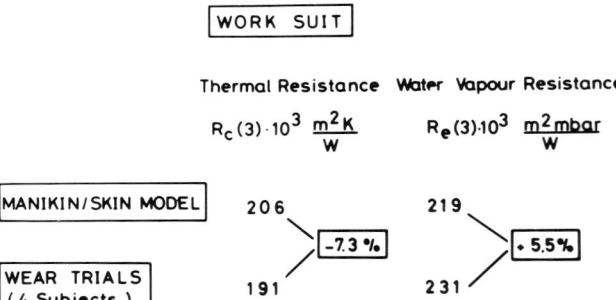

Figure 6.15.    *The correlation between manikin measurements and the results of wear trials with subjects.*

protective clothing. By placing a wind machine in front of the manikin (see Figure 6.13) this influence can be measured quantitatively. As can be seen from Figure 6.16, the thermal insulation for instance of a work suit decreases by approximately 50% if the wind speed in the surrounding air is raised from 0·3 to 10 m s$^{-1}$. This is partly due to the removal of the air layer adhering to the clothes' outer surface. Measurements have shown that with low wind speed this air layer contributes up to approximately 30% of the thermal insulation a work suit offers its wearer. The other part of the loss of thermal insulation with higher wind speed is due to the air blown right through the fabric into the microclimate near the body.

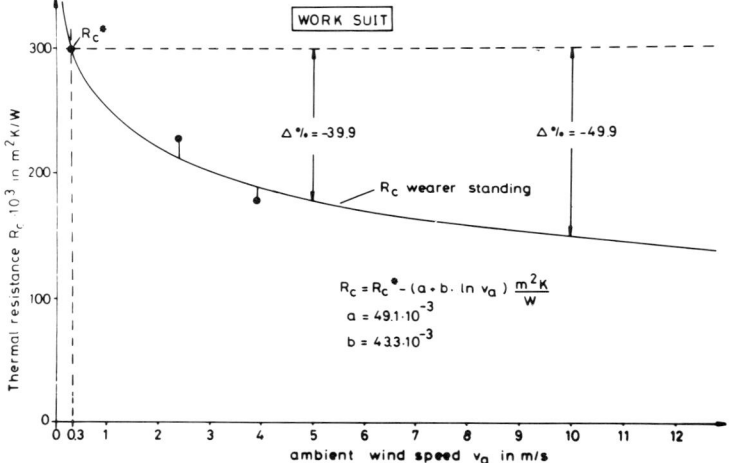

Figure 6.16.    *The correlation between the thermal insulation of a work suit and ambient wind speed.*

## 4. Predictive model for the evaluation of the physiological performance of clothing ensembles

Converted by a physiological evaluation model employing thermo-dynamic principles (Mecheels and Umbach, 1977b) the specific data of a clothing ensemble measured with the manikin and skin model express its wear properties. The formulae for the calculation of wear comfort in this evaluation model have been deduced empirically from the results of numerous wear trials with human subjects. In these wear trials conducted in climatic chambers (Level 3 of the system of analysis, Figure 6.1) people perform a defined programme of physical work under fixed ambient climatic conditions, for instance walking on a treadmill or, as in Figure 6.17(*a*) pedalling on a bicycle ergometer. Physiological data of the test subject—rectal temperature, skin temperature, heart rate, sweat production, temperature and humidity in the microclimate next to the person's skin—are measured by sensor elements (see Figure 6.17(*b*)) and collected on-line by a computer. The exhaled air is analysed for oxygen consumption and $CO_2$ production, yielding the person's metabolic rate. The subjective comfort sensation perceived by the person is quantified by means of multi-step ballots and questionnaires.

The evaluation of the results of these fundamental wear trials has shown a distinct correlation between the physiological data measured

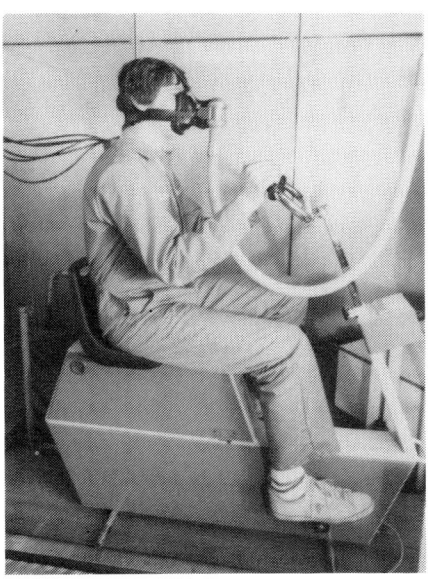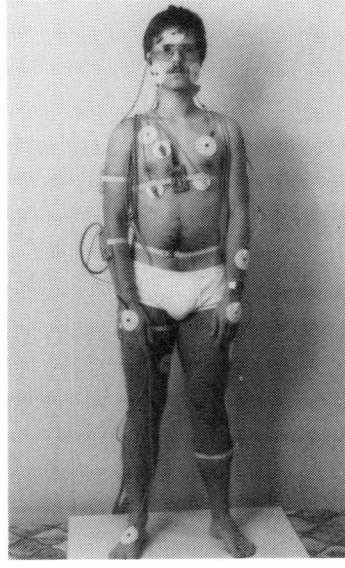

*Figure 6.17.   Controlled wear trials in a climatic chamber.* (a) *Subject on a bicycle ergometer and* (b) *with attached sensor elements (right).*

on the person's body and his comfort sensations perceived. Obviously—in contrast to a wide-spread belief—wear comfort is not an individually different and undefined varying sensation, but is directly caused by particular physiological quantities of the body. Because of this interaction it was possible to develop the predictive calculations described. Actually, the results they yield in predicting the thermophysiological behaviour of garments are in very good agreement with the wear comfort experienced by people during the garment's practical use.

The benefit of the predictive model for the garment manufacturer as well as for the user is the fact that—based on comparatively few laboratory tests—it describes a clothing ensemble's comfort characteristics under all possible wear and climatic conditions. The latter can be varied in the calculation model, based on the original set of laboratory tests with the manikin and skin model described. Thus, a universal evaluation of garments is possible which in its entirety could never be achieved by wear trials.

The principle of the predictive calculations is shown in Figure 6.18. In detail, considering the wearer's body stature and activity as well as specific comfort criteria, a garment ensemble's range of utility or comfort range can be determined. This comfort range is limited on the one hand by the minimum ambient temperature ($T_{a\ min}$), where the wearer performing a certain type of activity, expressed by his metabolic rate, is close to, but not yet feeling cold (mean skin temperature $T_s=32°C$) or close to, but not yet suffering from hypothermia (mean skin temperature $T_s=29°C$).

On the other hand, the comfort range is limited by the maximum ambient temperature $T_{a\ max}$, where the wearer is almost, but not quite sweating so heavily that he is feeling uncomfortable (skin wettedness or comfort factor $k=0.25$) or that he is suffering from hyperthermia (comfort factor $k_f=0.75$).

Because both $T_{a\ min}$ and $T_{a\ max}$ depend on the ambient humidity, the range of utility of a clothing ensemble is illustrated in a psychrometric chart, as in the example of a cold-weather protective suit, shown in Figure 6.20. This suit, combining a multilayer fabric construction with a sophisticated system of adjustable ventilation openings (Umbach, 1981; see Figure 6.19), possesses on the one hand a high thermal insulation, by which it can be worn under medium heavy work (metabolic rate $M=250$ W) at temperatures around $-16°C$ if the ventilation openings are closed. On the other hand, when they are opened, the good moisture transport as well as buffering properties resulting from a bellows effect caused by body movements make the suit wearable at temperatures around $+16°C$ without its wearer—again performing medium heavy work—sweating so much that his physical performance is inhibited. This suit model is an example of how a garment ensemble

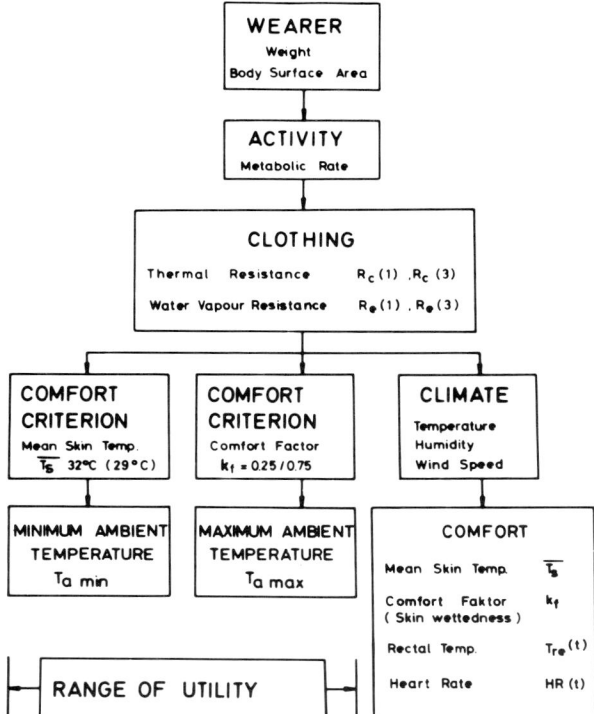

*Figure 6.18.   The principle of predictive calculations used for the evaluation of the comfort characteristics of clothing ensembles.*

can be furnished with an outstandingly broad range of utility, by optimizing ventilation by suitable pattern.

Obviously, from a psychrometric diagram like Figure 6.20 the garment maker or the user can deduce whether a particular item is actually meeting the climatic requirements dictated by the garment's field of application. Special effects such as wind-chill influence can be considered. When the tests with the manikin are performed at different wind speeds (see Figure 6.13), the predictive model directly shows how the lower limit of the range of utility is afflicted. For example, for a work suit the minimum ambient temperature ($T_{a\ min}$) shifts from approximately $-3°C$ to $+8°C$, if the ambient wind speed increases from 0·3 to 4 m s$^{-1}$ (see Figure 6.21).

Also from the psychrometric diagram it can be seen how constructional changes, e.g., in pattern or garment compositions, affect the clothing ensemble's range of utility. This provides a valuable means for optimizing the physiological performance of clothing.

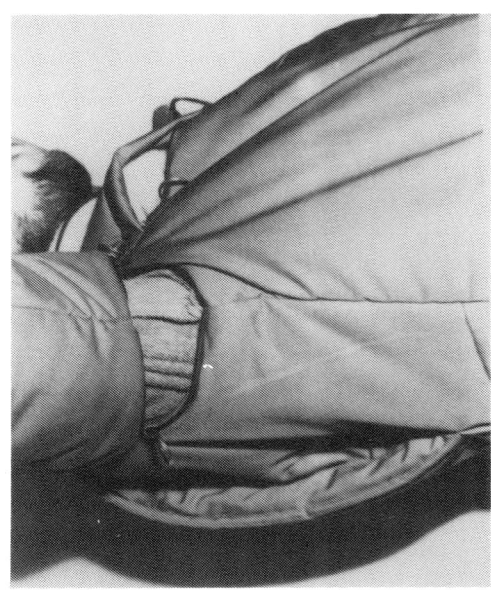

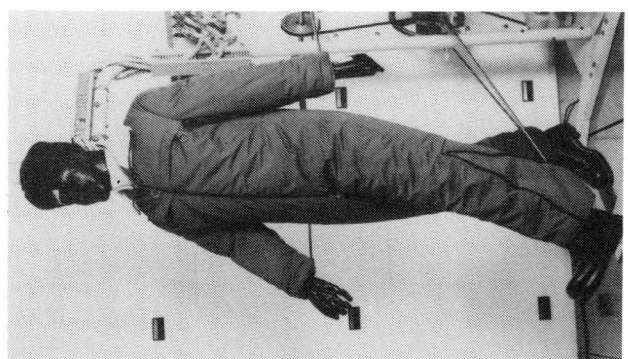

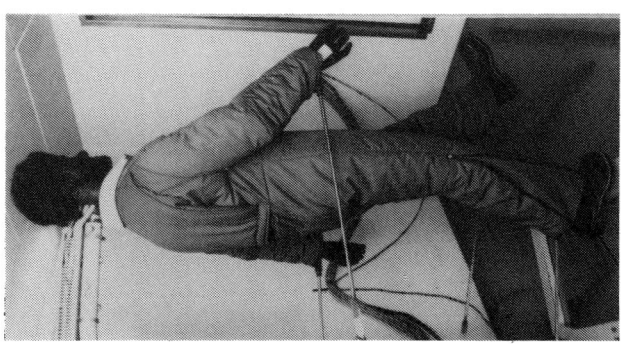

*Figure 6.19. A cold-weather protective suit with zipper-adjustable ventilation openings (left and centre), and openings underneath the armpits (right).*

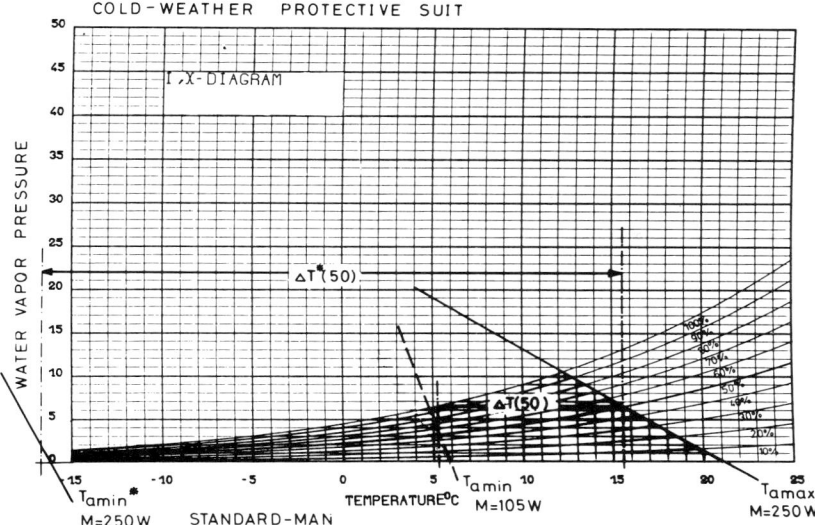

*Figure 6.20. The range of utility of the cold-weather protective suit shown in Figure 6.19. The wearer is performing medium-heavy work.*

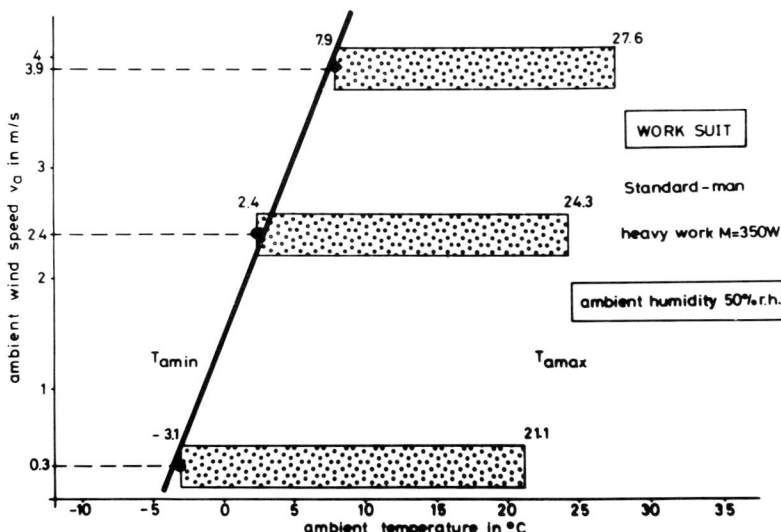

*Figure 6.21. The influence of outer wind speed on a work suit's range of utility.*

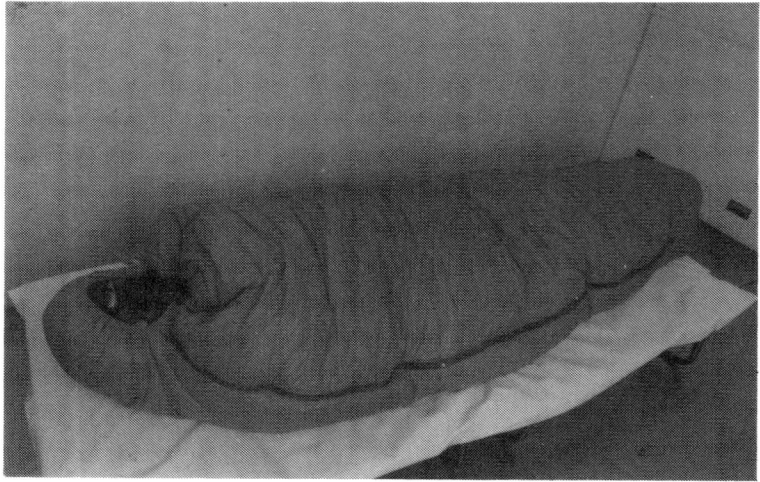

*Figure 6.22. The evaluation of a sleeping bag's range of utility with a manikin (top), and measuring the tolerance time for a life-preserving immersion suit (bottom).*

Furthermore, this evaluation system is not restricted to clothing only. As can be seen from Figure 6.22, the range of utility for example of sleeping bags can also be determined.

The usefulness of the predictive model is not exhausted with the calculation of the range of utility. The wear comfort of a clothing ensemble under particular climatic and activity conditions can also be judged using the model. According to Figure 6.18, the wearer's mean skin temperature ($T_s$) and rectal temperature ($T_{re}$) can be calculated.

These quantities show whether the wearer is feeling too cold or whether s/he is already suffering from hypothermia. Thus, tolerance times can be deduced, showing the longest possible time lapse for which clothing near its very limit of utility can protect its wearer from health problems. The example of the life-preserving immersion suit pictured in Figure 6.22 shows that the laboratory evaluation of garments described has proceeded into a region which could never be covered by wear trials with human subjects.

The time pattern of the rectal temperature ($T_{re}$) and heart rate (HR), calculated together with the skin wettedness (expressed in the comfort factor $k_f$) in a clothing ensemble, indicate whether the wearer is feeling too hot or whether s/he is already suffering from hyperthermia. This deduction is possible, because—as already stated—there exists a significant correlation between man's subjective comfort feeling and certain physiological data of the body. For instance, in Figure 6.23 the correlation between rectal temperature and the heat sensation perceived (judged according to an 8-step ballot) is shown.

With these correlations, universally established in wear trials with human subjects, predictive formulae for the wear comfort of clothing ensembles can be developed, yielding a comfort vote (*WC*) ranging between 1 (very good) and 6 (completely unsatisfactory):

Heat load: $WC = 1{\cdot}52\ T_{re} + 0{\cdot}3\ T_s + 3{\cdot}1\ k_f + 0{\cdot}02\ HR - 67{\cdot}85$

Heat loss: $WC = -5{\cdot}61\ \varDelta T_{re} + 0{\cdot}6\ \varDelta T_s + 1{\cdot}51$

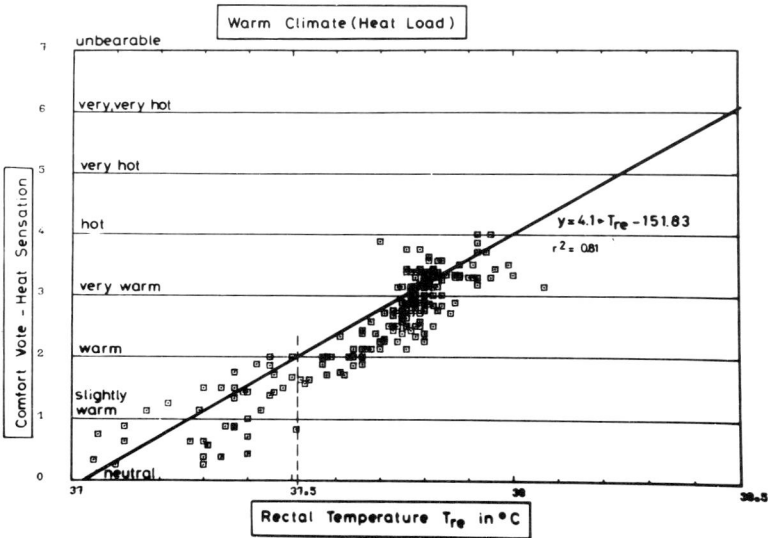

*Figure 6.23. The correlation between rectal temperature and heat sensation.*

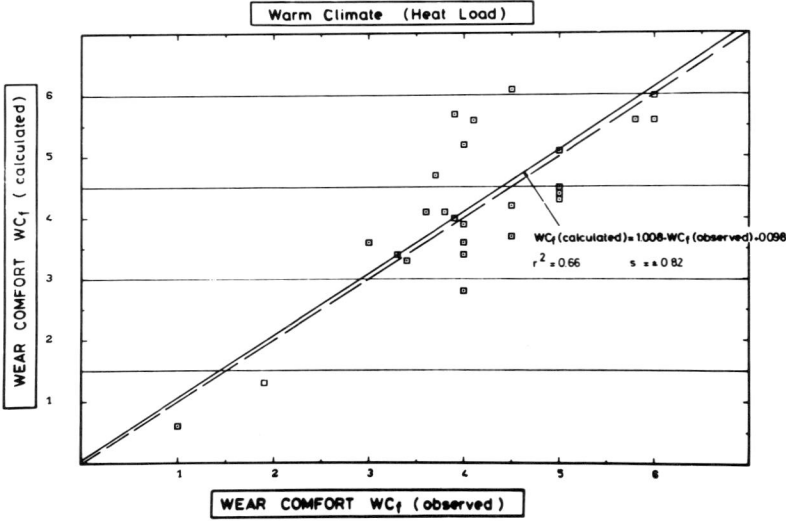

*Figure 6.24.   The agreement between predicted and actually perceived wear comfort of clothing ensembles.*

where $\Delta T_{re}$ and $\Delta T_s$ = decrease in rectal and mean skin temperature, respectively.

All the quantities included in these formulae can be calculated within the predictive model. As can be seen from Figure 6.24, there exists a good agreement between the calculated wear comfort votes and the subjective sensations of persons actually wearing the clothing.

For the textile and garment manufacturer it is of particular benefit that the predictive model can also be reversed. Thus, the thermal and water vapour resistances required by either the fabrics or the total clothing ensemble can be calculated, in order to yield good wear comfort under particular climate and activity conditions. These values derived from the predictive model can be readily checked by skin model and manikin tests. Thus, already at the very first stages of development the manufacturer can verify whether his products meet the physiological requirements demanded by the end-user.

## Acknowledgement

The test methods described are the results of research projects conducted with financial assistance from Forschungskuratorium Gesamttextil, Frankfurt am Main, with funds from the Arbeitsgemeinschaft Industrieller Forschungsvereinigungen (AIF), Cologne, the Deutsche

Forschungsgemeinschaft, Bonn, and the Commission of the European Communities.

## References

DIN 54 101, 1984, Determination of physiological properties of textiles. Draft edition, Apr (Berlin: Deutsche Institut für Normung).

Mecheels, J. and Umbach, K.H., 1977a, Thermophysiological properties of clothing systems. *Melliand Textilber.*, English edition, Dec 1976, 1142−1146, Jan 1977, 76−85.

Mecheels, J. and Umbach, K.H., 1977b, The psychrometric range of clothing systems. In *Clothing Comfort*, edited by N.R.S. Hollies and R.F. Goldman (Ann Arbor, MI: Ann Arbor Science).

Umbach, K.H., 1980, Measuring the physiological properties of textiles for clothing. *Melliand Textilber.*, English edition, Jun, 747−757.

Umbach, K.H., 1981a, Textile transmission. *Textile Asia*, **12**, 75−79.

Umbach, K.H., 1981b, Protective clothing against cold with a wide range of thermophysiological control. *Melliand Textilber.*, English edition, Apr.

# 7. Ventilation of Rainwear Determined by a Trace Gas Method

## W. A. Lotens and G. Havenith

## 1. Introduction

A well-known problem with the comfort of rainwear is the condensation of evaporated sweat inside the garment, which results in continually increasing wetness of the clothing. Under certain environmental and workload conditions this may be effectively counteracted by the use of vapour permeable materials, but in particular for hard work and in a cool environment this runs short in function (Havenith and Lotens, 1984). The alternative way for moisture dissipation is ventilation of the microclimate under the garment. A theoretical study was carried out (Lotens, 1987) to estimate the required ventilation, in addition to permeability, to avoid condensation in the garment. The conclusion was that for moderate work ventilation must amount to 450 l min$^{-1}$, depending on the permeability of the fabric.

Are such ventilation flows achievable? One could try to investigate this indirectly by conducting an experiment with various designs of rainwear. The resulting wetness would be an indicator then for the sufficiency of the ventilation. Since ventilation is a function of body motion (Vogt *et al.*, 1983) as well as wind (windchill!) this would involve a large number of conditions, in fact too large for a manageable experiment. Far better would be the use of a method that provides direct ventilation figures.

An obvious method would be the washing out of a trace gas, analogous to the dye dilution or heat dilution methods used to determine cardiac output. Such attempts have been made by Crockford *et al.* (1972). They exchanged the air under the garments for pure nitrogen and watched the oxygen level to restore the normal value of 21%. The time constant of this process can be converted to a ventilation rate.

At least two factors stain this basically elegant method. The first is that the washout process is not a simple first-order process and does not

162

reveal unique time constants. Seemingly, the ventilation changes during the measurement. The second is that only ventilation rates $(\text{min}^{-1})$ result, which are meaningless without a value for the ventilated volume, since the ventilation flow $(\text{l min}^{-1})$ is the product of ventilation rate and ventilated volume (l). Crockford and Rosenblum (1974) published a method to determine the volume, but this method is rather artificial and not free of experimental problems.

In this report attempts to improve Crockford's trace gas method are documented. The resulting instrumentation is described and the results of an investigation of limited size in the design of rainwear are reported. Finally, the obtained figures are compared to the aforementioned estimated requirements.

## 2. The trace gas method

Replication of Crockford's decay curve method showed that the decay curves are not of a simple exponential shape, but reflect at least two or three time constants. This is understandable since clothing really consists of several coupled compartments, with different magnitude of ventilation. Crockford *et al.* (1972) used an ambiguous method to avoid this problem. They sampled the concentration by a harness of tubings underneath the underclothing, but the inlet is under the rainwear. This indeed seems to simplify the decay curve, but it is now unclear what is actually measured: the ventilation of rainwear or that of rainwear and underclothing together. We decided to leave the matter of time constants and try another approach, the mass balance. This method is essentially simple. When a trace gas is constantly entered under the rainwear the concentration in the microclimate increases until equilibrium is settled. In that situation the total mass of trace gas washed out by ventilation meets the inflow:

$$\text{inflow} \times C_{\text{p}} = \text{vent} \times C_{\text{out}} \ (\text{g min}^{-1}) \tag{1}$$

where inflow is the flow of pure trace gas $(\text{l min}^{-1})$; vent is ventilation of the microclimate by fresh air $(\text{l min}^{-1})$; $C_{\text{out}}$ is trace gas concentration in the microclimate $(\text{g l}^{-1})$; and $C_{\text{p}}$ is the concentration of pure trace gas $(\text{g l}^{-1})$. Thus the ventilation may be determined from the measurement of inflow and $C_{\text{out}}$.

Preliminary experiments with distribution of pure trace gas showed that the mixing was far from ideal. Clouds of the heavier trace gas seemed to drop out of the jacket, not only producing a noisy signal, but causing an erroneous mean value as well.

For this reason the trace gas was next premixed with air, to bring the concentration and specific density down to near the expected value in

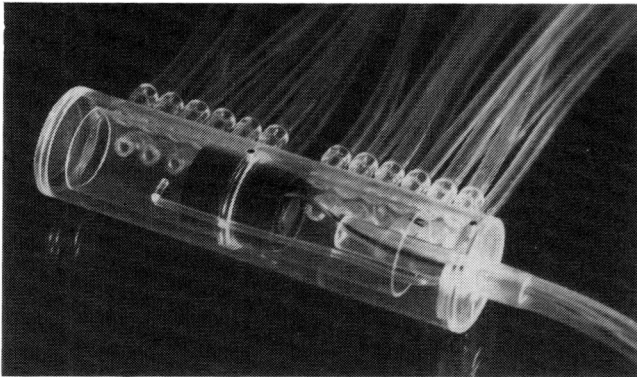

*Figure 7.1.   Distribution (right) and sampling (left) harnesses with mixing chambers and a miniature fan.*

the microclimate. This method gave satisfactory results. The required air was taken from the microclimate to avoid forced ventilation due to the method. In this way a circulation was established into which the trace gas was injected. The next trial was with distribution and sampling harnesses connected to the circuit (Figures 7.1 and 7.2). A miniature fan was used for the circulation. It showed, however, that the performance of the fan was too small to prevent flow fluctuations during body motions. Consequently a more powerful fan had to be used, but that one had to stand on a table instead of being carried. Two tubes (about 1 cm diameter) connected the fan with the harnesses. This set-up is the current product of this development.

The design of the harness is such that the trace gas mixture is distributed evenly over the body. To this purpose the number and length of the tubes were carefully adjusted to body size and surface area. The pressure–flow relationship for the various tubes has been determined by means of an empirical formula, covering our measurements on a wide range of tubes:

$$V = \frac{d^{2 \cdot 3} \, (3 \cdot 3 - \log L) \, p^{0 \cdot 54}}{572} \tag{2}$$

where $V$ is flow ($1 \ s^{-1}$); $d$ is diameter (mm); $L$ is length (cm); and $p$ is pressure (cm $H_2O$). The tubes of the distribution system were all of the same diameter (4 mm inner size) and the pressure is the same for all tubes. The only factor relevant for the distribution of the flow is thus the term $(3 \cdot 3 - \log L)$ in equation (2). In Table 7.1 the resulting relative flow is compared to the relative body surface area, showing that there is a close correspondence. Head, hands and feet are considered to be outside the rainwear.

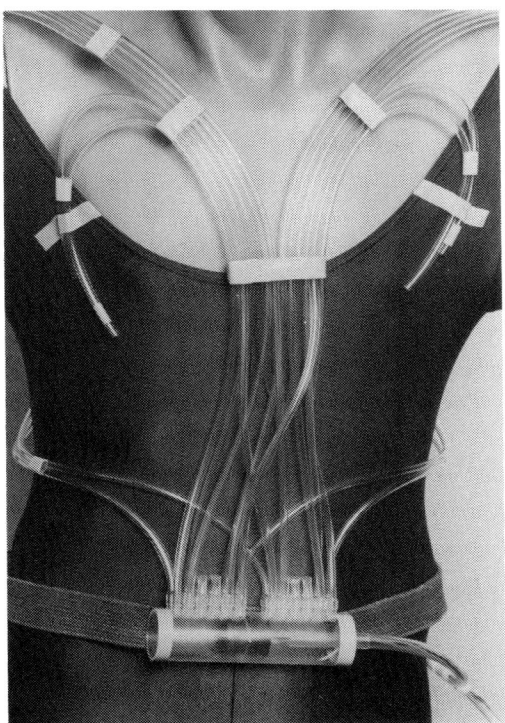

*Figure 7.2.* The harnesses as worn by a manikin.

*Table 7.1.* Dimension of the distribution harness and relative flow compared to relative body surface area.

| Compartment | Number of tubes | Length (cm) | Relative flow | Relative area (%) |
|---|---|---|---|---|
| **Upper body** | | | | |
| Upper arms | 2 | 80 | 7·5 | 7 |
| Lower arms | 2 | 105 | 7 | 7 |
| Back | 2 | 55 | 8 | |
| Chest | 2 | 55 | 8 | |
| | | | | 36 |
| **Lower body** | | | | |
| Buttocks | 2 | 50 | 8·5 | |
| Belly | 2 | 40 | 9 | |
| Upper legs | 4 | 50 | 17 | 18 |
| Lower legs | 2 | 90 | 14 | 13 |
| | +2 | 105 | | |
| | | | 79 | 81 |
| Hands | — | | | 5 |
| Feet | — | | | 7 |
| Head | — | | | 7 |
| | | | | 100 |

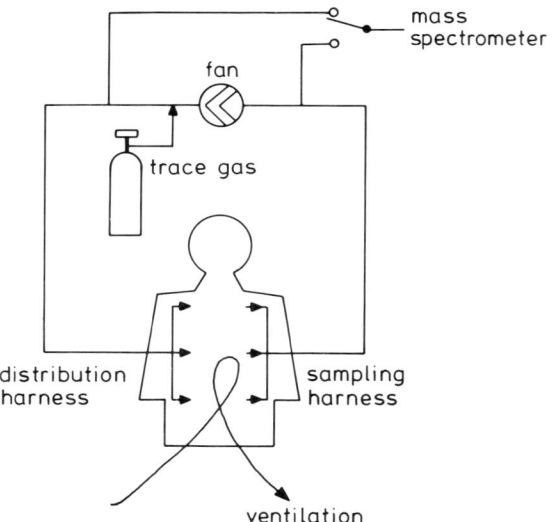

*Figure 7.3.   Scheme of the trace gas method. The trace gas is injected from a container into circulating air. The mixture is mixed again with outside air due to ventilation of the microclimate.*

The sampling harness is similar to the distribution harness. Tubes of both harnesses are mounted pairwise at the body, with a separation of about 7 cm. This is enough to allow adequate mixing. The total system is separated in a lower and an upper part, for the measurements of jackets and trousers separately. Head, hands and feet are excluded. Figure 7.3, finally, shows a scheme of the whole set-up. The actual measuring device is a mass spectrometer, which is a versatile and sensitive instrument.

The trace gas used so far is $N_2O$, an anaesthetic gas. Since it is rarefied down to levels lower than 100 ppm in the room air, there is no danger involved. Theoretically it would be good to use a trace gas with the same diffusion coefficient as water vapour. This requires a gas with a molecular mass close to that of water. Unfortunately all masses in that range show a considerable background level on the mass spectrometer, which masks the signal. Mass 20 (neon) would be feasible, but it is an inert gas and has quite different diffusion characteristics to water vapour. An alternative would be the use of $CH_4$, measuring the radical $CH_3^+$ as mass 15, but methane is highly explosive. For this reason we decided to keep on using the available $N_2O$. As long as convection processes dominate pure diffusion, which is usually the case, no errors will be introduced.

## 3. Theoretical analysis

Equation (1) is a simplification of what really goes on. In the first place an inflow of trace gas would push some air out of the garment and should be corrected for. And in the second place a garment is not one well mixed balloon but a structure of more or less well mixed compartments. Many compartments will thus be ventilated with microclimate air of the other compartments, instead of with fresh air. Both these arguments are taken into the account with equation (3):

$$(\text{inflow} + \text{circ})C_{\text{in}} + \text{vent}C_{\text{e}} = (\text{vent} + \text{inflow} + \text{circ})C_{\text{out}} \quad (3)$$

where inflow is flow of pure trace gas $(1 \text{ min}^{-1})$; circ is circulating flow over the fan $(1 \text{ min}^{-1})$; vent is ventilation $(1 \text{ min}^{-1})$; $C_{\text{in}}$ is concentration in the distribution harness $(\text{g } 1^{-1})$; $C_{\text{out}}$ is concentration in the sampling harness $(\text{g } 1^{-1})$; and $C_{\text{e}}$ is concentration in the immediate environment $(\text{g } 1^{-1})$.

Figure 7.4 shows the flow system. Equation (3) represents the mass balance for the part of the garment under investigation. The terms at the left of the sign of equality represent the sources of trace gas and those at the right the drains. When equation (3) is solved for vent, equation (4) results:

$$\text{vent} = \frac{(\text{inflow} + \text{circ}) \ (C_{\text{in}}/C_{\text{out}} - 1)}{1 - C_{\text{e}}/C_{\text{out}}} \quad (4)$$

Since the calculation of vent is dependent on $C_{\text{e}}$, the actual ventilation (air exchange) is only known when all concentrations inside the garment $(C_{\text{e}})$ are known. However, it is not so much the air exchange as the mass transfer that is of interest. This mass transfer $M$ $(\text{g min}^{-1})$ is:

$$M = \text{vent} \ (C_{\text{out}} - C_{\text{e}}) = (\text{inflow} + \text{circ}) \ (C_{\text{in}} - C_{\text{out}}) = \text{vent}_{\text{eff}} \times C_{\text{out}} \quad (5)$$

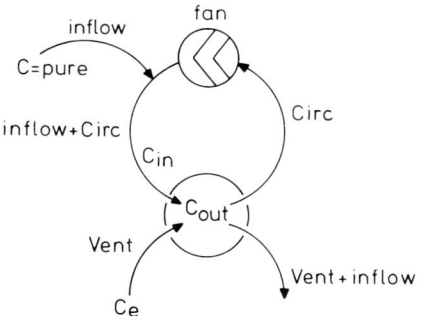

*Figure 7.4.   Flows and concentrations of trace gas for a part of a garment.*

By means of equation (5) effective ventilation vent$_{eff}$ is defined as the ventilation that would give the same mass transfer as vent if the ventilation had taken place with fresh air. By substituting (4) in (5) vent$_{eff}$ can be calculated as:

$$\text{vent}_{eff} = (\text{inflow} + \text{circ})\,(C_{in}/C_{out} - 1) \tag{6}$$

For the determination it is apparently sufficient to measure the flows (constant during the experiment) and the ratio of inlet and outlet concentration. Thus calibration of the mass spectrometer is not required. This facilitates experimentation.

An important question is, whether summation of the various body parts is allowed. Experimentally, the average sample concentration is determined by:

$$\overline{C_{out}} = \frac{\Sigma \, \text{circ}_i \, C_{out\,i}}{\Sigma \, \text{circ}_i}$$

and the circulating flow by

$$\text{inflow} + \text{circ} = \Sigma \, (\text{inflow}_i + \text{circ}_i)$$

where $i$ denotes the $i$th compartment. The average effective ventilation is then, according to (6):

$$\overline{\text{vent}_{eff}} = \Sigma(\text{inflow}_i + \text{circ}_i) \left( \frac{C_{in} \Sigma \, \text{circ}_i}{\Sigma \, C_{out\,i} \times \text{flow}_i} - 1 \right) \tag{7}$$

The summation of the effective ventilation of the various compartments, however, is:

$$\text{vent}_{eff} = \Sigma \, \text{vent}_{eff\,i} = \Sigma \, (\text{inflow}_i + \text{circ}_i)\,(C_{in}/C_{out\,i} - 1) \tag{8}$$

Expressions (7) and (8) are only identical when all $C_{out\,i}$ are equal, in other words, when there are no concentration differences under the garment. This is generally not so, but the even distribution of trace gas is very helpful in this respect. A numerical analysis shows that concentration differences of a factor of 10 over large body parts may cause serious calibration errors (underestimation by a factor of 3) but for concentrations not further than a factor of 2 apart, errors of the order of 10% result. In view of the variability in the data between subjects, fit of clothing and adjustment of fastenings this is not a major problem. The option is still open to measure the concentration at the various body parts separately, but this laborious method is not yet justified.

## 4. *Validity test*

The validity of the method was tested by means of forced ventilation of a jacket. Care had to be taken that the distribution of the forced air flow was area weighted, since only in that situation does the real ventilation compare with the measured effective ventilation. The area weighting was approximated by a simple tubing system, which was connected to an air pump via a gas flow meter.

The measurements were first done on a static manikin. Figure 7.5 (left frame) shows that a linear relationship results between forced and effective insulation, with a slight overestimation of the forced ventilation. The initial 10 l min$^{-1}$ effective ventilation, without forced ventilation, represents the diffusion through seams, zipper, apertures, etc. The deviation from the identity may well be explained by errors in the area weighting of the forced air flow. This weighting becomes less critical when body motion is introduced, pumping the air under the jacket back and forth, thus improving the mixing. This was tried with a marching subject (Figure 7.5, right frame) wearing the jacket well tied up. Again, a linear relationship results, with a slight underestimation of the forced ventilation this time. Indeed the increased mixing changed the calibration. In neither case, however, are the deviations large. Taking into account that ventilation of clothing is highly variable due to fit, tightness of apertures, body motion, wind, etc., the agreement between forced and effective ventilation is satisfactory. Therefore, the enhanced trace gas method is adequate both in a qualitative and a quantitative sense.

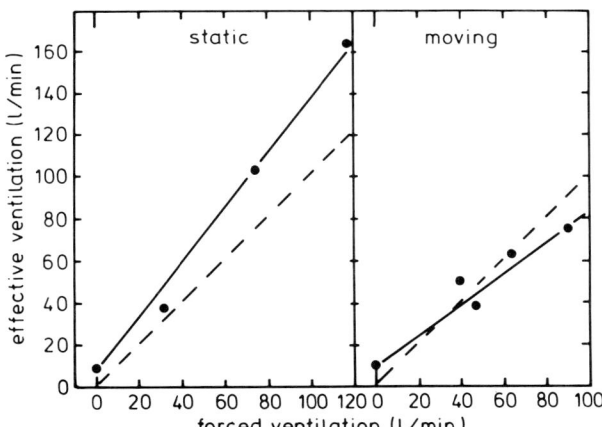

*Figure 7.5. Measured effective ventilation vs. forced ventilation of a jacket. The left frame refers to a static manikin, the right frame to a moving subject.*

## 5. *Measurements on rainwear*

The feasibility of the method was tested on an experimental garment, designed by the Fibre Research Institute, TNO. This garment is a conventional two piece rainsuit made out of air impermeable fabric, provided, however, with a number of vents that can be opened or closed at will (Figure 7.6). In an earlier study (Lotens, 1984) it was shown that the design of vents with special aerodynamic properties (either due to wind or to pumping by body motion) is difficult, if not impossible. The major factor seemed to be the space under the garment, the design of the vent did not matter so much. For this reason in the experimental suit of this study the simplest possible design was chosen: a hole with a mesh.

Five subjects participated in three experiments with the following aims:

1. To investigate the combined effect of three speeds of walking (0, 2·5 and 5 km h$^{-1}$) and three wind speeds (0, 2 and 6 m s$^{-1}$) on ventilation.

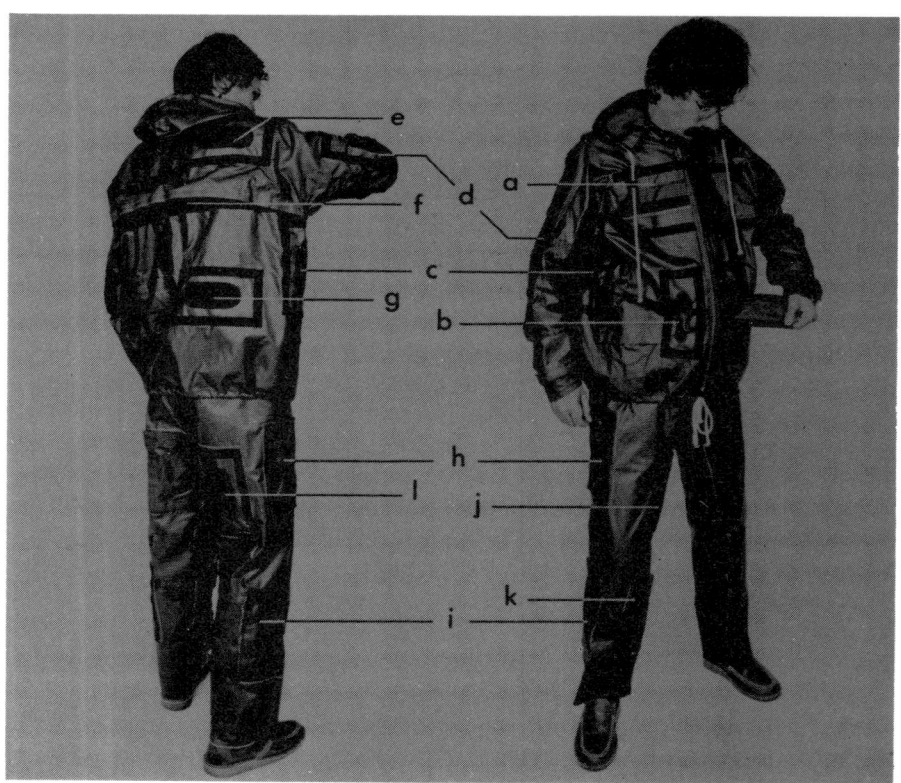

*Figure 7.6. The experimental rain suit with adjustable vents. The characters refer to Table 7.2.*

2. To investigate the ventilation effect of the seven locations of vents on the jacket and the five locations on the trousers.

3. To investigate the ventilation effect of a spacer (about 4 cm wide) under the jacket, with and without the vents opened.

During the experiments the circulatory flow was constant at $26.4\ l\ min^{-1}$ and the trace gas flow (inflow) was $0.1\ l\ min^{-1}$. This resulted in an inlet concentration of $0.4\%$ and a sample concentration in the magnitude of $0.01-0.1\%$. These concentrations are easily detectable with the mass spectrometer.

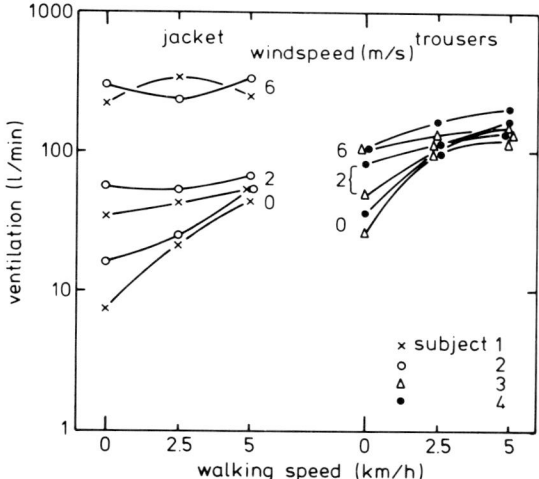

*Figure 7.7.* *Ventilation of jacket and trousers as a function of walking speed and wind speed. Vents were all closed.*

Figure 7.7 shows the results of *Experiment 1*. Two subjects were used for the measurements for the jacket and two others for the trousers, with all vents closed. The subjects walked on a treadmill in a wind-tunnel with the wind directed to their front. When the subjects were standing in still air ventilation was quite low, $10\ l\ min^{-1}$ for the jacket and $30\ l\ min^{-1}$ for the trousers. This basic ventilation is likely to be due to diffusion. A pilot experiment with a jacket of semipermeable material showed ventilations over $50\ l\ min^{-1}$ under the same conditions, reflecting the additional diffusion through the fabric.

Ventilation increases due to motion, but even more due to wind. For wind speeds over $2\ m\ s^{-1}$ the motion effect drowns in the wind effect. This is less so for the trousers which are already ventilated by the wind that is produced by the slinging motion of the legs. For the purpose of dissipation of moisture, the motion pump is much more adequate than the wind. Violent motion is usually associated with high sweat production and due to the increased ventilation, the dissipation increases with the demand. Wind, however, is uncorrelated with the demand for dissipation and therefore often undesired. From this point of view a ventilation of 60 l min$^{-1}$ for the jacket and 150 l min$^{-1}$ for the trousers is the best achievable for this design. The two subjects did not noticeably differ in their experimental results.

Table 7.2 lists the average effects of the vents in *Experiment 2*. Vents were always open at both sides of the garment and the data thus pertain to two vents in symmetrical position. The increase in ventilation ranges roughly from 15–65 l min$^{-1}$ for the jacket and 40–50 l min$^{-1}$ for the trousers. The majority of the vents are in the range 40–60 l min$^{-1}$. Exceptions are vents b and c, located at the stomach and under the arms respectively. An obvious explanation for their poor function is their location to the already ventilating bottom of the jacket. The highest ventilation is found for vents e and f, both located at the back. The other vents are all in the same range.

Table 7.2. *Average effect of the vents for walking at 5 km h$^{-1}$ in a 2 m s$^{-1}$ wind. Averaged over three subjects for the jacket and two subjects for the trousers. The location of the vents is indicated in Figure 7.6.*

| Jacket | Ventilation (l min$^{-1}$) | Trousers | Ventilation (l min$^{-1}$) |
|---|---|---|---|
| Closed | 56 | Closed | 149 |
| Vent h open | 102 | vent h open | 190 |
| b | 69 | i | 198 |
| c | 71 | j | 191 |
| d | 98 | k | 194 |
| e | 119 | l | 197 |
| f | 110 | | |
| g | 103 | | |

Just a few data were obtained with combinations of vents. In all cases the combination provides less ventilation than the sum of the two separate vents. This seems logical because opening another vent would cause ventilation of already ventilated microclimate air. The data do not allow analysis of vents that are close to each other in contrast to well separated vents. Following the above reasoning this should make some difference.

The effect of the spacer (*Experiment 3*) is shown in Table 7.3. The spacer should probably be regarded as an enhancing rather than an additive factor, since no additional avenues of ventilation are opened. The spacer merely enables more circulation under the jacket. The

*Table 7.3.    Ventilation (l min⁻¹) effect of the spacer worn under the jacket. Average over three subject.*

|  | Without spacer | With spacer |
|---|---|---|
| Closed | 56 | 125 |
| Vent a open | 102 | 174 |
| c | 71 | 111 |
| f | 110 | 156 |

magnifying factor is rather varying for the different designs, ranging from 1·4 to 2·2. The effect is largest for the closed suit, where the ventilation is most restricted.

## Discussion

Crockford *et al.* (1972) measured the ventilation rates of various designs of rainwear. A single raincoat offered ventilation rates of about 6 min⁻¹ while sitting in a 2 m s⁻¹ wind. The next best design was a duck suit with bibbed trousers, which may give less ventilation than our waist-high trousers. They found exchange rates of about 3 min⁻¹ for a hauling task in a 1·65 m s⁻¹ wind. In a later publication (Crockford and Rosenblum, 1974) they estimated the volume of the garment at 23 l, multiplying to a ventilation of ~70 l min⁻¹. This seems somewhat less than our results, although the task and clothing are too different to be certain about this. Opening wrist and ankle cuffs made a difference of about 25 l min⁻¹ which is again a bit less than we found for the various vents (during walking).

Shivers *et al.* (1977) used the same method as Crockford. They measured ventilation rates for women's raincoats of 4 min⁻¹ during rest and 8 min⁻¹ during walking in quiet air. In particular, during rest there was a marked effect of belting—*increasing* the ventilation rate. This must be due to the decreased microclimate volume. No difference was found between a set-in sleeve and a low dolman sleeve. The lack of data about the microclimate volume prevents comparison with our results, but they compare roughly to Crockford's raincoat.

Both these studies prove how important it is to know the ventilation in absolute flows instead of as ventilation rates. In this respect, the current method is an improvement.

The trace gas method may also be applied to permeable clothing, but care has to be taken in interpreting the data. Many fabrics will pass the trace gas in a similar way as they pass water vapour, by convection and diffusion. Hygroscopic materials, however, may pass water in other ways as well, for instance by wicking and on a molecular basis in hygroscopic films. These processes will not be simulated by the trace gas method.

## 6. Functional design of rainwear

We have shown above that a two-piece garment, provided with vents in both the jacket and the trousers and with a special provision to keep a space under the jacket may give a ventilation of:

| | |
|---|---|
| jacket | 170 |
| trousers | 190 |
| total | 360 l min$^{-1}$ |

This might increase a bit with an increasing number of vents. It is not likely, however, that a garment with many vents is really waterproof. Experience shows that any seam that is not adequately taped forms a leak during sustained rain. The more complicated the construction of the clothing, the higher the risk that leaks show up. Another problem is that rainwear may be used in such a variety of postures, depending on the application. In particular, military rainwear may be used in situations where rain will penetrate through the vents. Spacers are not available with commercial rainwear, to our knowledge. They would certainly increase the pack volume and therefore rarely be applied. For these reasons the figure of 360 l min$^{-1}$ must be regarded as an optimistic maximum ventilation for rainwear. Is this a sufficiently large quantity? Figure 7.8, reproduced from Lotens (1987), gives the theoretical requirements for a moderately hard working person. These requirements are dependent on the vapour permeability of the fabric. Lower limits are presented for the ventilation and permeation together, that will prevent condensation of vapour on the inner face of the rainwear, according to the calculations.

The permeation is expressed in the permeability index $i_m$ (Woodcock, 1962). The best rainwear materials available have an $i_m$ of 0·16, but many good materials fall in the range 0·03–0·06. Lower $i_m$ values may be found in less sophisticated (and less expensive) materials.

Figure 7.8 shows that in the cold (when condensation is more likely to occur than in the heat) even optimal ventilation is not sufficient to keep the rainwear of a moderately hard working person free of condensation, but in cool or warm environments ventilation probably can do. When ventilation is insufficient, permeation is required as well. According to Figure 7.8, for moderately hard working people fair permeability together with optimal ventilation is sufficient. Permeability alone, on the other hand, is sufficient only with the best materials. Rainwear made out of still good fabric may require up to 150 l min$^{-1}$ of ventilation. For harder work, the requirements increase to a level that cannot be met by design and fabric technology.

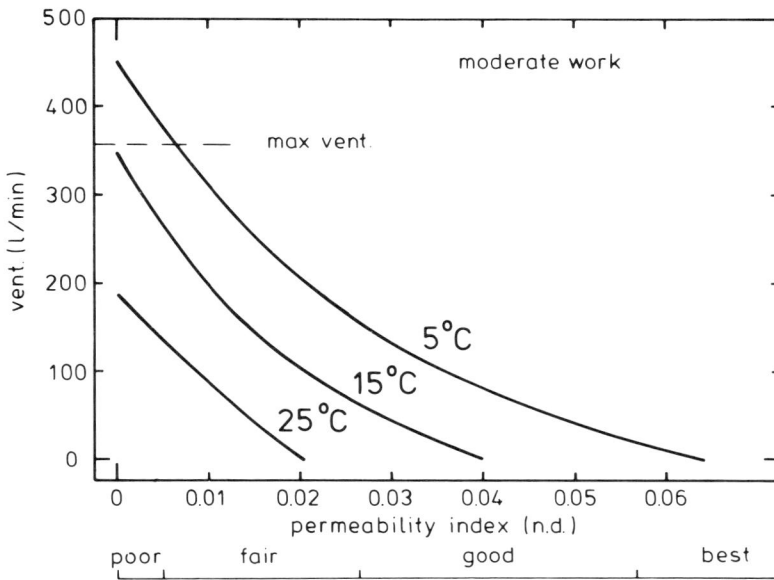

*Figure 7.8. Lower limits for the permeability of the fabric and the ventilation necessary to provide enough vapour dissipation to prevent condensation on the inner surface of the rainwear (Lotens, 1987).*

## 7. Conclusions

The enhanced trace gas method presented here is a convenient method. When effective ventilation is the variable of interest, the experimental procedure is fast and uncomplicated. A further advantage is that the result is expressed in l min$^{-1}$ in contrast to the ventilation rates in previously published methods. No assumptions about the actual process of ventilation are required, but care must be taken that the trace gas concentrations under the garment do not vary too much over the body, because averaging would not be allowed then.

The experimental suit clearly showed ventilation due to the motion pump, but this effect is drowned in the ventilation due to external wind for wind speeds over 2 m s$^{-1}$. The trousers are less sensitive to wind than the jacket. Vents may increase the ventilation (at 5 km h$^{-1}$ walking, 2 m s$^{-1}$ wind) by some 40–60 l s$^{-1}$, but somewhat less for vents low on the jacket. The effect of vents may be enhanced by a factor of 1·4–2·2 by introducing an air gap under the garment (4 cm wide) depending on the ventilation already present. Altogether a ventilation of 360 l min$^{-1}$ is an optimistic maximum for carefully designed rainwear.

This ventilation would often be sufficient to keep the rainwear of moderately hard working persons free of condensation. For harder work permeability of the fabric is also required.

## Acknowledgement

This investigation was supported by the Commission of the European Communities.

## References

Crockford, G.W. and Rosenblum, H.A., 1974, The measurement of clothing microclimate volumes. *Clothing Res. J.*, **2**, 109−114.

Crockford, G.W., Crowder, M. and Prestidge, S.P., 1972, A trace-gas technique for measuring clothing microclimate air exchange rates. *Brit. J. Industr. Med.*, **29**, 378−386.

Havenith, G. and Lotens, W.A., 1984, What actually, is the advantage of permeable over impermeable rainwear? *Report of the Institute for Perception*, 1984–6.

Lotens, W.A., 1984, Second semi-annual progress report on project TEX 23 NL (N). *Report of the Institute for Perception*, 1984−38.

Lotens, W.A., 1987, Balancing vapour permeability of rainwear against ventilation; a theoretical study. *Report of the Institute for Perception*, November.

Shivers, J.L., Yeh, K., Fourt, L. and Spivak, 1977, The effects of design and degree of closure on microclimate air exchange in light weight cloth coats. In *Clothing Comfort*, edited by N.R.S. Hollier and R.F. Goldman (Ann Arbor, MI: Ann Arbor Science Publishers), pp. 167−181.

Vogt, J.J., Meyer, J.P., Candas, V., Libert, J.P. and Sagot, J.C., 1983, Pumping effects on the thermal insulation of clothing worn by human subjects. *Ergonomics*, **26**, 963−974.

Woodcock, A.H., 1962, Moisture transfer in textile systems, part 1. *Text. Res. J.*, **32**, 628−633.

# 8. Simple Relationships among Current Vapour Permeability Indices of Clothing with a Trapped-air Layer

**Takahumi Oohori, Larry G. Berglund and A. Pharo Gagge**

## 1. Introduction

The permeation ratio of clothing is particularly important for clothing studies, because it is not easy and sometimes expensive to measure evaporative insulation of clothing directly. If the permeation ratio is known in advance, one can calculate the evaporative insulation of clothing from the thermal insulation and the permeation ratio of clothing.

The purposes of our study are to analyse and compare three current indices of evaporative heat exchange through clothing, namely, $i_{cl}$ for clothing, $i_a$ for outer-air and $i_m$ for combined clothing and air layers; and to incorporate explicitly the trapped air between skin and clothing into heat and mass transfer models.

## 2. Dry and evaporative heat transfer

We begin the present study with an analogy between dry and evaporative heat transfer from skin to ambient air. Heat transfer coefficients are described in two different formats, namely a resistance format and the Burton/Nishi conductance format. By using a resistance format, the dry heat transfer coefficient is expressed as the reciprocal of the sum of air ($I_a$) and clothing ($I_{cl}$) insulations:

$$\text{Dry}/\Delta T = 1/(I_a + I_{cl}) \qquad (\text{W K}^{-1}\text{ m}^{-2}) \qquad (1)$$

Similarly, the evaporative heat transfer coefficient is expressed in terms of evaporative insulations for air ($I_e$) and clothing ($I_{cle}$)

$$E_{sk}/\Delta P = 1/(I_e+I_{cle}) \qquad \text{(W m}^{-2}\text{ Torr)} \qquad (2)$$

On the other hand, Burton introduced a thermal efficiency factor for dry heat exchange ($F_{cl}$). Nishi also introduced a permeation efficiency factor for evaporative heat exchange ($F_{pcl}$) analogous to Burton's $F_{cl}$.

$$F_{cl} = I_a/(I_a+I_{cl}) \qquad \text{(N.D.)} \qquad (3)$$
$$F_{pcl} = I_e/(I_e+I_{cle}) \qquad \text{(N.D.)} \qquad (4)$$

In terms of $F_{cl}$ and $F_{pcl}$, the dry and evaporative heat transfer coefficients can be expressed in simple conductance formats, namely,

$$\text{Dry}/\Delta T = h\ F_{cl} \qquad \text{(W m}^{-2}\text{ K)} \qquad (5)$$
$$E_{sk}/\Delta P = h_e\ F_{pcl} \qquad \text{(W m}^{-2}\text{ Torr)} \qquad (6)$$

## 3. Lewis relations and permeation ratios

All current indices and models of evaporative cooling while clothed are based on the Lewis relation. The Lewis number is defined as the ratio of mass transfer coefficient by evaporation to heat transfer coefficient by convection (no radiation), which is constant at sea level, at 2·2:

$$\text{Lewis number} = h_e/h_c = 2\text{·}2 \quad \text{(K Torr}^{-1}) \qquad (7)$$

The Lewis relation can be extended to those for outer-air layer, for clothing layer and for combined clothing and air layers. The effective Lewis number for each layer is defined as the ratio of dry to evaporative heat transfer coefficients. Three non-dimensional permeation ratios, $i_a$, $i_{cl}$ and $i_m$, are defined as the ratios of the Lewis number of each layer to that of an equivalent non-radiative air layer, namely, 2·2 (see Table 8.1).

Note that each ratio is not always constant. For example, $i_a$ and $i_{cl}$ vary with air movement and type of fabrics, respectively and $i_m$ does with both.

## 4. Relationships between $i_m$, $i_{cl}$ and $i_a$

The three permeation ratios $i_m$, $i_{cl}$ and $i_a$ are related by two simple relationships. First, the reciprocal of $i_m$ is the average of the reciprocals

*Table 8.1.* *Lewis numbers and permeation ratios for air, for clothing, and for combined clothing and air.*

| | Heat transfer coefficient | | Lewis number $(=B/A)$ K Torr$^{-1}$ | Permeation ratio $(=B/2·2\,A)$ ND |
|---|---|---|---|---|
| | Dry $(=A)$ W m$^{-2}$ K$^{-1}$ | Evaporative $(=B)$ W m$^{-2}$ Torr$^{-1}$ | | |
| Air | $h(=h_r+h_c)$ | $h_e$ | $h_e/h$ | $i_a=h_e/h$ |
| Clothing | $h_{cl}$ | $h_{cle}$ | $h_{cle}/h_{cl}$ | $i_{cl}=h_{cle}/2·2\,h_{cl}$ |
| Combined clothing and air | $hF_{cl}$ | $h_eF_{pcl}$ | $h_eF_{pcl}/hF_{cl}$ | $i_m = h_eF_{pcl}/2·2\,hF_{cl}$ |

of $i_a$ and $i_{cl}$, weighted by Burton's $F_{cl}$ and $(1-F_{cl})$, respectively, which is an impermeability relation:

$$1/i_m = F_{cl}/i_a+(1-F_{cl})/i_{cl} \qquad \text{(N.D.)} \qquad (8)$$

Second, $i_m$ is also the weighted average of $i_a$ and $i_{cl}$ using Nishi's $F_{pcl}$ and $(1-F_{pcl})$, which is a permeability relation:

$$i_m = F_{pcl}\,i_a+(1-F_{pcl})/i_{cl} \qquad \text{(N.D.)} \qquad (9)$$

In these equations, it is clear that $i_m$ is dependent on both air movement and clothing insulation.

Figure 8.1 shows numerical relationships between $i_{cl}$ and $i_m$ as air movement varies from still air to $2·5$ m s$^{-1}$. The clo value is kept constant ($0·6$ clo). $i_{cl}$ values are plotted on the abscissa and $i_m$ on the ordinate. A low value of $i_{cl}$ means impermeable clothing; a high value means permeable. For impermeable clothing, $i_m$ is a little bit higher than $i_{cl}$. On the other hand, for permeable clothing, $i_m$ is lower than $i_{cl}$, and this difference becomes larger as air movement decreases.

Figure 8.2 also shows relationships between $i_m$ and $i_{cl}$ as clothing insulation varies over 0 to 2 clo values. Air movement is kept constant at still air. All curves in this figure pass through a common point, $i_{cl}=i_m=i_a=0·39$. This specific identity point is independent of clothing insulation and varies only with air movement or the ratio $h_c/h$. For permeable clothing, $i_m$ is always lower than $i_{cl}$ and decreases with decreasing clo value.

## 5. *The trapped-air effect*

So far, it has been assumed that all clothing factors—$I_{cl}$, $I_{cle}$ and $i_{cl}$—include trapped-air layers between skin and clothing. In a case of

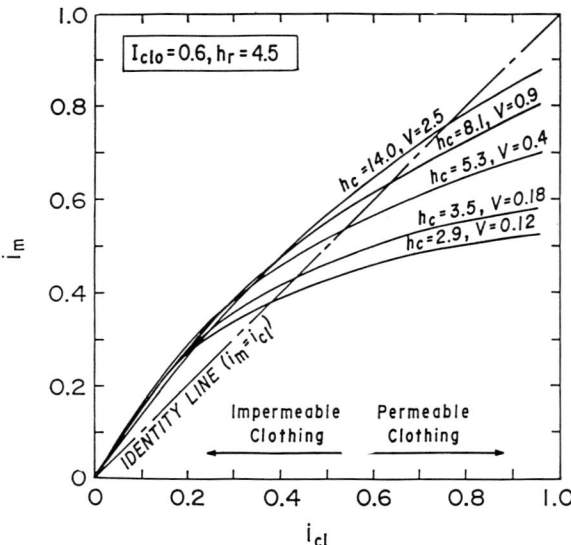

*Figure 8.1.    The influence of air velocity (V) on the $i_m$–$i_d$ relationship, where* h = 2·9
*(V = 0.12 m s⁻¹) and* $h_c$ = 8·6 V^{0.53} *(otherwise) W m⁻² K.*

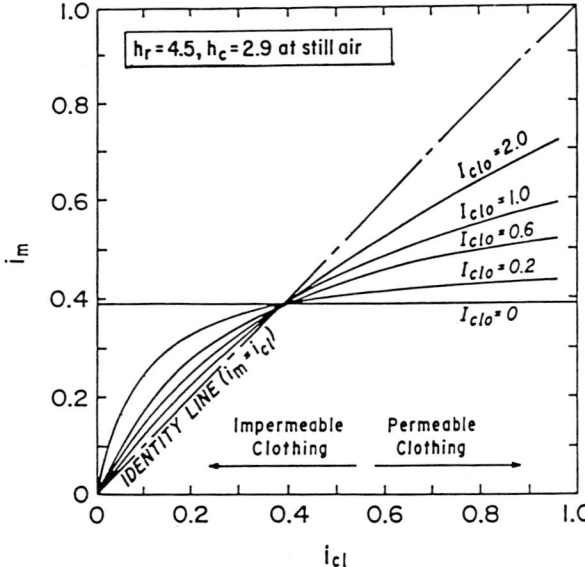

*Figure 8.2.    The influence of clothing insulation ($I_{clo}$) on the $i_m$–$i_{cl}$ relationship.*

low activity, such as sedentary, with one-layer of clothing this trapped-air effect may be explicitly incorporated into heat and mass transfer models. In this case, the effective thermal insulation of clothing ($I_{cl}$) is expressed as the sum of intrinsic insulations of fabric itself ($I_{cl}\star$) and trapped-air layer ($I_{at}=1/(h_{rt}+h_{ct})$):

$$I_{cl} = I_{cl}\star + I_{at} \qquad (\text{m}^2 \text{ K W}^{-1}) \qquad (10)$$

Similarly, the effective evaporative insulation of clothing $I_{cle}$ is the sum of the evaporative insulations of the fabric itself ($I_{cle}\star$) and the trapped-air layer ($I_{et} = 1/2 \cdot 2 \ h_{ct}$):

$$I_{cle} = I_{cle}\star + I_{et} \qquad (\text{m}^2 \text{ Torr W}^{-1}) \qquad (11)$$

In a case of clothing with the trapped-air layer, the effective permeation ratio $i_{cl}$ can be expressed in terms of $i_{cl}$, $i_{at}$ and a weighting factor $F_{at}$:

$$1/i_{cl} = F_{at}/i_{cl}\star + (1-F_{at})/i_{at} \quad (\text{N.D.}) \qquad (12)$$

where $i_{cl}\star$ ($= I_{cl}\star/2 \cdot 2 \ I_{cle}\star$) and $i_{at}$ ($= h_{ct}/(h_{rt} + h_{ct})$) are the permeation ratios for the fabric itself and the trapped-air layer, respectively; $F_{at}$ ($= I_{cl}\star/(I_{cl}\star+I_{at})$) is the thermal efficiency factor of the trapped-air layer, namely, the ratio of the dry heat transfer through clothing with trapped-air and without trapped-air.

Figure 8.3 shows the relationships between $i_{cl}$ and $i_{cl}\star$ as the thickness of the trapped-air layer ($d$) varies from 0 to 6 mm. The conductive heat

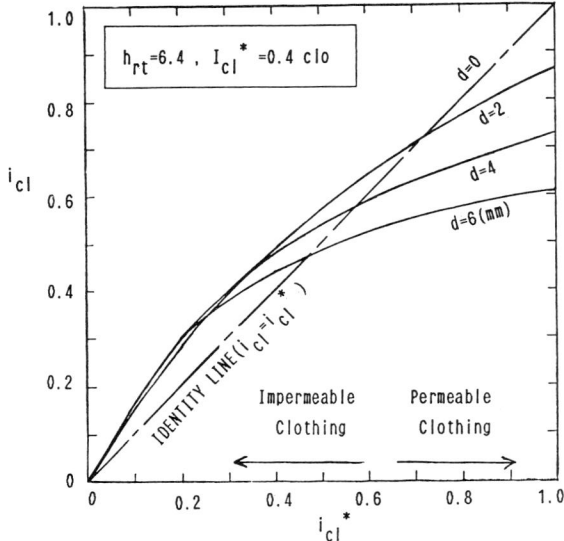

*Figure 8.3.* The influence of trapped-air thickness (d) on the $i_{cl}$–$i_{cl}\star$ relationship.

transfer coefficient through trapped-air $(h_{ct})$ is inversely proportional to the air thickness. On the other hand, the radiative component $h_{rt}$ and intrinsic clothing insulation value $I_{cl}^\star$ are kept constant, at 6·4 and 0·4, respectively. For impermeable clothing, the effective permeation ratio $i_{cl}$ is a little bit higher than $i_{cl}^\star$ and independent of air thickness. However, for permeable clothing, the ratio decreases as air thickness increases.

## 6. Conclusions

We have developed simple relationships between the current indices of evaporative heat exchange through clothing and obtained new heat and mass transfer models which explicitly include the trapped-air effect.

The conclusions of our paper are summarized as follows:

1. Three permeation indices—$i_m$, $i_a$ and $i_{cl}$—are related in terms of Burton's $F_{cl}$ or Nishi's $F_{pcl}$, namely,

$$1/i_m = F_{cl}/i_a + (1-F_{cl})/i_{cl}$$
$$i_m = F_{pcl}\, i_a + (1-F_{pcl})i_{cl}$$

2. For impermeable clothing, $i_m$ is a little bit higher than $i_{cl}$, and $i_m$ is independent of air movement and clo value.

3. For permeable clothing, $i_m$ is lower than $i_{cl}$, and this difference becomes larger as air movement or clothing insulation decreases.

4. For clothing with a trapped-air layer, $i_{cl}$ is more dependent on air thickness for permeable fabric than impermeable fabric, and $i_{cl}$ decreases as trapped-air increases for permeable fabric.

### Acknowledgement

The authors wish to thank Professor Y. Nishi for his critical reading of the paper and his helpful comment.

### References

ASHRAE, 1985, *ASHRAE Handbook of Fundamentals* (New York: American Society of Heating, Refrigerating, and Air-Conditioning Engineers, Inc.).

Burton, A.C. and Edholm, O.G., 1955, *Man in a Cold Environment* (New York: Hafner).

Lotens, W.A. and Van der Linde, E.J.G., 1983, Insufficiency of current clothing description. In *International Conference on the Biophysical and Physiological Evaluation of Protective Clothing*, Lyon, 4–8 July.

Nishi, Y. and Gagge, A.P., 1970, Moisture permeation of clothing—A factor governing thermal equilibrium and comfort. *ASHRAE Trans.*, **76**, 137–145.

Oohori, T., Berglund, L.G. and Gagge, A.P., 1984, Comparison of current two-parameter indices of vapor permeation of clothing—As factors covering thermal equilibrium and human comfort. *ASHRAE Trans.*, **90**, 85–101.

# 9. The Measurement of Clothing Air Exchange and its Role in Clothing Design

## G. W. Crockford

## 1. Introduction

The role of the clothing micro-environment air exchange in achieving thermal comfort for the wearer and minimizing thermal stress was probably recognized several thousand years ago by man when he started covering a substantial part of the body with clothing. Belding (1949) showed that the clothing worn by the Eskimo and Bedouin, for example, has well-developed design features to ensure micro-environment air exchange. In the case of the Eskimo's clothing the air exchange is controlled by draw strings which increase and decrease air exchange as required.

Air, and with it water vapour, enters and leaves the clothing micro-environment space as a result of diffusion through permeable fabrics, as a mass transfer across the fabric produced by wind and the pumping action of body movements, and as a flow of air in and out of openings due to wind pressures and body movement. The rate at which air moves through the micro-environment space is therefore determined by:

1. The air permeability of the fabrics.
2. The design of the garment (*a*) in providing openings and (*b*) in the development of a pumping action.
3. Body movements.
4. Wind speed.
5. The volume of the micro-environment.
6. The presence of restrictions to the movement of air through the spaces in the clothing, e.g., a belt.

The volume of the micro-environment will influence air exchange by providing less resistance to air movement with the increases in cross-sectional area that are likely to accompany the larger volumes and by

184

influencing the volume of air available for exchange. For a given air exchange rate, the larger the volume of air participating the greater the total volume exchanged and hence the greater the potential for the removal of heat and water vapour from the wearer. For this reason, when a comparison of garments with differing micro-environment volumes is being made, it is necessary to measure the volume of the micro-environment.

Forty years after Belding's review, clothing ventilation remains a major design problem. When asked to improve the design of fishermen's protective clothing it became apparent that the fisherman's main problem was losing metabolic heat in a controlled manner and, with an impermeable garment fabric, this meant ventilation of the micro-environment. The fishermen's garments that had evolved over the years, the smock and duck suit, enabled a high rate of air exchange to be achieved in the former and apparently an acceptable rate of exchange in the latter, although both were made from impermeable material to provide protection against water and fish oils. The impermeable garments appeared to act as wind breaks and the fishermen wore clothing suitable for the prevailing weather conditions under the wind break. We knew that if the fishermen were to be offered alternative working garments which incorporated improved health and safety features, the new garments would have to have air exchange rates similar to those of the conventional garments. At the time, mid 1960s, it was not possible to determine the air exchange of a garment except by wearing it and making a subjective assessment or by physiological studies. However, the air exchange under an impermeable garment appeared to be so important to the thermal comfort of the garment that a method of measuring it was required and preferably one that was quick and easy. The first attempts (Crockford, 1977) in retrospect were not quick or easy. They involved measuring the air temperature of the micro-environment and its water vapour pressure and the use of carbon dioxide as a trace gas. The carbon dioxide was introduced into the micro-environment by placing the subject in a plastic bag, securing it round his neck and then feeding carbon dioxide in. The bag was then removed from the subject and samples of micro-environment air taken at intervals using a 50 ml syringe attached to plastic tubes leading into the clothing ensemble. Although crude, differences could be identified between garments and between the skin and outer layer of a garment. The technique was therefore developed and the rate at which oxygen levels returned to atmospheric values after being depressed by feeding nitrogen into the micro-environment were monitored. Because the fishermen's clothing was impermeable it proved possible to introduce the nitrogen into the micro-environment via a network of tubes and to sample continuously using a similar system of tubes (Crockford *et al.*, 1972).

The technique proved to be very precise and it was not unusual to have dilution curves which followed the theoretical curve almost exactly.

Physical activity of any magnitude is accompanied by the production of sweat, at least after a latent period during which the body is warming up. Fanger (1970) gives the relationship between work rate and sweat production (latent heat loss) for the condition of thermal comfort as

$$\bar{E}_{SW} = 0{\cdot}42\ A_{DU}\ (H/A_{DU} - 50)\ (kcal\ h^{-1})$$

and for the skin temperature when thermally comfortable as:

$$\bar{t}_s = 35{\cdot}7 - 0{\cdot}032\ H/A_{DU}\ (°C)$$

where $A_{DU}$ is the surface area of the nude body (m$^2$) and H is the metabolic heat production.

If the subject is cold or hot, less or more sweat will be produced than is indicated by the equation. If the equation is presented as a graph and the metabolic heat load divided into the sensible and latent heat fractions the relationship between work rate and volume of sweat produced can be easily seen. Once the rate at which sweat is being produced is known, the volume of air required to evaporate it can be calculated making some assumptions about the psychrometric state of the ambient air, the extent to which the air can be loaded with water vapour from the sweat and the temperature at which the air leaves the clothing. In this way an estimate can be derived for the volume of air required to remove the metabolic heat in the desired ratio of sensible to latent heat for thermal comfort.

For example, hard work (465 W) requires the evaporation of 3·7 g of sweat a minute which, at ambient air temperatures of 5°C and 15°C and 60% RH, requires 300 and 470 l min$^{-1}$ of air, respectively, if it leaves the garment at 26°C and 60% RH. The sensible heat would require 700 and 1400 l min$^{-1}$, respectively, if thermal comfort were to be maintained. Clearly these micro-environment air flows cannot be achieved by using permeable fabrics or the pumping effect. Wind or forced ventilation provide the only methods available, apart from enduring discomfort or taking the clothes off. Fortunately the clothing designer does not always have to deal with high work rates and in the colder environments the evaporative capacity of the air is high. Garments can therefore be designed which provide sufficient thermal insulation and achieve adequate air exchanges for at least the lower work rates and levels of sweating.

The difficult design problems arise when the garments have an overriding protective or safety function, and that is when the designer must identify and try to achieve the best compromise possible between comfort and protection.

## 2. Review of past work

The insulation of garments and complete clothing ensembles can be measured on heated manikins, some of which are able to move (Seppanen *et al.*, 1972; Olesen and Nielsen, 1972; Goldman, 1974; Mecheels and Umbach, 1976; Olesen *et al.*, 1982; Olesen and Madsen, 1983). Clothing insulation values can also be determined on subjects and in recent years such experiments have become very sophisticated and able to determine the influence of air movement and activity on the total insulation value of the garments, the individual components, the air boundary layer and the clothing itself (Nishi *et al.*, 1975; Holmer and Elnas, 1981; Vogt *et al.*, 1983, 1984; Nielsen and Olesen, 1984; Nielsen *et al.*, 1985). In the above studies it is assumed that the reduction in the insulation value of the clothing that occurs between the wearer being static and the wearer moving and working is due to the bellows or pumping effect of the body's movements. The air which is being pumped in and out of the garment thus contributes another pathway for the loss of sensible and latent heat from the body. The value of this approach is that it considers both the physics of the clothing and the wearer's physiology and so provides quantitative physical and physiological data and an understanding of the physical processes and design principles. Amongst the disadvantages associated with the use of subjects for investigating clothing are the high cost, expensive facilities, high level of technical skill, inter- and intra-subject variability, and the length of time involved in even simple studies.

Vogt *et al.* (1983), in referring to the influence of the pumping effect on the thermal insulation of clothing, state "the measurement and control of the pumping effect seems to be the key for further advances in the field of thermal insulation of clothing ensembles". As has already been shown above the "pumping effect" can be defined in terms of its effect on the thermal insulation value of a garment ensemble or it can be measured directly in terms of litres $min^{-1}$ of air flowing through the clothing micro-environment. Vogt *et al.* also remind us that in high air temperature environments the sensible heat carried into the micro-environment could exceed the increase in heat loss due to evaporation of sweat. The insulation value of the clothing would then appear to increase as micro-environment ventilation increased. This, firefighters are aware of, and so their clothing is designed to prevent the ingress of hot air. The ability of physiological procedures to dissect the complex relationship between garment design and air exchange is, however, limited and much faster techniques are required, such as the use of a trace gas.

The first direct measurements of the air exchange due to the pumping effect appear to have been made by Crockford (1969) whilst trying to develop clothing with built-in safety features for fishermen. In the published report of the technique (Crockford *et al.*, 1972) attention was drawn to its high resolving power. Also the sensitivity and reproducibility of the technique enables the influence of garment design, undergarments, sizing, fit, wind velocity, body movements, posture and work routines and fabric properties on air exchange to be investigated and quantified as an air exchange rate for both the garment as a whole and for its different parts. The term 'exchange rate' was used to replace pumping or bellows effect as the movement of air between the clothing micro-environment and the environment is due to air movement produced by wind, diffusion through permeable fabrics and warm air rising from the clothing and escaping from openings as well as the bellows effect generated by body movement.

The technique for measuring air exchange is very simple and has a number of advantages which are worth comparing with the physiological method.

1. The results obtained are not influenced by the subject's physiological state, e.g., fitness or state of acclimatization.
2. The exchange rate is quantified.
3. It can look at parts of a garment.
4. It is fast.
5. It is inexpensive.
6. It does not require expensive facilities.
7. It does not require highly skilled staff.
8. It enables the designer to get to grips with those aspects of garment design which determine air exchange.

It is important to recognize, however, that it is a physical technique which only answers questions on the air exchange characteristics of a garment. But, as this particular feature is connected by physical laws to the heat loss characteristics of the garment, it is clearly relevant and important to its physiological impact on the wearer. It can therefore be used to reduce or avoid unnecessary physiological work. For example, new garment designs for chemical protection can be compared with the garment in current use. If the new garments restrict micro-environment air exchange compared with the current one, there is a very real possibility that they will be rejected as uncomfortably warm or because sweat accumulates in them. The trace gas technique enables the designer to measure and ensure his garments maintain adequate air exchange for the purpose for which they are intended.

The question arises, why not measure the water vapour pressure next to the skin directly? A good design is one which maintains a low vapour pressure and a poor one is where the vapour pressure is high when tested under similar conditions. This has in fact been done. Fujitsuka and Ohara (1977) studied water vapour pressure gradients across clothing ensembles from the skin to the ambient air. They showed that the pattern of vapour pressure gradients from the skin to the ambient air was not greatly influenced by the external humidity but that the absolute values were. The pattern of vapour pressure gradients was, however, influenced by the textiles used for the coat and underwear. Holmer and Elnas (1981) have also measured the water vapour pressure gradient between the skin and ambient air using an adaptation of the technique developed by Crockford *et al.* (1984). Using a multi-point sampling harness they withdrew air samples from the microenvironment next to the skin and then analysed the air for oxygen. However, they did not put in a trace gas but assumed that any decrease in the oxygen concentration would be due to the increase in the partial pressure of water vapour. The higher the rate of sweating or the lower the air exchange rate the lower the oxygen concentration would become. The technique is slightly more complex in that the water vapour must not be allowed to condense in the sampling tubes or oxygen analyser. By measuring at the same time the water vapour pressure of the ambient air, the water vapour pressure gradient across the garment is determined. Holmer and Elnas also measured continuously the rate at which the subjects evaporated sweat and with partitional calorimetry were able to determine both evaporative and convective heat transfer simultaneously. The data presented in their paper indicate that once the subject is in equilibrium the technique could pick up the influence of wind speed and work tasks on the vapour pressure gradient across the garment, although they did not attempt this.

Berglund and Cunningham (1983) measured the dew point at a number of different skin sites using miniature dew point sensors. They were able to identify that the thermal comfort of the ensembles tested could be enhanced most by increasing the ventilation and permeability of the torso garments. It appears that the miniature dew point sensors provide a method for quantitatively evaluating garments in terms of their ability to maintain low vapour pressures next to the skin.

The measurement of water vapour pressure in the micro-environment undoubtedly aids the evaluation of clothing but it does appear to depend on the use of sweating subjects. This inevitably leads to problems with subject variability and lengthy experiments. The techniques therefore appear to be more suited to the physiology laboratory rather than the clothing development laboratory. Also, if they are to be used effectively, the trace gas technique may still be required in order to make the best use of these vapour pressure techniques.

## 3. Technical aspects

The basic trace gas technique developed by Lotens and Havenith (see Chapter 7, this volume) appears to be a very important development, if they can get round the use of a mass spectrometer. The work of the Simon Fraser University group is also reported in this volume and will not be discussed in detail (Mekjavic and Sullivan, 1988).

When the trace gas technique was originally developed it was in the context of fishermen's protective clothing which are impermeable garments enclosing micro-environments of many litres. The air exchange takes place at openings and is relatively slow. This type of garment ensemble is ideal for the trace gas technique. A close fitting garment such as a T-shirt on the other hand is not suitable, although a T-shirt does appear to influence the dilution curve if compared with no shirt. The reason is that the micro-environment volume is small and the exchange is very rapid.

The trace gas technique is based on a number of assumptions which include:

1. That the oxygen and water vapour molecules behave in the same way.
2. That the dilution of the sampling system is very fast compared with that of the micro-environment.
3. That the air exchange next to the skin is the important exchange rate and not one between the clothing layers.
4. That the volume of sample extracted does not influence the exchange rate or suck air down from higher levels in the clothing.
5. That the nitrogen is distributed evenly and starts off at the same value at all sampling points.

These assumptions may not always be valid in that, in practice, liquid sweat can move to layers where the ventilation rate could be higher than it is near the skin. More heat would then be removed than is indicated by the exchange rate. The work of Fujitsuka and Ohara (1977) on water vapour pressure gradients indicates that air exchange may well proceed at different rates in adjacent layers of a clothing ensemble. The dilution curve of the sampling system depends on its construction and 'dead space' volume and it is well to keep this as small as possible and to draw more sample through than is required by the oxygen analyser. Whether the sample volume influences the exchange rate will depend on the relative values of the sample and local micro-environment volume. To ensure equal distribution of nitrogen throughout the micro-environment when testing permeable garments with high rates of exchange, the subject can be enclosed in a plastic bag up to the neck, into which the nitrogen is fed, instead of using a pipe distribution system as previously described. When ready to start the work task the bag is pulled down quickly and the subject steps out of it. To ensure that the sampling tubes are working properly and none of them are blocked, a U tube manometer is used to measure the negative pressure in the sampling tube.

With such a device, when the subject is dressed, each sampling tube can be checked in turn by placing a finger over its end and observing the increase in suction pressure as a result of blocking one of the tubes. The negative pressure in the sampling line can also be determined and checked during the run. One big contributor to the delay in response of the system is the tube connecting the sampling harness to the analyser. The length of this tube running between the subject and analyser is inevitably determined by the work routine the subject is following and its diameter by the power of the pump.

In our latest work we have connected the output of the oxygen analyser to a micro-computer which determines the slope of the dilution curve between set oxygen values. Three curves are analysed and presented as a mean with standard deviation. This facilitates easy comparison of garments and work routines whilst the experiments are proceeding.

## 4. Information provided by the trace gas technique

The question arises as to how much the investigator can learn about fabrics and garment design using the technique. To date the question normally asked by the investigator when using the technique is, how do the top coats, sleeping bags, skirts, etc. compare or rank in terms of air exchange rates? Very often the comparison is between garments with similar micro-environment volumes and ranking the rates of exchange is sufficient. If the micro-environment volume is known, the clothing ventilation index can be determined, so enabling garments enclosing different volumes of air to be compared (Birnbaum and Crockford, 1978). This is important when investigating designs differing in this respect. Other aspects are the pumping action developed, which can be designed in or out, the contribution of fabric permeability, the vulnerability of air exchange to sizing, fit and wind. If the volume of air moving through the micro-environment is determined by measurement its sensible and latent heat carrying capacity can be estimated.

## 5. Garment micro-environment volume

The method originally developed for measuring micro-environment volumes is lengthy and demanding on the subject. The fully dressed subject is enclosed up to the neck in a plastic suit and as this stage of a study always appears to coincide with a heat wave subjects are not easy to come by. The improved method developed by Sullivan and colleagues and presented in this volume is therefore very welcome. One aspect of the procedure that has not attracted much attention is its ability to define the compressional resistance of a clothing ensemble. The volume enclosed by a garment and hence its insulation value is very sensitive to pressure. The response of a garment ensemble to wind pressure can therefore be assessed at the same time as its volume.

## 6.  The value of the technique to the designer of work wear and protective clothing

The clothing designer/scientist/technologist has for many years had detailed information on the heat transfer and other physical properties of the fabrics available. Such information has been used in the development of a design and to predict with some accuracy what happens when the garment is placed on a standing wearer. The problem of determining what happens when the wearer sets about his normal tasks in a working environment is traditionally done using physiological studies and field trials. One very important aspect of performance where the garment can go wrong is in its ventilation characteristics. This can now be measured so easily that new garments should not go forward to the laboratory or field trials without meeting the exchange rates required or at least having them measured.

One way of determining a reference value for the exchange rate ($R$) or clothing ventilation index ($Q$) is to use the value of the garment already in use. This $R$ and $Q$ value can then be considered as a minimum performance standard which the new garments must meet. The extension of the use of such a performance standard to all work wear and protective clothing would help the user and protective clothing consultant considerably when trying to identify new and hopefully better garments for a work situation and should be considered seriously by the work wear and protective clothing industry. Its use would also stimulate research into the role of design in achieving thermally acceptable garments.

## 7. Future research

One of the aims of future research must, I believe, be the original one, namely, to develop a low cost, quick, reliable, reproducible method of defining the air exchange characteristics of a garment which can be used by the clothing manufacturing industry as well as by research laboratories. Also necessary is the development of standard procedures to facilitate inter-laboratory work and methods of defining a garment's ventilation. Following from this, it should be possible to move towards the setting of garment performance standards in terms of the clothing ventilation index.

Another aim should be to investigate design features relevant to the development and control of air exchange (Vokac *et al.*, 1973) and the influence of body movement and wind on the air exchange. The latter is most important for the development of work wear and protective clothing for use in hot environments. The natural movements of the body cannot move air through the micro-environment in sufficient quantity to deal with high heat loads. The only way to do it is by building vents into the garment and blowing air through the micro-environment. This can be done as Pirastu (1981) demonstrated using a physical model but is by no means straightforward, as Crockford *et al.* (1984) found out when investigating the influence of garment vents on thermal stress in wind speeds of up to 6 m s$^{-1}$. The vents worked on a physical model and on a subject wearing an impermeable coverall but not on subjects wearing a permeable coverall. However, the air flows required are so high that the trace gas technique will be of little use with such design problems and dew point sensors, napthalene balls and sweating manikins may provide the only way forward.

# References

Belding, H.S., 1949, In *Physiology of Heat Regulation and the Science of Clothing*, edited by L.H. Newburgh (Philadelphia, PA: Saunders), pp. 351−366.

Berglund, L.G. and Cunningham, D.J., 1983, Skin wettedness and discomfort estimates from dew point measurements under clothing. *International Congress of Medical and Biophysical Aspects of Protective Clothing*, Lyon, July.

Birnbaum, R.R. and Crockford, G.W., 1978, Measurement of clothing ventilation index. *Applied Ergonomics*, **9**, 194−200.

Crockford, G.W., 1969, Trawler fishermen's protective clothing. *Ergonomics*, **12**, 767.

Crockford, G.W., 1977, Trawler fishermen's protective clothing. In *Human Factors in Work, Design and Production*, edited by J.S. Wiener and H.G. Maule (London: Taylor & Francis), pp. 65–100.

Crockford, G.W., Crowder, M. and Prestidge, S.P., 1972, A trace gas technique for measuring clothing microclimate air exchange rates. *Brit. J. Industr. Med.*, **29**, 378−386.

Crockford, G.W., Sen, R.N. and Spencer, J., 1984, The design of work wear and protective clothing for use in the tropics. *Proceedings of the International Symposium on Applied Physiology & Ergonomics*, Calcutta, India, 1983, edited by R.N. Sen and H. Chattopadhyay.

Fanger, P.O., 1970, *Thermal Comfort* (New York: McGraw-Hill).

Fujitsuka, C. and Ohara, K., 1977, Studies on water vapour pressure gradient from external air through clothing to skin in relation to external humidity and clothing conditions. *J. Human Ergol.*, **6**, 75−85.

Goldman, R.F., 1974, Clothing design for comfort and performance in extreme thermal environment. *Trans. N. Y. Acad. Sci.*, **36**, 531–544.

Holmer, I. and Elnas, S., 1981, Physiological evaluation of the resistance to evaporative heat transfer by clothing. *Ergonomics*, **24**, 63–74.

Kakitsuba, N., Michna, H. and Mekjavic, I.B., 1986, Clothing surface area as related to body volume and clothing micro-environment volume. *Ann. Physiol. Anthropol.* **5**, 164.

Lotens, W.A. and Havenith, G. 1988, Ventilation of rainwear determined by a trace gas method. In this volume, pp. 162–176.

Mecheels, J. and Umbach, K.H., 1976, *Melliand Texilber.*, **57**, 1029–1032.

Mekjavic, I.B. and Sullivan, P.J., 1988, Constant wear thermal protection garments for helicopter pilots. In this volume, pp. 240–263.

Nielsen, R., Olesen, B.W. and Fanger, P.O., 1985, Effect of physical activity and air velocity on the thermal insulation of clothing, *Ergonomics*, **28**, 1617–1631.

Nishi, Y., Goinzales, R.R. and Gagge, A.P., 1975, Direct measurement of clothing heat transfer properties during sensible and insensible heat exchange with thermal environment. *ASHRAE Trans.*, **81**, 183–199.

Clesen, B.W. and Madsen, T.L., 1983, Measurements of the thermal insulation of clothings by a moveable manikin. *International Congress of Medical and Biophysical Aspects of Protective Clothing*, Lyon, July.

Olesen, B.W. and Nielsen, R., 1972, Thermal insulation of clothing measured on a moveable thermal manikin and on human subjects, Technical University of Denmark, Technical Report no. 7206/00/914.

Olesen, B.W., Slivinska, E., Madsen, N.L. and Fanger, P.O., 1982, Effect of body posture and activity on the insulation of clothing. Measurements by a moveable thermal manikin. *ASHRAE Trans.*, **88**, 791–805.

Pirastu, R., 1981, Occupational hygiene in the tropics: A problem of convective heat loss in the presence of protective clothing. MSc Report (Occupational Hygiene), 1981, London School of Hygiene & Tropical Medicine, University of London.

Seppanen, O., McNall, P.E., Munson, D.M. and Sprague, C.H., 1972, Thermal insulating values for typical clothing ensembles. *ASHRAE Trans.*, **78**, 120–130.

Sullivan, P.J., Mekjavic, I.B. and Kakitsuba, N., 1988, Determination of ventilation indices of helicopter pilot suits. *Ann. Physiol. Antropol.* **5**, 141.

Vogt, J.J., Meyer, J.P., Candas, V., Libert, J.P. and Sagot, J.C., 1983, Pumping effect on thermal insulation of clothing worn by human subjects. *Ergonomics*, **26**, 963–974.

Vogt, J.J., Sagot, J.C., Meyer, J.P. and Candas, V., 1984, Basic, effective and resultant clothing insulations. *ASHRAE Trans.*, **90**, 1091–1098.

Vokac, Z., Kopke, V. and Keul, P., 1973, Assessment and analysis of the bellows ventilation of clothing. *Text. Res. J.*, **43**, 474–482.

# 10. A New Immersible Thermal Manikin

## Ed A. Smallhorn

## 1. Introduction

A thermal manikin test system is a means for evaluating the thermal insulation of thermal protective clothing. The system described below consists of a hollow aluminium manikin equipped with temperature sensors and electric heaters, connected to a computer system.

In the operational mode, the manikin is dressed in the suit to be tested and exposed to the required environment (dry or wet). The computing equipment then controls the heaters and measures the electrical power required to maintain the skin of the manikin at the set temperature. This power is equivalent to the heat escaping through the suit due to the temperature difference across it. The power and temperature difference are then used, along with the known surface area of the manikin, to calculate the thermal resistance offered by the suit.

## 2. Methods

The scheme of the thermal manikin test system is illustrated in Figure 10.1. The manikin is fitted inside the test garment. The combination of the aluminium shell of the manikin and the output of the heaters inside it provide for an approximately uniform temperature over the manikin surface. The manikin surface temperature is sensed by sensors embedded in the manikin shell and passed to the control module.

The thermal manikin is cast from an aluminium alloy and is divided into 13 sections with thermal breaks made of Delrin between each section. The head, arms and legs are removable to allow for maintenance. The shoulder and hip joints are single axis rotatable to enable various test positions to be utilized. There is provision for a removable hook in the top of the head for dressing and storage. A flat face-plate provides for the installation of the connectors for all wiring services to

195

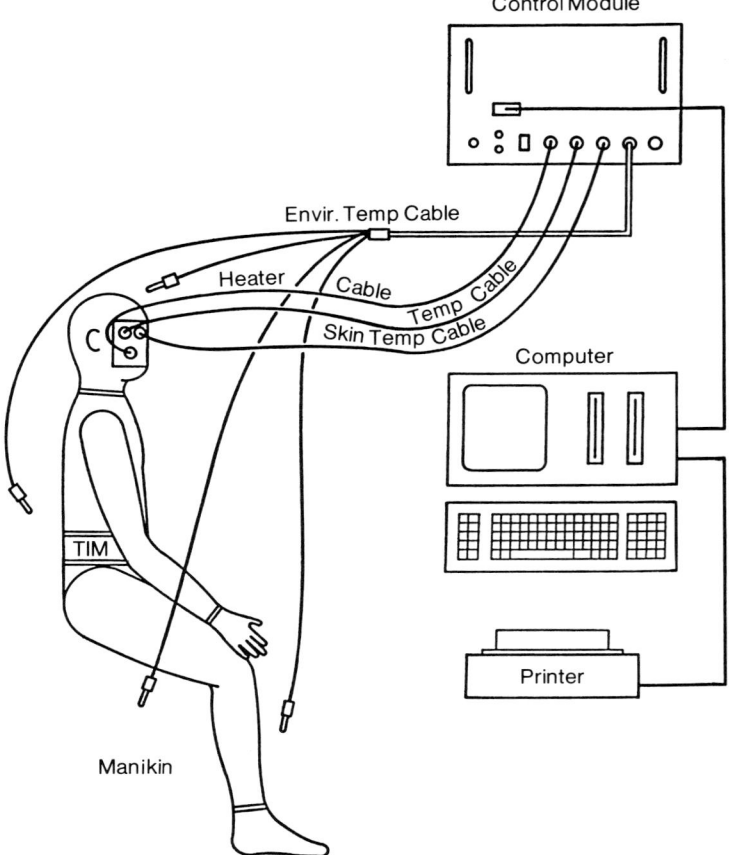

*Figure 10.1. A schematic of the thermal manikin test system.*

the manikin. The manikin is fresh water compatible inside, sea water compatible outside and watertight to a depth of 6 feet (180 cm).

Two three-wire RTDs are embedded in the aluminium shell of each section. In addition, four RTD sensors monitor ambient temperature in the close proximity of the manikin.

The control module consists of a programmed data acquisition system (MicroMac 5000 system with two 4010 analog input expander boards) and heater relays (Crydom MS-24H Solid State Relay Board). The data acquisition system receives data from the temperature sensors on the manikin and controls the heater relays, thus maintaining the manikin surface temperature constant. The ambient temperature and power supplied to the manikin heaters are also monitored by the control module.

The heating hardware produces heat in each of the 13 sections of the manikin. The heaters provide a total power input of 1 kW to the manikin and are of a power density, size and number in each manikin section that ensures the uniform surface temperature of each section. Heat input to each manikin section is controlled separately. Two types of heater are used: wire heaters and cartridge heaters. The wire heaters are 26 gauge type E thermocouple wires bonded to the inside of the manikin shell. Wire heaters are installed in the chest, back, abdomen

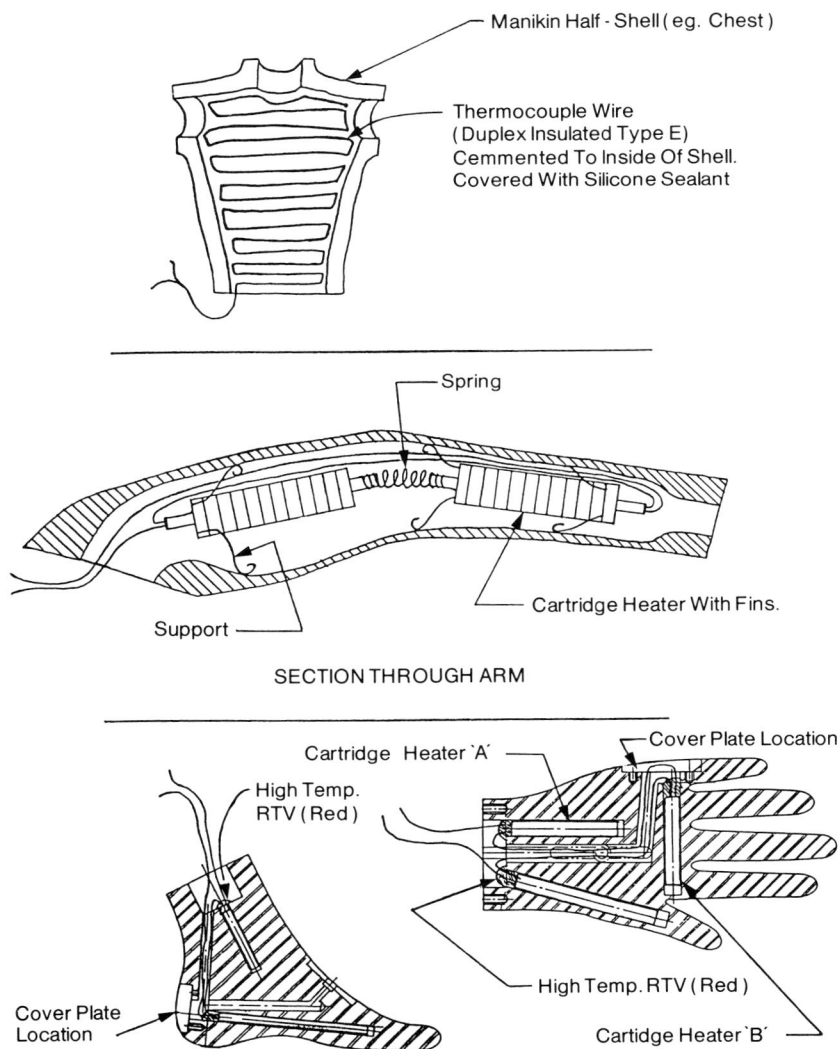

*Figure 10.2. Examples of the placement of wire and cartridge heaters in the manikin.*

and buttock sections. Cartridge heaters are factory-made cylindrical heaters, which are embedded in the aluminium shell of the manikin, as in the foot and hand sections, or suspended in the air space inside the shell, as in the head and leg sections (see Figure 10.2). All data are processed by an Eagle PC Spirit XL personal computer.

During the test procedure, the manikin is suited with the garment under test and is allowed to warm up in the test environment, either air or water. The heaters are controlled to bring the user selected manikin sections to the desired set temperatures. The screen display during the warming phase is presented in Table 10.1. The pertinent user input data is displayed at the top and the user selected manikin sections are listed in the left column. Only temperature data are displayed during the initial warming phase. Once the surface temperature stabilizes, the control system calculates the regional insulation provided by the test garment. The criterion used to determine that steady state has been reached is that the surface temperatures of all user selected manikin sections must be within 0·2°C of the set temperature. The sectional and overall insulation is displayed as shown in Table 10·2. Near the end of the test, the insulation values should approach constant values and the 'change/ min.' (of insulation) values should approach zero.

## 3. Results

Over 300 hours of test time has been logged with the thermal immersion manikin over a period of 6 months. A parametric study revealed that a test duration of 4 hours ensures that the manikin has achieved thermal equilibrium and produces reliable results. For water immersion

*Table 10.1.   Information displayed by the control system during the warming phase.*

```
TEST NO: 123-A                   TEST TITLE: TUTORIAL SAMPLE RUN
SUIT: HELLY-HANSEN TYPE E-305 SURVIVAL SUIT. HOOD, GLOVES & BOOTS ON.
UNDER GARMENTS: HELLY-HANSEN DEEP-PILE THERMAL UNDERWEAR. 1 LAYER ON LIMBS, 2
    TORSO
ENVIRONMENT: AIR; 50 DEG C; 86% R.H. ; 6MPH WIND FROM FRONT
SET POINT TEMP:  35              ENVIRONMENT TEMPERATURE:  20.34(   1.5)
        DATE:  1/ 6/ 86                  TIME: 14:19:05.06
START TIME: 14:05:34.47   * * WARM-UP PERIOD * *    ELAPSED TIME, MIN:  13.5
    SECTION:          SKIN TEMP (C):   DEV'N FROM SP:       SPREAD:
RIGHT HAND              35.04            0.04               1.04
LEFT HAND               35.07            0.07               0.40
RIGHT FOOT              29.21        -   5.79               0.40
LEFT FOOT               28.96        -   6.04               0.05
RIGHT ARM               27.29        -   7.71               1.91
LEFT ARM                28.48        -   6.52               1.02
RIGHT LEG               23.71        -  11.29               0.38
LEFT LEG                23.52        -  11.48               2.00
ABDOMEN                 22.34        -  12.66               2.59
BUTTOCKS                22.39        -  12.61               0.93
CHEST                   26.23        -   8.77               0.97
BACK                    25.78        -   9.22               0.45
HEAD                    26.87        -   8.13               0.54
```

Table 10.2.  Information displayed during the test of insulation.

```
SUIT: HELLY-HANSE TYPE E-305 SURVIVAL SUIT. HOD, GLOVES & BOOTS ON
UNDER GARMENTS: HELLY-HANSEN DEEP-PILE THERMAL UNDERWEAR; 1 LAYER ON LIMBS, 2
   ON TORSO
ENVIRONMENT: AIR; 5 DEG c; 86% R.H. ; 6 MPH WIND FROM FRONT
SET POINT TEMP: ˙35                  ENVIRONMENT TEMPERATURE:  20.3    ( 2.909668)
        DATE:  1/ 6/ 86                       TIME: 15:12:09.53
START TIME: 15:11:01.13        * * TEST PERIOD * *     ELAPSED TIME MINS):    1.1
```

| SECTION: | SKIN TEMP (C): | SPREAD (C): | POWER(W): | INSUL. (Clo): | CHANGE/Min: |
|---|---|---|---|---|---|
| RIGHT HAND | 35.06 | 1.00 | 6.93 | 0.6619 | 0.1074 |
| LEFT HAND | 35.32 | 0.24 | 11.31 | 0.4025 | -0.0194 |
| RIGHT FOOT | 35.00 | 0.21 | 0.85 | 7.5602 | 7.2226 |
| LEFT FOOT | 35.00 | 0.26 | 8.09 | 0.7784 | 0.7439 |
| RIGHT ARM | 35.99 | 1.59 | 1.99 | 5.6932 | 5.4123 |
| LEFT ARM | 35.62 | 0.05 | 2.31 | 4.6756 | 4.4224 |
| RIGHT LEG | 34.93 | 0.30 | 187.87 | 0.1595 | 0.0046 |
| LEFT LEG | 35.39 | 1.61 | 30.52 | 1.0336 | 0.9815 |
| ABDOMEN | 35.02 | 2.57 | 16.88 | 0.3041 | 0.2859 |
| BUTTOCKS | 35.02 | 0.87 | 16.95 | 0.3126 | 0.2939 |
| CHEST | 35.01 | 0.81 | 32.57 | 0.4472 | -0.2461 |
| BACK | 35.03 | 0.71 | 38.56 | 0.4218 | 0.1336 |
| HEAD | 35.53 | 0.43 | 4.75 | 2.7513 | 2.5660 |
| OVERALL: | | | | .4369 | .1522 |

tests, a 1 hour warm-up period is required, for a total manikin running time of 5 hours.

Tables 10.3a and 10.3b represent the file data of two suit tests. The results in Table 10.3a indicate a difference in insulation between the right and left arms due to a slight leak in the left arm of the suit. Similarly, the chest section shows a high insulation, as the manikin was in a semi-horizontal orientation, so that the air in the suit collected in the chest area. In addition, the chest was partially awash and the manikin was wearing a life vest, which adds to the chest insulation. Table 10.3b is a test of the same suit without the life vest and with a lighter undergarment. Thus the overall insulation is lower (0·89 clo). Note also that in this test the left arm of the suit did not leak.

## 4. Discussion

Insulation values were affected when sections with large thermal barriers were run independently, due to the heat transfer between adjacent sections. For reliable results, it is necessary to run all sections adjacent to the section involving large thermal barriers or to establish the effect of the barrier heat transfer. However, this is not necessary in the majority of tests evaluating immersion suits.

The ocean environment proved to be quite a significant factor in producing different results from tests conducted in a pool with the same suit. This is probably due to the effect of wind and sea state on the insulation performance of a suit. In addition, the manikin exhibited

*Table 10.3.  Example of tests conducted on the same suit. (a) In addition to the suit, the manikin is also wearing a life vest. There is a leak in the left arm of the suit as evidenced by the difference in insulation values between the left and right arm. (b) The manikin is without the life vest and lighter undergarment. There are no leaks in this suit.*

(a)

```
TIME: 21:27:18.82        MINUTES SINCE START OF TEST:   240:04
ENVIRONMENT TEMPERATURE:
          INSTANTANEOUS: 17.1        AVERAGE OVER TEST TIME: 17:08
```

| SECTION | SKIN TEMP. Degrees C | TEMP.DIFF (Deg.C) INSTANT. | AVERAGE | POWER CONS. (WATTS) | INSULATION (CLO) |
|---------|------|------|------|------|------|
| Right Hand | 30.07 | 13.07 | 12.99 | 6.16 | 0.6679 |
| Left Hand | 30.09 | 12.86 | 13.01 | 6.79 | 0.5961 |
| Right Foot | 30.08 | 13.01 | 13.00 | 7.01 | 0.8198 |
| Left Foot | 30.04 | 12.89 | 12.96 | 4.98 | 1.1312 |
| Right Arm | 30.09 | 12.93 | 13.01 | 7.78 | 1.2261 |
| Left Arm | 30.10 | 13.05 | 13.02 | 13.88 | 0.6174 |
| Right Leg | 30.09 | 12.97 | 13.01 | 28.23 | 1.0593 |
| Left Leg | 30.07 | 12.95 | 12.99 | 24.92 | 1.1166 |
| Abdomen | 30.05 | 12.96 | 12.96 | 2.43 | 1.8883 |
| Buttocks | 30.04 | 12.96 | 12.96 | 5.86 | 1.2287 |
| Chest | 30.08 | 13.01 | 13.00 | 5.52 | 2.3673 |
| Back | 30.03 | 12.92 | 12.95 | 12.25 | 1.1343 |
| Head | 30.12 | 12.94 | 13.04 | 12.20 | 0.9357 |

```
     Total Power (W) For All Sections:   137.995
     Total Area  (Square Meters):   1.736
     Overall Insulation Resistance (CLO):  1.0548
```

(b)

```
TIME:  17:50:39.41       MINUTES SINCE START OF TEST:   240.0485
ENVIRONMENT TEMPERATURE:
          INSTANTANEOUS: 17.1        AVERAGE OVER TEST TIME:  17.1
```

| SECTION | SKIN TEMP. Degrees C | TEMP.DIFF (Deg.C) INSTANT. | AVERAGE | POWER CONS. (WATTS) | INSULATION (CLO) |
|---------|------|------|------|------|------|
| Right Hand | 30.09 | 12.90 | 13.00 | 6.21 | 0.6629 |
| Left Hand | 30.10 | 12.88 | 13.01 | 5.85 | 0.6919 |
| Right Foot | 30.07 | 12.90 | 12.97 | 6.60 | 0.8689 |
| Left Foot | 30.05 | 13.01 | 12.95 | 6.20 | 0.9082 |
| Right Arm | 30.06 | 12.98 | 12.97 | 8.87 | 1.0716 |
| Left Arm | 30.14 | 12.92 | 13.04 | 8.29 | 1.0361 |
| Right Leg | 30.05 | 12.93 | 12.96 | 37.65 | 0.7913 |
| Left Leg | 30.06 | 13.02 | 12.96 | 35.14 | 0.7899 |
| Abdomen | 30.04 | 12.92 | 12.95 | 3.35 | 1.3697 |
| Buttocks | 30.04 | 12.93 | 12.94 | 8.01 | 0.8972 |
| Chest | 30.08 | 13.02 | 12.99 | 5.99 | 2.1764 |
| Back | 30.03 | 12.94 | 12.94 | 17.78 | 0.7807 |
| Head | 30.13 | 13.14 | 13.03 | 12.64 | 0.9031 |

```
     Total Power (W) For All Sections:   162.566
     Total Area (Square Meters):   1.736
     Overall Insulation Resistance (CLO):   .8935
```

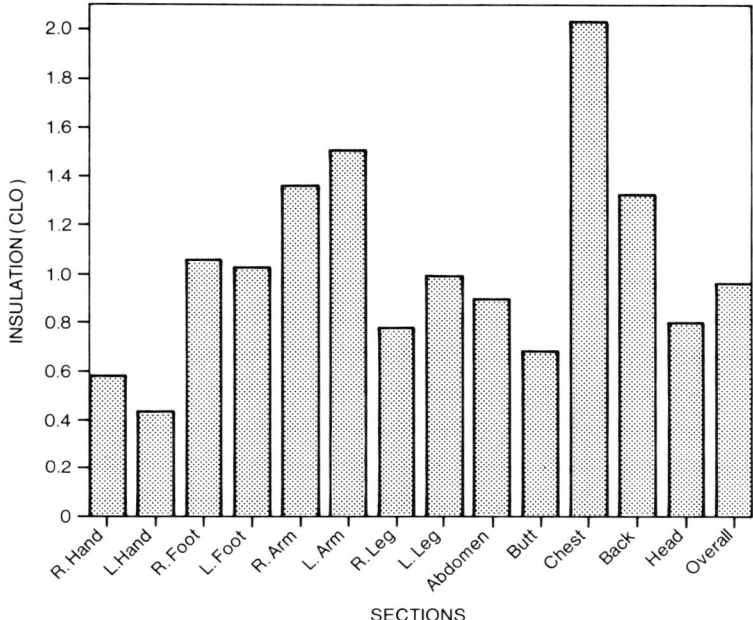

*Figure 10.3. An example of the test conducted on a protective garment, indicating regional insulation.*

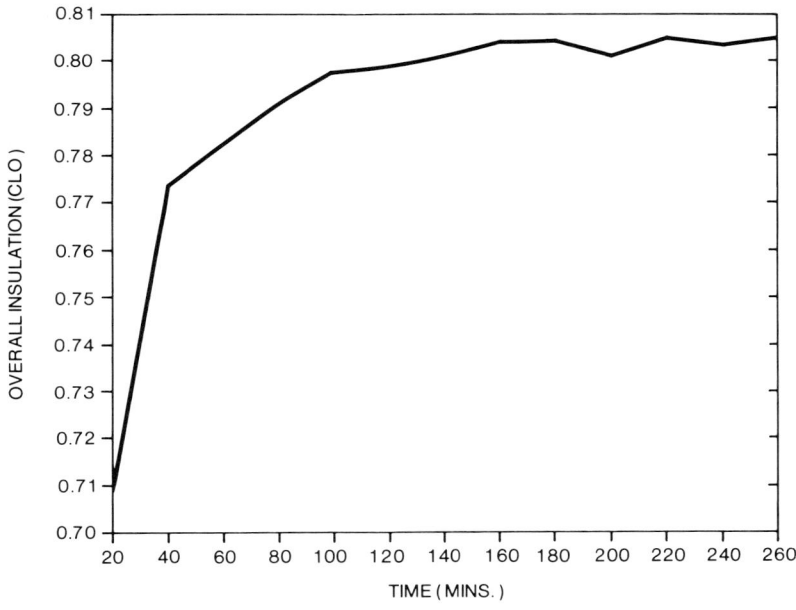

*Figure 10.4. An example of the test conducted on a protective garment, indicating the overall insulation with time of exposure.*

sensitivity to orientation in the water, emphasizing the significant role of hydrostatic pressure, and thus suit compression, on insulation values.

The result of the operational experience with the manikin has been the development of a test protocol and standard test report format which ensures consistent results and concise, clear output with a minimum of time and labour (see Figures 10.3 and 10.4). Thus the 156 thermal manikin test system is a useful tool for both manufacturers and agencies governing the formation of standards for protective garments.

## Acknowledgement

The first task in the development of the CORD thermal manikin test system, called TIM (thermally instrumented manikin), was a visit to the RAF Institute of Aviation Medicine at Farnborough, England. With Dr Allan's enthusiastic and expert advice, CORD and the Nova Scotia Research Foundation were able to design a versatile thermal manikin system, suitable for testing a range of protective garments: survival suits, commercial diving suits and outdoor recreational wear.

# PART III
# COLD WATER IMMERSION

# 11. A Technical Basis for the Development of Thermal Performance Standards for Immersion Protection

**J. R. Allan**

## 1. Introduction

Recent attempts to specify the thermal performance of immersion clothing have usually been based on the concept that there is a minimum allowable level of body core temperature below which survival is at risk. The core temperature selected has varied between 35°C and 33°C and it is important to note that these are not core temperatures normally associated with unconsciousness or cardiac arrhythmias but they represent the range of views as to the practical lower limit to which it is acceptable to allow core temperature to fall in a survival situation. Clearly there is no very precise evidence on which to base this limit.

Notwithstanding the commonality of objective, namely the setting of a lower limit on the fall of core temperature, there have emerged two sharply different approaches to the method of demonstrating compliance with the limit set. One involves direct measurements of body cooling in human subjects and the other is based on mathematical model predictions.

## 2. Human subject experiments

The International Maritime Organization (IMO, 1974) have based their proposals on a maximum fall of core temperature of 2°C, i.e., to about 35°C. They apply this limit to each of two standards of immersion protection, one providing 6 hours' protection in water at 0−2°C and the other providing 1 hour's protection in water at 5°C. To test compliance IMO recommend the use of human subjects exposed to calm, stirred water at the prescribed temperatures for the prescribed durations. Measurements of rectal temperature are used to demonstrate the fall in body core temperature achieved. This approach has been criticized by Allen and Hayes (1984) on the following grounds.

## Use of calm water

Poor suit design can lead to serious leakage and such weaknesses may not be apparent in calm water where the neck closure may be above the water level when the suit is worn with an adequate lifejacket by a largely passive subject. Even in the slightest sea, serious leakage may occur and lead to an important loss of insulation. Allan *et al.* (1985) have shown a loss of 30% of initial clothing insulation for a leak of 500 g. This may be less serious where built-in insulation consists of inherently water-proof material (e.g., neoprene foam). It has also been demonstrated that heat losses can be higher in waves (Steinman *et al.*, 1985).

## The use of human subjects

The use of human subjects may appear attractive as a direct measure of the performance of an immersion suit but can give rise to a number of problems such as:

1. High financial costs.
2. Methodological difficulties.
3. Ethical considerations.

### Cost

Human subject experiments are always highly expensive, particularly when they require initial medical examinations and investigations, and the presence of a skilled and experienced (in this field) physician throughout the tests. Such costs can be prohibitive for the purposes of the development testing of an immersion suit. Presumably any suit design once approved would have to be re-submitted in the event of any modifications. This would place a serious disincentive on manufacturer's efforts to seek improvements in their designs on a continuous basis.

### Methodological difficulties

Measurements of skin and deep body temperature under immersion conditions require meticulous care in the use of instrumentation and the interpretation of results. More difficult again is the problem of subject selection and the influence that this can have on the results obtained. It is well known that individuals differ significantly in their responses to cold water due not only to differences in age, fitness and physical characteristics such as body fat, but also due to other less tangible and less well understood characteristics such as experience of cold water immersions. The practical point is that it would easily be possible to select a panel of subjects falling within the rather vague physical characteristics contained in the IMO proposals that would be atypically cold tolerant or cold susceptible. This would leave obvious scope for errors in terms of the overall objective of the tests which is to give approval only to effective suits.

### Ethical considerations

Human experimentation involves an element of risk, even if small, and places serious responsibilities and obligations upon those in charge. It would be difficult to deny that exposure of human subjects to immersion in water at $0-2°C$ for periods of 6 hours does involve risk. If it did not then the stringent requirements for medical supervision, preliminary examination and continuous monitoring, for example by electrocardiograms (ECG), would not be necessary. The requirement for a continuous ECG monitor is presumably based on an expectation, however small, of cardiac arrhythmias and a need to detect them at an early stage. Many authorities may feel it necessary to provide facilities for emergency cardiac defibrillation. There is also the question of how much confidence can be placed in the system of temperature monitoring which is, after all, the subject's sole protection against totally unacceptable falls in deep body or skin temperature. A failed thermistor could lead to serious consequences.

   All these considerations will, of course, apply to any human experimentation in cold water, and those who undertake such work do so with great care and circumspection. However, it is open to question whether such experimentation is justifiable for use in the routine qualification testing of an immersion suit when an alternative approach is available.


# 3. Mathematical model predictions

An immersion suit protects by providing thermal insulation to a survivor immersed in water. It usually does this by excluding water and thereby maintaining a layer of insulating air within and between the layers of other clothing worn beneath the immersion suit. Some immersion suits have built-in insulation which may continue to provide a measure of protection even if the immersion suit leaks. Thus the basic characteristic of the suit which will determine the rate of heat loss of a survivor is the level of thermal insulation it provides, with or without specified additional clothing, and the extent to which that insulation is protected by the exclusion of water.

Whatever the results may be of a particular human subject immersion, those results are determined, insofar as the suit is concerned, by the insulation and water-excluding characteristics. It has been suggested, therefore, that it would be equally effective, more generally applicable, easier to regulate and much cheaper if we were to specify and require the measurement of these characteristics rather than to require the direct measurement of body cooling in human subjects (Allan and Hayes, 1984).

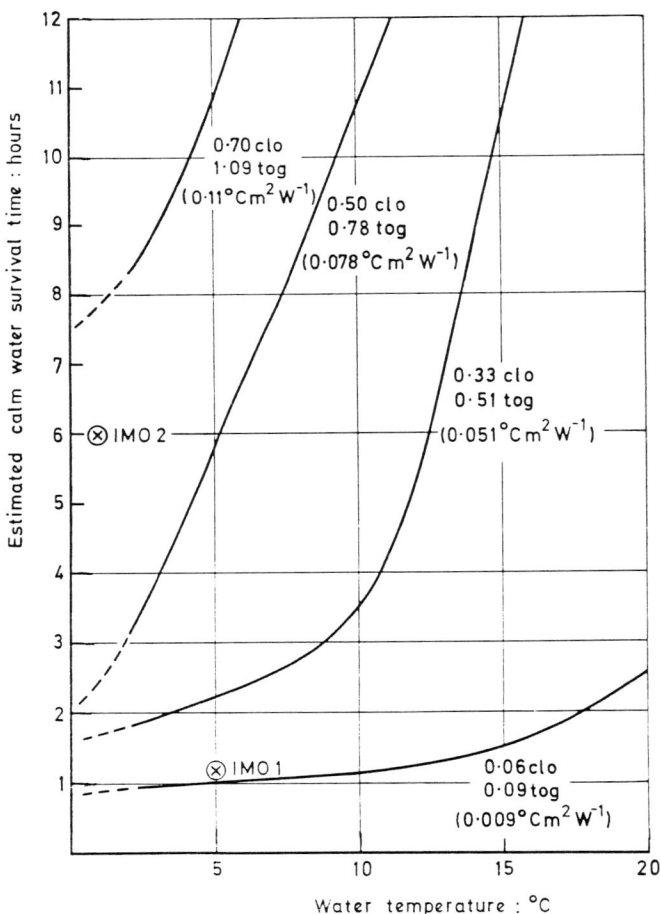

*Figure 11.1.   Predicted survival time against sea temperature for four levels of immersed clothing insulation.*

The general approach of using direct measurements of insulation together with a modelling technique for predicting body cooling has been described by Nunneley and Wissler (1980), and set out in the context of North Sea Helicopter Operations by Allan (1983). The technique has also been adopted as a standard by the five Nation "Air Standardization Co-ordination Committee" (Air Standard 61/40). Further development of the method, making allowances for differing thicknesses of subcutaneous fat, has been described by Nunneley *et al.* (1985).

The above descriptions of the practical application of thermal modelling show how it may be used to develop relationships between water temperature, clothing insulation and likely survival time for groups of subjects with differing subcutaneous fat thickness. An example for thin individuals, at approximately the 10th percentile for mean weighted skinfold thickness, is shown in Figure 11.1. This graph also shows the design points providing equivalence with the IMO requirements for 6 hours exposure to 1°C water and 1 hour exposure to 5°C. Of course the validity of such a predictive technique is only as good as the mathematical descriptions of physical and physiological responses incorporated in the model.

Modelling problems, however, are not the only difficulties with the predictive approach and I now come to another difficult area which concerns the physical measurements of clothing thermal resistance. A method is required to measure the insulation provided in water by any particular immersion protection clothing assembly. Conventional guarded hot-plate methods present difficulties because they do not allow for the trapped air layers in multi-layer assemblies, for hydrostatic compression with displacement of air during immersion, or for the damaging effects of any water leakage into the clothing.

## 4. Thermal manikins

To overcome these difficulties various types of thermal manikin have been devised of which the original form was a copper man and consisted of a hollow, single section copper figure with a single wire heater attached to the inside surface and controlled by temperature sensors also attached to the inner surface. When the dressed copper man is at temperature equilibrium in water, the heat loss through the clothing is equal to the heat input by the heater which is measured. Also measured is the temperature of the copper man and of the surrounding water. The clothing insulation is then derived from

$$I = \frac{(T_m - T_w)\,A}{H} \qquad (°C\ m^2\ W^{-1})$$

Where $T_m$ = surface temperature of copper man in °C; $T_w$ = water temperature in °C; $A$ = surface area of copper man in m$^2$; and $H$ = heat input in Watts.

Multi-section manikins, in which each section of the body is thermally isolated with a separate temperature sensor and heater, were originally built to simulate real human beings rather than simply to measure clothing insulation. They can, however, be used for the latter purpose and they have the distinct advantage of giving separate measurements for external insulation in different regions—for example over the trunk or limbs. These individual measurements may then be used in the more sophisticated mathematical models leading to improved predictive accuracy since these models allow for the differing physiological responses in different body regions. Multi-section manikins are obviously preferable but are also highly expensive and difficult to obtain.

In an effort to build a much simplified manikin we have studied a water-filled fabric manikin with built-in heaters and a water circulation system designed to produce even surface temperatures. This device is illustrated diagrammatically in Figure 11.2. The heat input is not controlled to achieve a target temperature but is simply fixed by switching on an appropriate number of heaters. The dressed manikin is

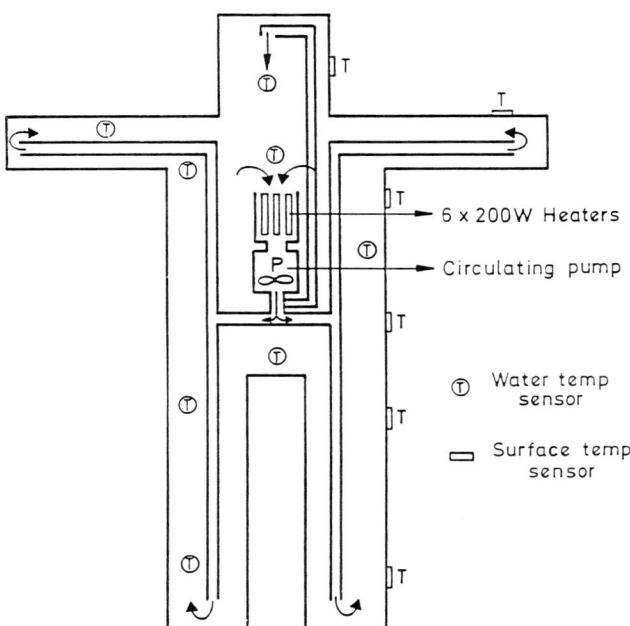

*Figure 11.2.    The circulation diagram for Hydroman.*

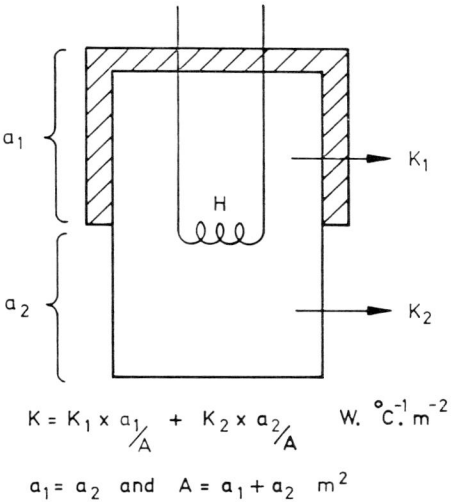

$$K = K_1 \times \tfrac{a_1}{A} + K_2 \times \tfrac{a_2}{A} \quad \text{W. } ^{\circ}\text{C.}^{-1} \text{m}^{-2}$$

$$a_1 = a_2 \quad \text{and} \quad A = a_1 + a_2 \quad \text{m}^2$$

*Figure 11.3. A simple cylinder manikin.*

then immersed in a stirred water pool and left to equilibrate. The temperature of the manikin and of the water is then measured which, together with the heat input and known surface area, allows calculation of the clothing insulation. 'Hydroman', as we call him, is thus extremely simple but suffers from the same difficulties as other single section manikins when insulation differs greatly from area to area. I will now come to these problems.

Let us consider the case where a partial coverage wet suit covers only the trunk and arms of the survivor. This may be represented by a simple cylinder with insulation placed over the upper half only, as in Figure 11.3. The conductance through the bare area, $k_2$, is much higher than the conductance through the insulated area $k_1$. The total conductance, $K$, is then equal to the sum of the parallel conductances through the two adjacent areas $a_1$ and $a_2$:

$$K = k_1 + k_2 \qquad (\text{W } ^{\circ}\text{C}^{-1} \text{ m}^{-2})$$

If we introduce a surface area factor then:

$$K = k_1 \times \tfrac{a_1}{A} + k_2 \times \tfrac{a_2}{A} \qquad (\text{W } ^{\circ}\text{C}^{-1} \text{ m}^{-2})$$

Where $a_1$ = surface area covered with material having conductance $k_1$; $a_2$ = surface area covered with material having conductance $k_2$; and $A$ = total surface area. In a multi-section manikin each of the expressions

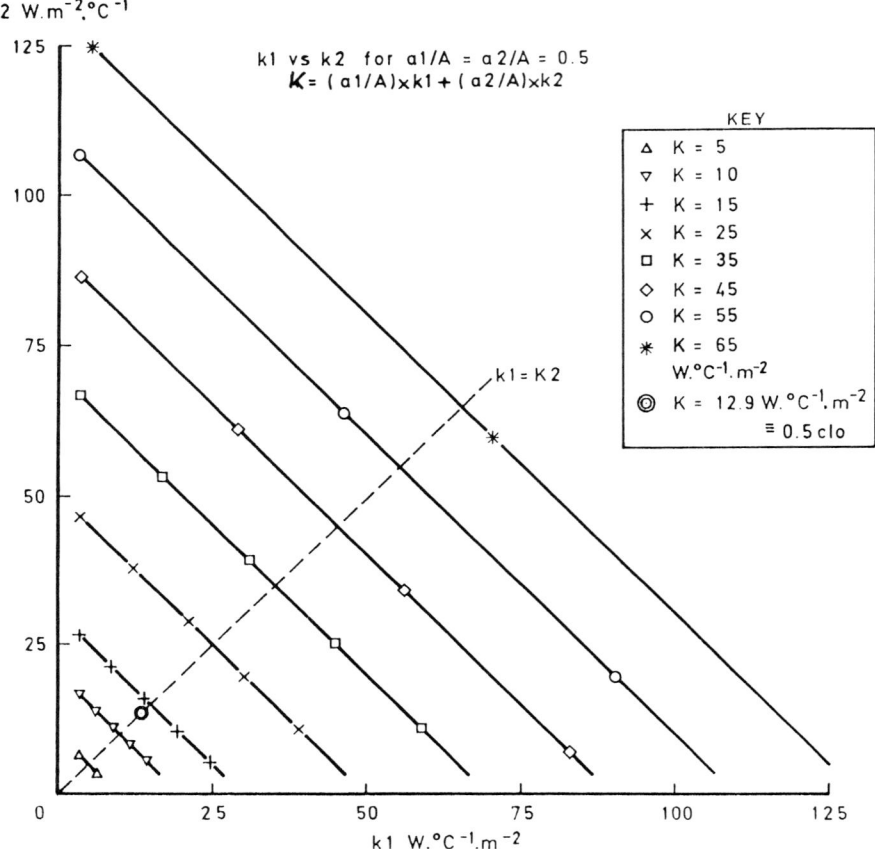

Figure 11.4.    *Plots of* $k_1$ *against* $k_2$ *for the simple cylinder manikin in Figure 11.3 for values of* k *from 5 to 65* $W°C^{-1} m^{-2}$.

$(k_1 \times a_1/A)$, etc. is calculated separately from the individual section measurements of heat input, temperature and area. However, in a single section manikin, only an overall figure for $K$ is measured and this tells us nothing at all about the differences in insulation $1/k$ between the covered and bare areas. If the areas $a_1$ and $a_2$ are equal, for each single value of $K$, $k_1$ and $k_2$ can vary from 0 to 2 K, as shown in Figure 11.4. If we do a similar exercise for insulation $(1/K)$ we get the result shown in Figure 11.5. In this way it may be shown, for example, that a suit giving an overall figure for insulation of 0·5 clo (0·78°C m$^{-2}$ W$^{-1}$) could have insulation over the trunk as high as 2·5 clo while insulation over the legs was only 0·25 clo or, somewhat unusually, the other way round. If the areas $a_1$ and $a_2$ are not equal then the overall influences of changes in conductance in one region will depend on the proportion of

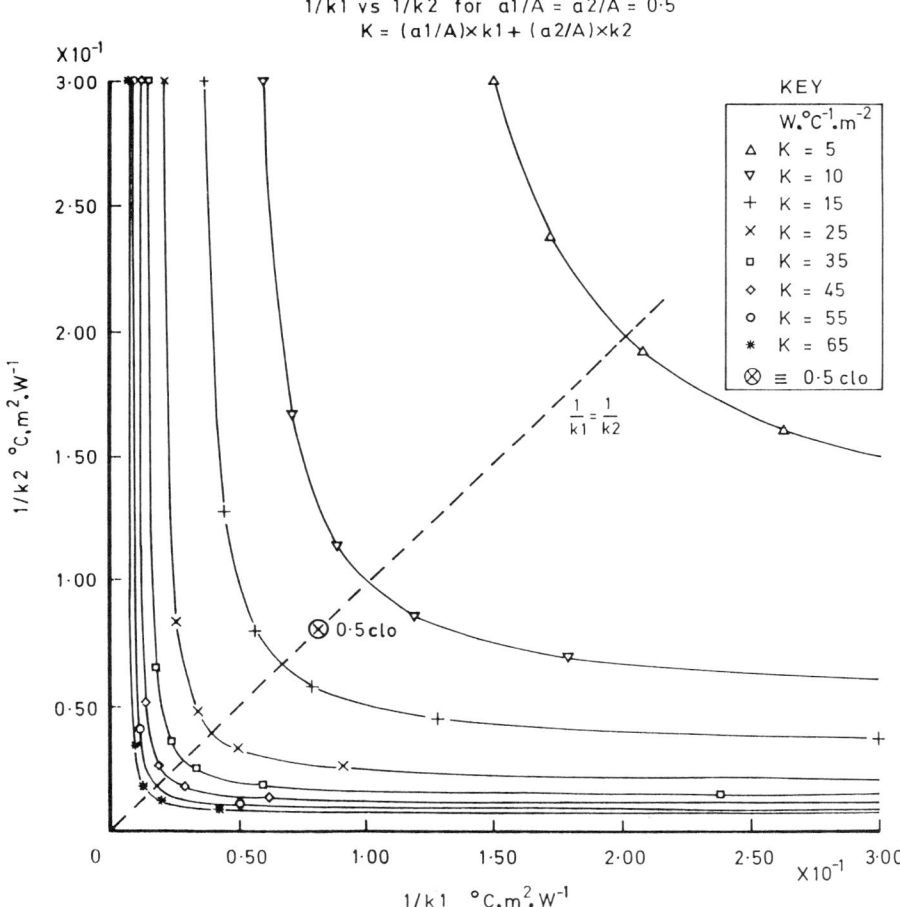

*Figure 11.5. Plots of $1/k_1$ against $1/k_2$ for the simple cylinder manikin in Figure 11.3 for values of k from 5 to 65 $W\,°C^{-1}\,m^{-2}$.*

the total area it represents. This is illustrated in Figures 11.6 and 11.7 for $k$ and $1/k$ respectively.

In Figure 11.8 are plotted model predictions of arterial temperature change (Wissler model) in water at 5°C for three combinations of insulation over the legs and elsewhere. If we take the surface area of the legs (and feet) as 0·8 $m^2$ and the remainder of the body as 1·0 $m^2$ these three combinations each give the same area-weighted overall conductance (equivalent to 0·5 clo)—and this is the figure that would be measured with a single-section copper man or 'hydroman'. The predicted time to arterial temperature $(T_{ar})$ 34°C is about 145 minutes for the evenly distributed insulation and almost doubled for the case with

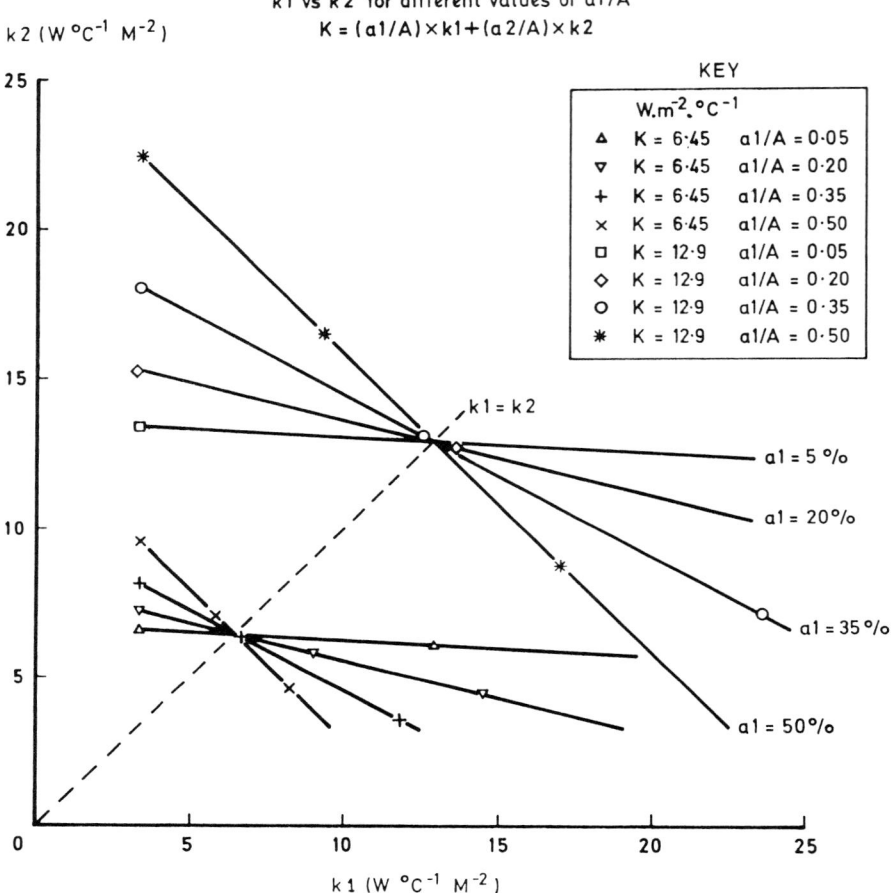

*Figure 11.6.    Plots of $k_1$ against $k_2$ for the simple cylinder manikin in Figure 11.3 for values of k of 6·45 and 12·9 $W\,°C^{-1}\,m^{-2}$ and values of $a_1/A$ from 0·05 to 0·5.*

very high insulation over the trunk. Note also that the Figure also illustrates another hazard of modelling, or any other attempt to set strict core temperature limits, in that the higher trunk insulation produces a prolonged thermal equilibrium of some 3·5 hours at $T_{ar}$ 33·9°C!

In practice it is virtually impossible to provide insulation in water as high as 2·5—3·0 clo and Figure 11.9 shows the results associated with an actual immersion suit of the type which protects the trunk and arms only. Such a garment is in use by Shell UK Ltd as a 'shuttle jacket' for short duration helicopter trips. Representative insulation figures for such a suit in water are 0·06 clo for the bare legs and 0·5 clo for the protected trunk, arms and head. An overall figure based on the sum of

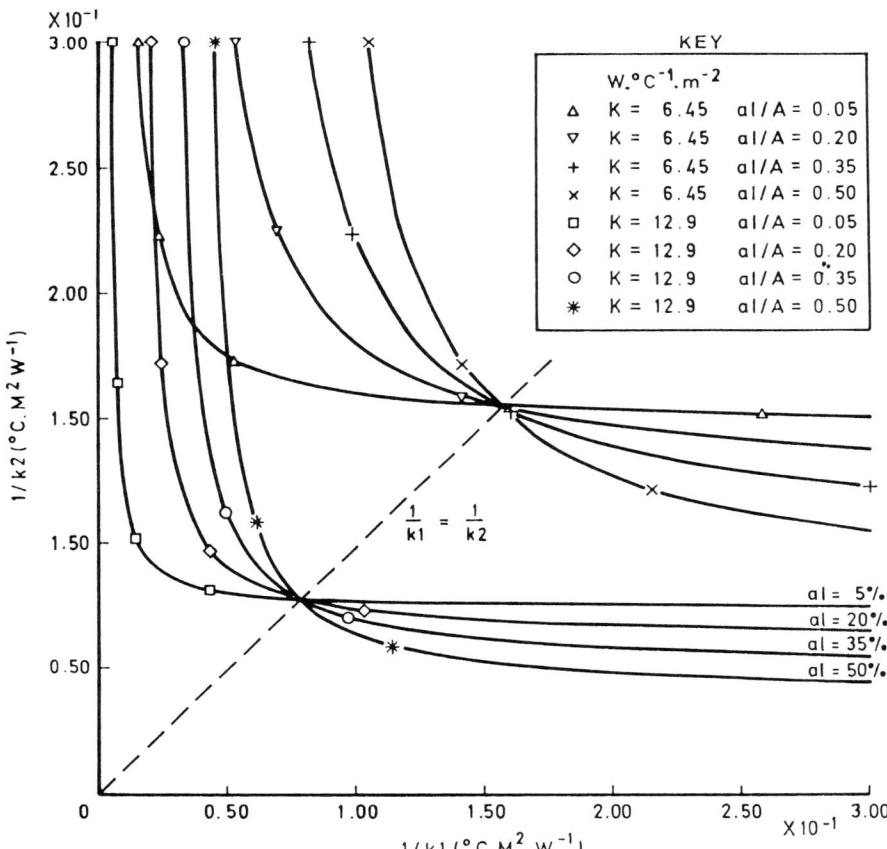

*Figure 11.7.   Plots of $1/k_1$ against $1/k_2$ for the simple cylinder manikin in Figure 11.3 for values of k of 6·45 and 12·9 $W °C^{-1} m^{-2}$ and values of $a_1/A$ from 0·05 to 0·05.*

the area weighted external conductances for each of these regions is approximately 0·18 clo and this would be the result measured with a single section copper man. In this case the 'survival' time prediction, based on the overall insulation figure of 0·18 clo, is some 22% shorter than the time based on the correct regional figures for the trunk and legs.

I now come to the practical implications of these difficulties. Several nations, including Japan, Canada and the UK, are developing immersion suit specifications to meet the IMO requirements which are based on prescribed levels of immersed insulation rather than a requirement for *ad hoc* human subject experiments. This is generally done by giving a single overall figure for required insulation and is based on the assumption that insulation is reasonably even over all body areas

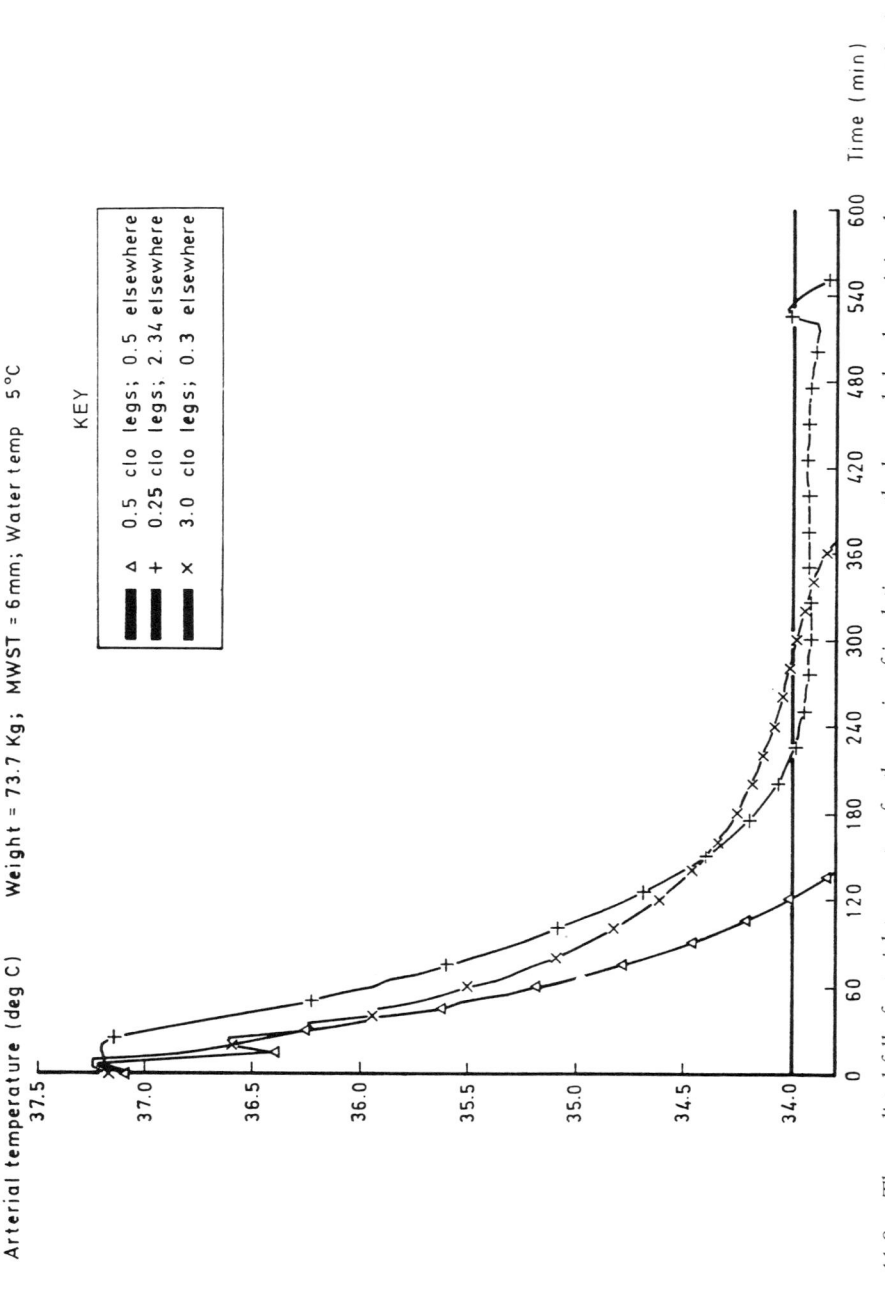

*Figure 11.8. The predicted fall of arterial temperature for three pairs of insulations over the legs and elsewhere giving the same area weighted 'mean' conductance of 12·9 $W °C^{-1} m^{-2}$ (equivalent to 0·5 clo insulation).*

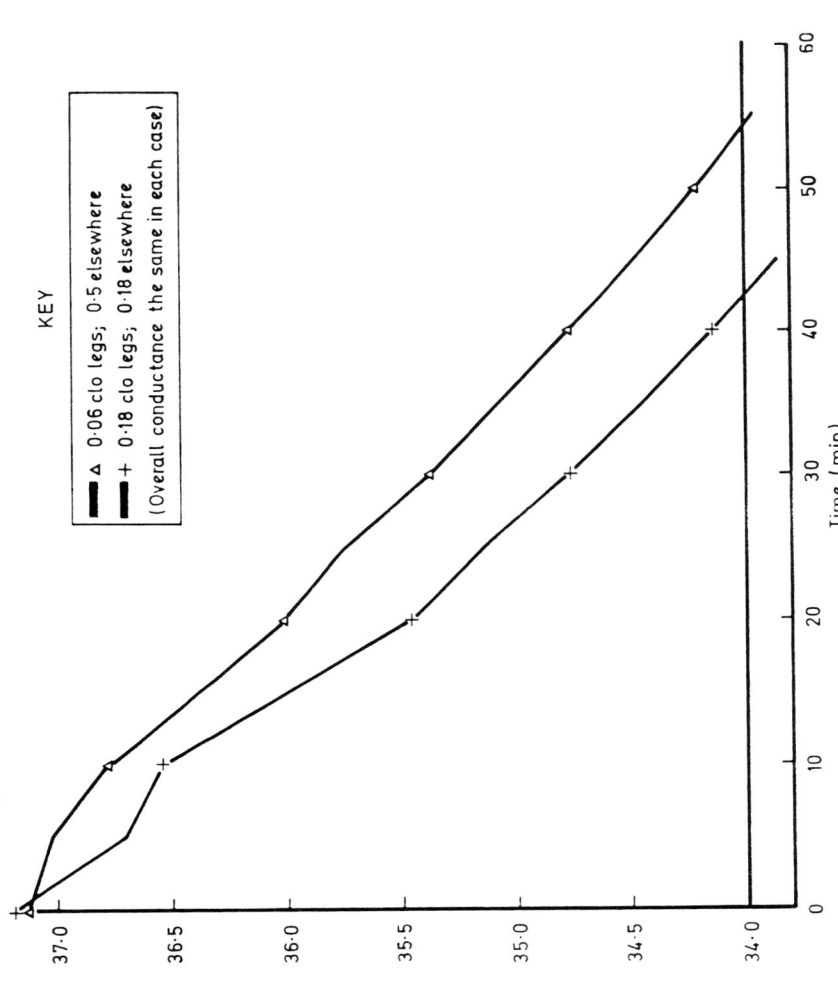

*Figure 11.9. Predicted fall in arterial temperature for a partial coverage wet suit (legs uncovered) when the area insulations are separately entered into the model compared with using an area weighted 'mean'.*

—'IMO' suits are required to cover all body areas. As yet there is no general agreement as to whether this overall figure should include all areas, including for example the head which is not generally immersed and the hands and feet which sometimes have separate cover. Such agreement will be an essential first step. If this approach is to be applied more widely so as to include other immersion protection such as partial coverage wet-suits then, for the reasons I have given, it will probably be necessary to specify regional insulations. Meanwhile it is important that those who make use of predictive graphs of the kind shown in Figure 11.1 are aware that the insulation levels refer to evenly distributed insulation. It may be that parallel conductances will result in pessimistic forecasts for suits with higher insulation over the trunk than the limbs. Whilst this may be unfair to some designs, it is clearly an error on the safe side.

## 5. Water leakage and flushing

A further area of difficulty in the use of manikin measurements of insulation concerns the method of allowing for the adverse effects of water leakage into dry suits or flushing beneath wet suits.

Hall and Polte (1956) have demonstrated a serious loss of insulation with leakage into dry suits, using a single section copper manikin. We have confirmed and extended Hall and Polte's findings using a multi-section aluminium manikin (Allan *et al.*, 1985) but with induced leaks up to 3 litres introduced over the trunk area (see Figure 11.10). Current national proposals for the testing of immersion suits built to IMO standards usually require the measurement of leak rates under realistic circumstances. Manikin measurements of insulation are then undertaken after the artificial introduction of water into the insulation in amounts commensurate with the measured leak rates. This procedure rather assumes that leakage will be evenly distributed which, in the case of small leaks, is often not the case. Here we are up against the same difficulty as with regional differences in insulation, in that a dramatic increase in local conductance in the area of the leak will have a disproportionately large effect on 'overall' insulation if measured on a single section manikin or, for that matter, calculated from measurements on a multi-section manikin. The difficulty is that the manikins do not react to sudden changes in local external conductance in the way that human bodies do. The effect of changes in external conductance caused by regional leakage in a suit will not be independent of consequent changes in the local tissue conductance beneath caused by vasoconstriction. For this reason the results of Hall and Polte and our own on the effects of leakage may be somewhat pessimistic but again the errors will be on the safe side.

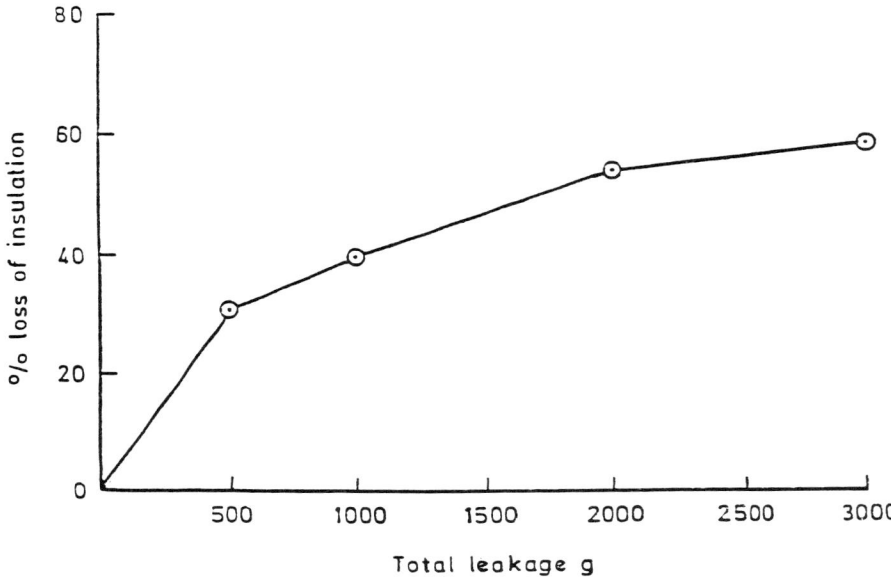

*Figure 11.10.   The loss of insulation plotted against leakage into a dry system (adapted from Allen et al., 1985).*

Similar problems will arise in the use of manikins in wet suit assessments because of the variable effect of flushing. Wolff *et al.* (1985) have shown that the flushing of water beneath a loose fitting wet suit can halve the effective external insulation. Such flushing is induced by body movements and by breathing which of course do not occur in most present day manikins. I expect these modifications will shortly appear!

By concentrating so much on the difficulties associated with manikin measurements of insulation I may have given the impression that they should be abandoned. I do not believe this to be the case but they must certainly be applied with care. Indeed, when compared with the problems associated with *ad hoc* human subject experiments and the potential errors in that approach to the thermal performance specification and testing of immersion suits, the manikin problems are less severe, more known and probably safer.

### References

Allan, J.R., 1983, Survival after helicopter ditching: A technical guide for policy-makers. *Int. J. Aviat. Safety*, **1**, 291–296.

Allan, J.R. and Hayes, P.A., 1984, The specification and testing of the thermal performance of immersion suits. *Aircrew Equipment Group Report No. 512* (Farnborough: RAF IAM).

Allan, J.R., Higenbottam, C. and Redman, P.J., 1985, The effect of leakage on the insulation provided by immersion-protection clothing. *Aviat. Space Environ. Med.*, **56**, 1107–1109.

Hall, J.F. and Polte, J.W., 1956, Effect of water content and compression on clothing insulation. *J. Appl. Physiol.*, **8**, 539–545.

IMO, 1974, *International Convention for the Safety of Life at Sea* 1974, Chapter III (revised) and Assembly Resolution A521(13) entitled 'Recommendations on testing of life-saving appliances.' (International Maritime Organization).

Nunneley, S.A. and Wissler, E.H., 1980, Prediction of immersion hypothermia in men wearing anti-exposure suits and/or using life rafts. Advisory Group for Aerospace Research and Development. Report No. AGARD-CP-286: A1-1 to A1-3.

Nunneley, S.A., Wissler, E.H. and Allan, J.R., 1985. Immersion cooling: Effect of clothing and skinfold thickness. *Aviat. Space Environ. Med.*, **56**, 1177–82.

Steinman, A.M., Nemiroff, M.J., Hayward, J.S. and Kubilis, P.S., 1985, *A Comparison of the Protection against Immersion Hypothermia Provided by Coast Guard Anti-exposure Clothing in Calm versus Rough Seas*, US Dept of Transportation, US Coast Guard, Office of Research and Development, Washington, DC 20593. Report No. CG-D-17-85.

Wolff, A.H., Coleshaw, S.R.K., Newstead, C.G. and Keatinge, W.R., 1985, Heat exchanges in wet suits. *J. Appl. Physiol.*, **58**, 770–777.

# 12. The Physiological Basis for the Development of Immersion Protective Clothing

**Philip Hayes**

## 1. Introduction

There are fundamentally two approaches that may be taken to define the thermal specification of a protective garment. One approach is to produce a set of physiological design criteria or end-points which a test subject wearing the garment must maintain for a prescribed duration and severity of exposure.

A second approach specifies a required level of immersed protection (usually a thermal resistance) and 'predicts' that an individual wearing the assembly would show an acceptable range of responses under the specified conditions. The 'prediction' standard is often presented as a model and requires many sources of reliable data for its construction and validation.

At first sight the two approaches seem not dissimilar but the difference is more than a contrivance when one examines the ethics and expense of testing a suit according to one approach or the other. However, the purpose of this review paper is not to discuss the relative financial and ethical merits of each approach but to examine the basic physiological assumptions and precepts which underlie each.

## A 'physiological design criteria' approach

A good example of this approach is shown by the design goals for the thermal protection of divers, updated by Webb in 1980. The essential requirements for the thermal protection suits were:

1. Maximum net body heat loss (change of enthalpy) should be not greater than 13 kJ kg$^{-1}$ body weight.
2. The core temperature should not be lower than 36°C or show a decrease of 1°C from its initial temperature, whichever is lower.

3. Mean skin temperature should not be lower than 25°C and no individual skin temperature should be lower than 20°C except for the hand, which may be as low as 15°C.

In effect, the suit was to prevent any loss of performance of the man in the water as a result of a decrement in body core or local skin temperature. Closer examination shows the inadequacy and inconsistency of the set of specifications. Theoretically, it requires many different thicknesses of suit to accommodate the many different body types and their severally different degrees of differences in heat loss rate. A fatter person under a set of specified conditions may require 2 mm of neoprene to maintain his core temperature above 36°C, the thin man may need 20 mm! Also, one must question if it is really the set temperature (of 36°C) or the quantity of heat loss from body stores (13 kJ kg$^{-1}$) that produces any observed decrement in performance. In the event of rapid body cooling the rectal temperature may lag behind the arterial or brain temperature and presumably the 13 kJ kg$^{-1}$ provides a limit for net heat loss consistent with a 1°C fall in the theoretical 'core'. Lastly, one must consider whether the minimum skin temperatures quoted in the specific tests are feasible and realistic. Is it practical, for example, to try to maintain the toes above 20°C in water at 0°C?

A similar approach to suit specification was taken by the International Maritime Organization, SOLAS Convention (IMO, 1974). A suit would 'pass' if it prevented body (rectal) temperature from falling below 35°C under a given set of conditions during a prescribed period. Tests were ordered to stop if any local skin temperature fell below 10°C. The IMO requirements are not so stringent as those for divers, as the former are to ensure survival and the latter good performance. However, the same principle is apparent in both requirements and prompts two further questions. First, how are survival and performance related to the defined 35°C and 36°C temperatures (or any other 'core' temperature for that matter), and second, should one stipulate a minimum temperature for local skin cooling when wearing what is, after all, a survival suit? Before attempting to answer these questions, an alternative approach to suit specification is described.

## A 'modelling' approach

Details of this approach were described by Allan (1983) but in essence the specification of the suit is based on measurements of immersed insulation derived from either a thermal manikin or heat flux transducer system and the use of a validated model predicting survival time relative to the insulation. An example of the most recent development of such survival curves generated by a mathematical model (Wissler, 1984) is shown in Figure 12.1. The assumptions on which the model is

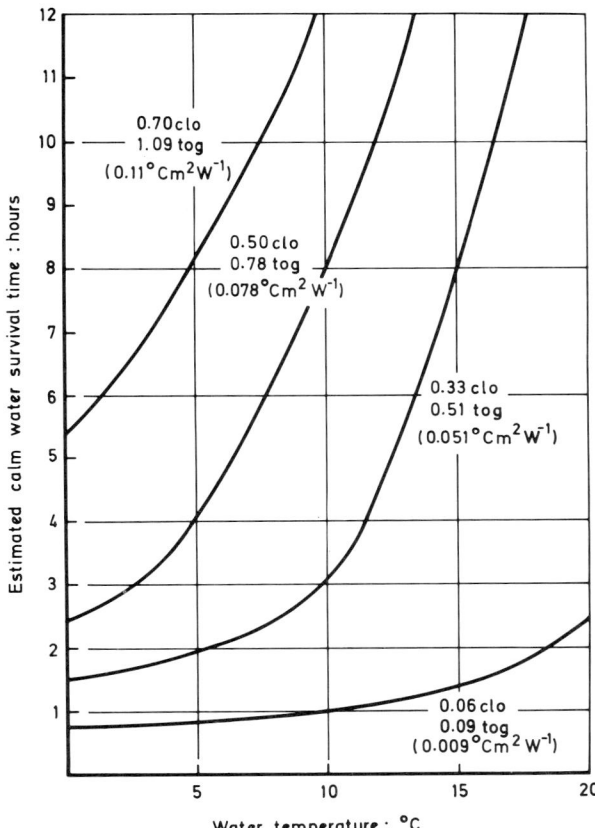

*Figure 12.1. A set of survival curves generated by a mathematical model of thermoregulation. "Survival" time is the time for arterial temperature to fall below 34°C.*

based and the physiological basis for the algorithms that are contained within it are worthy of careful examination. The curves shown in Figure 12.1 pertain to the 10th percentile man (USAF anthropometric data) by weight and mean skinfold (fat) thickness (MST). Survival time is taken as the time taken for the body core (arterial) temperature to fall below 34°C. The sea state is presumed calm and the thermal resistance of any protective clothing worn remains constant during immersion.

The accuracy of the predicted survival time cannot be tested directly using humans, but the ability of the model to describe less extreme cases of cooling of volunteers during physiological experiments can give an indication of the model's worth. Figure 12.2 demonstrates a number of individual experiments (16) involving body cooling in cold water. A range of thin and fat individuals is represented and the

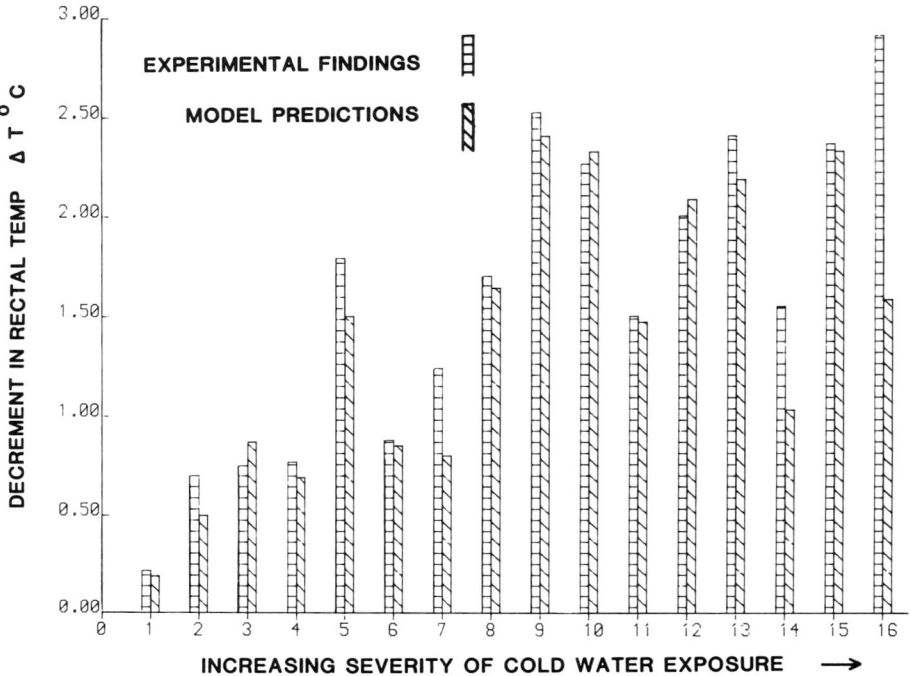

*Figure 12.2.   A comparison of the measured response to cold water immersion and that predicted from the mathematical model.*

predicted results of cooling from the model are compared with measured decrements in rectal temperature. Experimental and predicted end-points (for the same duration of immersion) are well correlated ($r=0.90$, $p \leq 0.001$) and the means (experimental 1·59°C, model 1·40°C) are barely significantly different ($p=0.041$, paired t-test).

The $x$ axis (Figure 12.2) represents (from left to right) increasing severity of cold exposure where the level of protection is reduced from winter aircrew assemblies (thermal resistance in excess of 1·30 togs) in 8°C water, down to nude immersions in water at 12°C and partially protected in 8°C. When comparing measured and model data using rates of body cooling, surface heat flux and rates of heat production, the greatest disparity between the two sets of data is found during severe exposures, for example nude immersion (case 16 in Figure 12.2) often involving thin individuals (case 14).

However, over the range of temperatures and level of protection which is of greatest interest, the model performs reasonably well when group data are analysed. Further development of the model is aimed at improving its predictive performance during cold exposure of 'individuals' of various body compositions; cold 'reactivities', varying clothing conditions (e.g., leakage), and sea state (e.g., waves).

A consistent finding when comparing model predictions with empirical data is the inability of metabolic (shivering) heat controllers to describe accurately observed data. In general terms, metabolic controllers used in a number of models (see Richardson, 1985 for review) are too conservative at low water (skin) temperatures and too sensitive at warmer levels. Real people tend to produce more heat in severe conditions than the models predict but less when well protected in warm water.

## Body temperature and control

Werner (1980) offers a convincing case for describing the thermoregulatory system as a "distributed parameter control loop" based on "adaptive spatial integration of temperatures plus local effector actuation". Such a concept does not require an explicit reference or set point and also can account for the occurrence of fever and circadian rhythm. However, the complexity of Werner's description is matched by the difficulties in attempting to model such a concept based on human data. The requirement for a measure of many deep body temperatures as well as a great proliferation of superficial and skin temperatures is beyond existing practical methods. As a consequence, simple manipulations of core and mean skin temperatures are offered as metabolic heat controllers, examples being that of Stitt *et al.* (1974), further modified for human subjects by Hayward *et al.* (1977). Hayes *et al.* (1985) used the same basic equation:

$$M = \infty(T_S - T^0{}_S)(T_C - T^0{}_C)$$

to try and fit data obtained from immersion experiments involving a range of water temperatures $(7-30°C)$, people, protection level and wave motion. $M$ = Energy expenditure, $W\ kg^{-1}$; $\infty$ = coefficient describing sensitivity of the system; $T_S$ = mean skin temperature; $T^0{}_S$ = reference mean skin temperature at which the metabolic response to a change in central temperature is effectively zero; $T_C$ = central core temperature; and $T^0{}_C$ = reference core temperature at which the metabolic response to a change in skin temperature is effectively zero.

The correspondence between measured and expected results using this algorithm were far from satisfactory. This type of equation does not describe obtained data very well and seemed a regressive step returning to 'reference points' and similar inventions from which to make predictions. A more profitable approach was to consider the considerable array of factors, other than absolute levels of temperature, which may modify the rate of heat production in the cold. The list is extensive and includes body mass and fat, sensitivity or reactivity to cold, habituation and local acclimatization, cold induced vasodilation, regional cold sensitivity, fatigue and physical fitness, wave motion and exercise. Some of these factors are discussed below.

All too often, rate of body cooling and survival time are examined from a perspective of reducing heat loss, little regard being paid to developing methods of increasing and maintaining heat production. The rate of heat production is as important as rate of heat loss since the balance between the two ultimately determines the duration of survival in accidental immersion.

The balance between heat loss and heat production is the net heat transferred and in the case of an overall loss (NHL) is assumed to be linearly related to the fall in core temperature ($\Delta T_r$) and consistent between individuals on a per unit mass basis (i.e., kJ kg$^{-1}$). This assumption is inherent in the design limitation of 'no greater heat loss being tolerated than 13 kJ kg$^{-1}$, mentioned earlier. In reality, the non-linearity of the NHL relationship is reflected in the poor correlation between the two measures shown in Figure 12.3 ($r=0.44$ long lukewarm immersion, $r=-0.13$ short cold immersion). These were collected from 18 men who underwent both a cold immersion (about 15°C) for 1 hour and a lukewarm immersion between 28 and 32°C for 4 hours, both in the nude. It is apparent that a fall of 1°C in body temperature may be equal to a $20-45$ kJ kg$^{-1}$ net heat loss. Alternatively, the stated criterion level of 13 kJ kg$^{-1}$ represents a fall in temperature of between 0.25 and 0.70°C. Further analysis of these data shows that the difference in rectal temperature between cold and lukewarm immersion is a function of the variation in heat production, not the heat loss (Padbury, 1984). Inter-individual variation, both in the $\Delta T_r$/NHL relationship and the rate of heat production (at each water temperature), is sufficiently large to suggest that a single value of 13 kJ kg$^{-1}$ is of little practical significance for any one person: it also appears to be somewhat conservative as a tolerable limit.

It has been shown that heat production may be a source of great variation in the outcome of immersion hypothermia—a more detailed discussion follows.

## 2. Factors affecting heat production

### Fat and mass

There is an abundance of anecdotes to suggest that people routinely exposed to a cold environment (e.g., divers, Arctic travellers) and fatter individuals, do not 'feel the cold' as others do. Initially, let us assume that there are two different mechanisms at work in each of the two

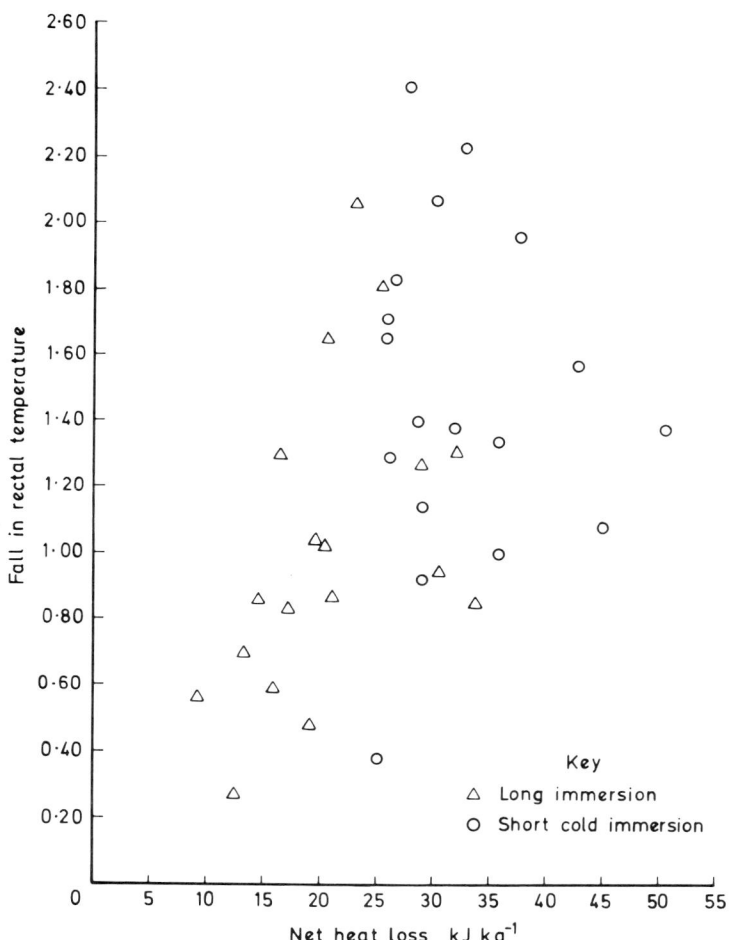

*Figure 12.3.  The relationship between the fall in rectal temperature at the end of immersion against the measured net heat loss. Long immersions are for 4 hours in water at 28–32°C and short cold immersions are approximately 15°C for 1 hour, n = 18 males.*

groups. Divers and their like may undergo some form of cold adaptive response whereas the role of additional body fat is principally related to a physical resistance to heat transfer. Scholander *et al.* (1958) presented evidence that young men could adjust to a chronic cold stress by raising their heat production, a process they termed metabolic acclimation. The other possibility (insulative acclimation), which would be obtained by a colder skin at a reduced heat production, was not observed. The latter possibility was observed by Skreslet and Aarefjord (1968) on a group of divers routinely exposed to Arctic waters in inadequate thermal protection. The divers went through successive stages of adaptation where, following a period of initial increase in metabolism there was a fall in rectal temperature believed to be caused by habituation of the CNS to the cold stimulus. Insulative acclimation was the last and final stage.

The concept of habituation, a reduced response to a sustained cold stress, is interesting and could provide a link between the lack of response measured in persons accustomed to cold (Hayward and Keatinge, 1969) and those who are fat (Cannon and Keatinge, 1960; Strong *et al.*, 1985; Wyndham *et al.*, 1968). Wyndham *et al.* (1968) suggested that fatter persons become habituated to a lower average skin temperature. Because of their greater thermal insulation the skin remains lower most of the time compared with thin men. Fat men therefore need to experience a greater fall in skin temperature in order to stimulate shivering to a level where there is a marked increase in heat production. Strong *et al.* (1985) developed equations relating heat production to core and mean skin temperatures ($T_S$) relative to the duration of cold immersion. The metabolic functions defined planes of thermogenic activity demonstrating much steeper slopes with respect to changes in $T_S$ if the subject were thin, than for the larger fatter individual.

The thermal insulative capability of fat is well established (Cannon and Keatinge, 1960; Rennie *et al.*, 1962; Sloan and Keatinge, 1973), but it would seem, from the observations noted above, that the amount of fat may also be a determinant of heat production. A large fat mass is usually synonymous with a large overall mass and a small surface area (SA) to mass ratio, which in turn is beneficial in resisting heat loss. Strong *et al.* (1985) found the reciprocal ratio, mass (kg)/SA(m$^2$), to be an important determinant of both vasomotor insulative activity as well as the metabolic response to cold water immersion.

## Regional sensitivity to cold

Van Someren *et al.* (1982) were able to reverse the body cooling accompanying water immersion at 29°C by further local cooling of the hands and feet in 12°C water. It appears that a uniform cutaneous temperature of 29°C does not provide sufficient stimulus to activate heat production processes even when core temperature is as low as 35·6°C. Additional cooling of hands and feet initiates an increased metabolic heat production sufficient to reverse the downward trend in the rectal temperature due to the cold immersion. The practical significance of regional thermal sensitivity was highlighted by Kaufman (1982) who considered various clothing assemblies for protection against cold weather exposure. Core and skin temperatures were analysed in persons exposed to different levels of cold, which showed that the distribution and level of the thermal insulation over the body may either compliment heat conservation processes or detract from them. Heavy insulation on the hands and feet may maintain greater comfort during acute cold exposures but over-insulation of the same area during chronic cold exposure may result in a decreased core temperature and depressed level of shivering. A graded system of insulation with greater protection of the torso than the distal regions appears to augment thermogenic and vasoconstrictor processes.

Both sets of experiments demonstrate the importance of a regional pattern of differing temperature (and presumably heat loss) in obtaining the best performance from a thermal protection and/or survival assembly. Subjective comfort is not necessarily the objective but increased heat production and a fully activated thermoregulatory system are. Suit design should reflect this requirement.

## Physical fitness and fatigue

Hayward and Keatinge (1981) found that the.ability of persons to stabilize body temperature in cold was related as much to their metabolic response as to their subcutaneous fat thickness. They considered the high muscle mass in relation to surface area and a high state of physical training to be important in promoting high rates of shivering heat production. Individuals who were more reactive to cold developed high heat production rates and often had low tissue conductance relative to their fat thickness. Recent observations indicate that endurance fitness does influence the cooling rate during immersion (Jacobs *et al.*, 1984), where increased fitness extended the time taken for the body core to fall by 1°C during immersion. In addition, Golden *et al.* (1979) had previously shown that peak oxygen uptake ($VO_2$), observed during shivering in cold water, was significantly correlated with the $VO_2$ max obtained during exercise, but not with skinfold thickness or rate of cooling.

Increased endurance fitness should extend the duration of shivering and delay the onset of shivering fatigue. The model producing the curves shown in Figure 12.1 uses the principle of a fixed 'quantity' of shivering potential which can be used to exhaustion at a slow or fast rate depending upon the severity of cold exposure. There is insufficient data available at present to describe the process more effectively, but without doubt the changes occurring in heat production and the distribution and level of blood flow at the cessation of shivering are of great importance to the survival estimates of subjects exposed to the cold temperature.

## 3. Performance at a lowered body temperature

The tacit assumption in the IMO specification (1974) is that performance (in this case the expectation of survival from hypothermia) is severly reduced when rectal temperature falls below 35°C. Alternatively, the model approach defines a limit for survival of 34°C (arterial temperature). In each case it seems unlikely that an imminent demise of an individual could be a consequence solely of hypothermia: it is more likely related to a combination of factors including loss of manual performance as well as a reduction in the level of performance of complex perceptual-cognitive tasks. In simple terms, a hapless victim is no longer able to find the strength or sense to prevent himself from drowning.

The detrimental effect of cold water exposure on the performance of tasks involving tactile sensitivity, grip strength and dexterity has been reasonably well established (Coppin *et al.*, 1978; Gaydos and Dusek, 1958; Provings and Morton, 1960). Depending upon the precise nature of the physical activity and the individual attributes, the limit of local skin cooling lies between 8 and 15°C. The lower end of this range also describes a limit at which non-freezing cold injury (NFCI) might be expected to occur during a period of as little as 6–8 hours (in sensitive individuals). During prolonged periods (48 hours) injury may result from exposure even when tissue temperatures are as high as 15°C. NFCI is a phrase used to describe a syndrome that results from damage to tissues which have remained for long periods in temperatures ranging between freezing (−0·55°C) and 15°C (Francis, 1984). The pathogenesis of NFCI has been reviewed by Francis and Golden (1985). They propose a mechanism for the onset and sequelae of NFCI based on a vicious circle of cooling and vasoconstriction in the presence of a high level of sympathetic tone that is not counteracted by CIVD (cold induced vasodilatation). The occurrence and magnitude of CIVD may be reduced by generalized body cooling (Keatinge, 1957) and may lead to a higher propensity to NFCI for those cooling in immersion suits, particularly in areas such as the feet which inevitably are less well protected. The feet are often the coldest part of the body during testing of immersion suits and a minimum limit of 8°C for their skin temperature appears both practically feasible and maintains an element of safety relative to the expected survival time and capability of most rescue services.

Coleshaw *et al.* (1983) demonstrated impaired memory registration and calculation speed at core temperatures between 34 and 35°C. In a test involving serial choice reaction times, Ellis (1982) reported that consistently large increases in error were attendant upon reductions in mean skin temperature rather than a fall in rectal temperature. Reaction times were unaffected. Vaughan (1975) examined a number of performance tasks during 6-hour tests diving in 2°C water. His results suggest a complex task-dependent effect of cold stress from deep body cooling and he concluded, in addition, that rectal temperature change was probably a poor index of cold stress for use as a correlate of any degradation in mental performance.

If there is a requirement to define a level of core temperature at which hypothermia is sufficiently developed to merit attention then 35°C is an accepted level (Royal College of Physicians of London, 1966). However, from this brief review of performance in cold we may still make only an intelligent guess about the body temperature at which life is seriously jeopardized in the open sea. An arterial temperature of 34°C may prove to be too conservative for many individuals but such a choice is commensurate with our provision of a survival policy that errs on the side of caution. Indeed the choice of a 10th percentile individual as an example of survival capability is also a low estimate of the general capability of the great majority of men to survive.

## 4. Body fat measurements

Although it is apparent that the rate and duration of heat production in man has a significant role in the overall survival equation, the level of subcutaneous fat thickness is also a major determinant of cold water survival time. Recently, an exceptional example of survival in cold water was reported by Keatinge *et al.* (1986) where the ability of the individual to swim for about 6 hours in Icelandic waters (5−6°C) was attributable to his size (125 kg, surface area 2·54 m²) and substantial fat thickness (mean weighted value 14 mm (determined by ultrasonic measurement), according to Hayward and Keatinge, 1981). In the majority of studies, since the early work of Baker and Daniels (1956) involving cooling and body fat, the subcutaneous fat thickness has been estimated with skinfold calipers. The simple addition of a number of skinfold caliper measurements is often used to provide an index of obesity (Keys and Brozek, 1953; Larsson *et al.*, 1984). Caliper thicknesses have also been measured in a number of workforces at risk from accidental or intentional cold water immersion, such as the RAF (Bolton *et al.*, 1974), USAF, North Sea offshore workers (Light and Dingwall, 1985) and US Navy divers (Beatty and Berghage, 1972), where a mean weighted skinfold (fat) thickness (MWST) may be related to a predicted survival time in the sea (Nunneley *et al.*, 1985). A recent study compared fat thickness measured by ultrasonics ('A-scan') to that measured by calipers on a group of young women (Volz and Ostrove, 1984).

However, do measurements made with ultrasound or calipers provide an accurate indication of the subcutaneous fat thickness?

Hayes *et al.* (1985) described a method of imaging human adipose tissue using a magnetic resonance imaging (MRI) method. MRI can be used to provide cross-sectional images of the human body at any level from which the fat and skin thickness can be measured directly. Magnetic resonance imaging was used as a standard for the quantitative assessment of subcutaneous fat thickness against which measurements made by calipers and ultrasonics were compared (Cohen *et al.*, 1986). Mean fat thickness from 12 sites per individual was calculated in a group of 21 males (range of % fat mass 9−26% by densitometry) and 20 females (% fat mass 11−37%) by each of the three methods. In males the overall mean fat thickness obtained by calipers (divided by 2 to halve the 'fold') was $6·1\pm0·6$ mm ($\pm$sem) and by ultrasound was $6·4\pm0·6$ mm. These means were not significantly different from each other but were significantly less ($p\leq0·001$) than the MRI reading of $9·5\pm0·9$ mm. Duplicate statistical analysis on the female data using ANOVA showed that calipers ($8·6\pm0·8$ mm) were less than ($p\leq0·001$) MRI ($11·5\pm1·1$ mm) were less than ($p\leq0·001$) the ultrasound ($12·9\pm1·2$ mm) (see Figure 12.4). All three methods were well correlated in

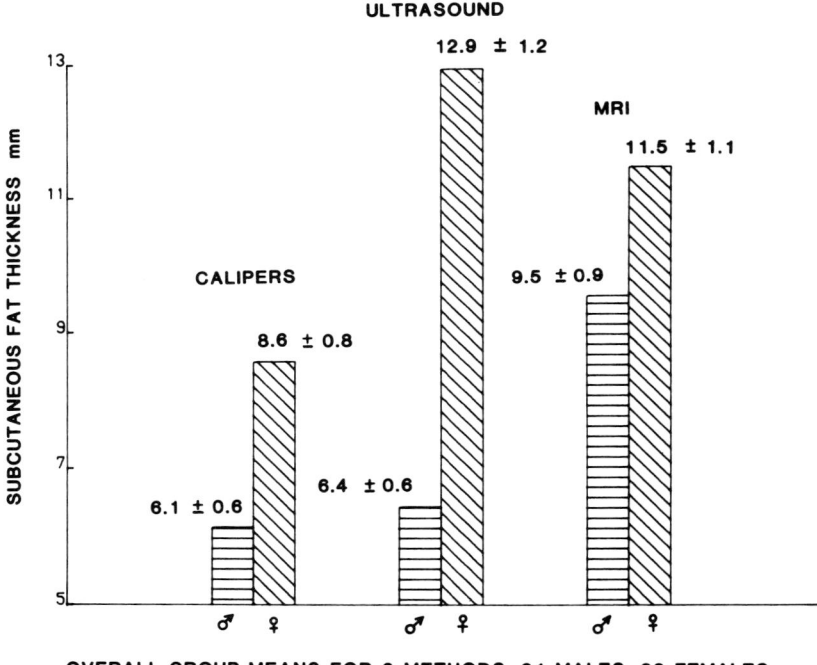

OVERALL GROUP MEANS FOR 3 METHODS, 21 MALES, 20 FEMALES

*Figure 12.4.   A comparison of mean subcutaneous fat thickness (12 sites) measured by three methods: calipers, ultrasound and magnetic resonance imaging on males (n = 21) and females (n = 20).*

males (all $r>0.86$, $p \leq 0.001$) but in females only the calipers and MRI are well correlated ($r=0.87$, $p \leq 0.001$).

Some explanation for these findings may come from the presence of something akin to fascial planes within the subcutaneous fat layer. In areas where the layer is thick the calipers may well be less than MRI because they can only 'pinch' off the top layer of fat whereas in reality more of the fat deposit is left below, closely adhered to the underlying structures. The occurrence of these layers produces multiple echoes on the ultrasonic trace and in combination with multiple echoes obtained from the muscle leads to difficulties in interpretation and unsatisfactory measurements from ultrasonics, particularly in the fatter females.

Earlier in this review it was suggested that the level of fat was instrumental in determining not only the level of insulation but also the magnitude of thermogenic and vasomotor response to cold. In keeping with this argument it might be anticipated that a more accurate assessment of fat thickness (by MRI) will eventually provide a better correlation with, and hence prediction of, thermoregulatory responses to cold immersion. This is indeed the case as shown by the data below (Table 12.1) collected by Sowood *et al.* (1986). Thirteen of the 21 males involved in the MRI fat study were immersed without clothing in water at 18°C and 24°C. Mean weighted surface heat flux (MWHT) was calculated according to the method of Bell *et al.* (1985). It was interesting to find that the average rate of heat flux (average MWHT), average rate of heat production and body heat conductance were better correlated with the MRI measure of fat thickness than those derived from either ultrasonics (U/S) or calipers (Cal).

*Table 12.1   Correlation data between different methods of measurement of subcutaneous fat and physiological responses to cold water immersion, n = 13, all correlations significant, p<0.05.*

| Water temperature | Average surface heat flux | | | Average rate of heat production | | | Average tissue heat conductance | | |
|---|---|---|---|---|---|---|---|---|---|
| | MRI | U/S | Cal | MRI | U/S | Cal | MRI | U/S | Cal |
| 18°C | 0.92 | 0.88 | 0.88 | −0.73 | −0.66 | −0.66 | −0.93 | −0.89 | −0.89 |
| 24°C | 0.79 | 0.72 | 0.77 | −0.89 | −0.78 | −0.85 | −0.82 | −0.76 | −0.81 |

In summary, it is apparent that in our sample of 41 people both sexes have a mean subcutaneous fat thickness measure by MRI imaging which is greater than has previously been measured with calipers. However, the MRI and caliper measurements are well correlated in both sexes (if not the ultrasonics), and one should be able to predict a true mean fat thickness (as if it were from MRI) from a set of caliper readings. Increasing the size of the caliper readings by a given amount (according to the relationship between the two methods★) will provide a better indication of fat insulation but will not provide a better correlation between fat thickness and the measures of physiological response described in Table 12.1: individual examination of body composition is required.

During human testing of immersion suits tremendous inter-individual variation can be anticipated, even within a given weight and height range. A description of the height and weight does not define a percentage body fatness or subcutaneous fat thickness from which one could anticipate or predict a specific response to cold immersion. Weight and caliper thickness should provide a method of standardizing individuals, but it has been shown that the caliper thicknesses can themselves be a source of error. Variation in response, as a consequence of body composition, is augmented by that attributable to inherent cold reactivity, local fluctuations in thermal resistance of clothing as well as physical fitness and cold habituation.

The estimation of fat using MRI methodology suggests that in future survey work involving measurement of the thickness of fat, the percentile levels will need to be recalculated and carefully defined in terms of the site(s), method of measurement and weighting coefficients used. Implications of the MRI work for thermal modellers are considerable as the morphological representation of the human body 'built' by the model programs will need to take into account greater local variation and magnitude of subcutaneous fat thickness. The relationships between fat and the thermoregulatory factors it influences require further examination together with validation of the models initiated to reflect these changes. There is little logic in applying mean fat measurements of one group of workers to those of another and one cannot readily transfer an estimate of survival capability used, say, by the USAF, to those who work in the North Sea. In addition, offshore waters are seldom calm, and it is unlikely that an immersion protection suit would remain at a constant level of thermal resistance for the duration of any accident. Inevitably there will be fluid invasion, either from the sea itself or from within (urine) and these will cause a gradual erosion of the level of insulation. Figure 12.5 is an attempt, using the Wissler model, to combine some of these features of the real environment such as leakage and wave motion: it is a set of survival curves aimed exclusively at the offshore workers in the North Sea and is based on the anthropometric data of Light and Dingwall (1985). Curve 1 represents a good immersion suit worn over working clothes (trousers, shorts, sweater) under ideal conditions (constant insulation, calm sea).

---

★ Corrected fat thickness $= 1.45 \left( \dfrac{\text{MWST calipers at 4 sites}}{2} \right) - 0.48$

$r = 0.92$

The four sites of caliper measurement are biceps, triceps, subscapular and suprailiac.

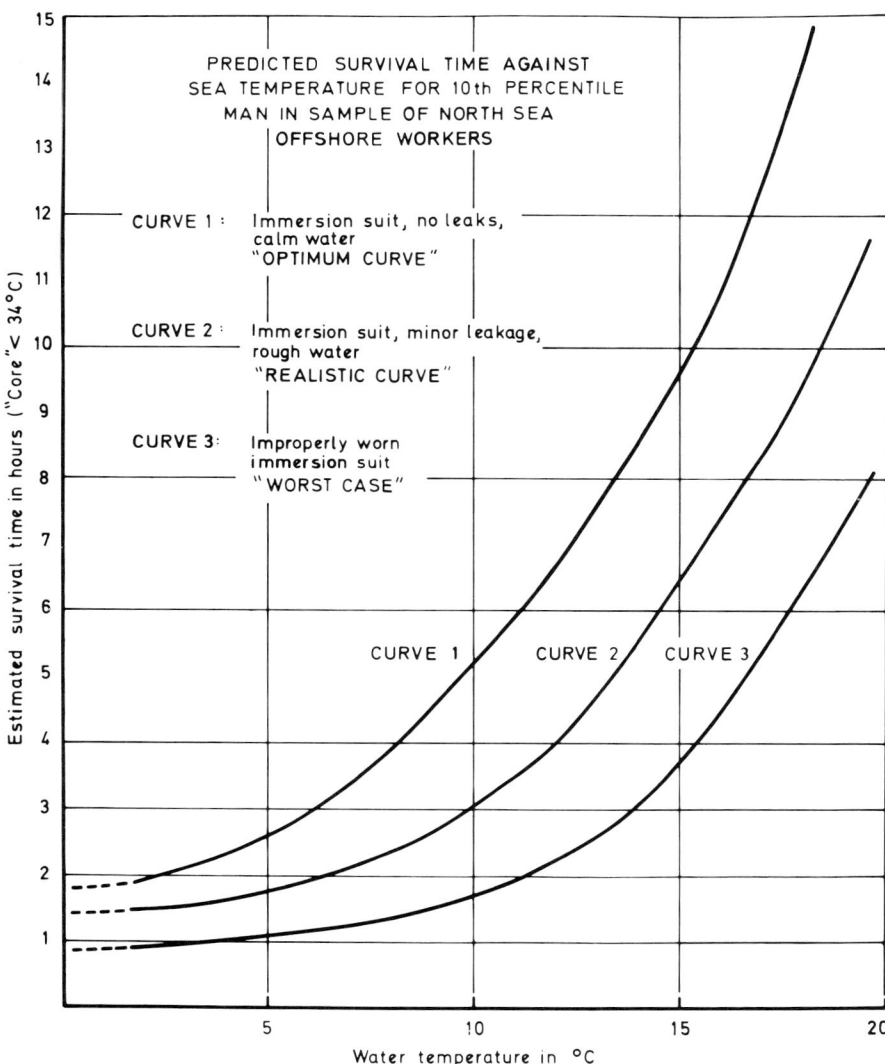

*Figure 12.5. Survival curves generated by a mathematical model for a sample of North Sea offshore workers. Fat thickness was measured by calipers. Curve 1 represents clothing insulation value of 0·51 (mean) togs; curves 2 and 3 are reduced according to leakage and activity.*

Curve 2 represents the situation where 200 g of water leak into the suit at the point of immersion, followed by a further leakage of 200 g h⁻¹ from a combination of the sea invasion and urine loss. Wave action is simulated by increasing the activity level and limb motion of the immersed victim and by reducing the insulation of the clothing protection. Curve 3 represents the case where an immersion suit is not fastened and most of the insulation is wetted during initial entry into the water. The range of survival times described between Curves 1 and 3 at any one temperature is an indication of how much variation one might find in a 'normal' thin man depending on a number of practical features. In the light of the discussions above we might also expect survival times remarkably different from those shown in Figure 12.5 even without any change in body composition. The inclusion of many other physiological variables awaits further research and future generations of models.

One immediate advance would be to develop the curves from a representation of a single fat level to a series of charts representing many different body builds. Inevitably the complexity of so many sheets of paper would outweigh practical use. Alternatively, there would be great practical advantage in establishing the survival time for any combination of the variables—water temperature, thermal insulation, body weight and fat thickness. It would be better still to have means to assess clothing requirements for a given set of conditions for a predetermined survival time.

We have attempted to produce a Mk II version of the Wissler model for use in cold water only, simple enough for calculator use.

Over 200 survival predictions have been performed using the model based on a range of body weights (wt), clothing assemblies (togs), fat thicknesses (MWST) and water temperatures (temp). An equation has been produced relating the variables wt, togs, MWST and temp to the survival time, defined as the time for the arterial temperature to fall below 34°C. Based on data where the survival time was <900 minutes ($n=94$) the following equation (Hayes and Belyavin, 1986) describes the relationship between the four variables and the survival time (ST) predicted by the Wissler program:

$$\ln(\text{ST}) = A + B \ln[C + (D \times \text{togs}) + \text{MWST}]$$

where $A = -8.971 + 0.560 \times \text{temp}$
$B = 3.951 - 0.124 \times \text{temp}$
$C = 13.537 - 0.795 \times \text{temp}$
$D = 38.227 - (0.278 \times \text{wt}) + 0.897 \times \text{temp}$

This equation accounts for 87% of the variance found in the data with a correlation of $r=0.98$ between the logarithm of the fitted and expected survival times. As more data is added the equation will be developed further.

## References

Allan, J.R., 1983, Survival after helicopter ditching: A technical guide for policy-makers. *Int. J. Aviat. Safety*, **1**, 291–296.
Allan, J.R. and Hayes, P.A., 1984, The specification and testing of the thermal

performance of immersion suits. *Aircrew Equipment Group Report No. 512* (Farnborough: RAF IAM).

Baker, P.T. and Daniels, F., 1956, Relationship between skinfold thickness and body cooling for 2 hours at 15°C. *J. Appl. Physiol.*, **8**, 409–416.

Beatty, H.T. and Berghage, T.E., 1972, *Diver Anthropometrics*. Experimental Diving Unit Report 10-72. Navy Experimental Diving Unit, Washington Navy Yard, Washington, DC 20390.

Bolton, C.B., Kenward, M., Simpson, R.E. and Turner, G.M., 1974, *An Anthropometric Survey of 2000 Royal Air Force Aircrew, 1970/1971*. FPRC/ 1327 Mod (Air Force Dept).

Cannon, P. and Keatinge, W.R., 1960, The metabolic rate and heat loss of fat and thin men in heat balance in cold and warm water. *J. Physiol.*, **154**, 329–334.

Cohen, J.B., Gemmell, H., Hayes, P.A., Smith, F.W. and Sowood, P.J., 1986, A comparison of magnetic resonance imaging, ultrasound and skinfold caliper techniques to measure subcutaneous fat thickness in humans. *J. Physiol.*, **380**, 61.

Coleshaw, S.R.K., Van Someren, R.N.M., Wolff, A.H., Davis, H.M. and Keatinge, W.R., 1983, Impaired memory registration and speed of reasoning caused by low body temperature. *J. Appl. Physiol.: Respirat. Environ. Exercise Physiol.*, **55**, 27–31.

Coppin, E.G., Livingstone, S.D. and Kuehn, L.A., 1978, Effects of handgrip strength due to arm immersion in a 10°C water bath. *Aviat. Space Environ. Med.*, **49**, 1322–1326.

Ellis, H.D., 1982, The effects of cold on the performance of serial choice reaction time and various discrete tasks. *Human Factors*, **24**, 559–598.

Francis, T.J.R., 1985, Non-freezing cold injury: a historical review. *J. Roy. Nav. Med. Serv.*, **70**, 134–139.

Francis, T.J.R. and Golden, F. St C., 1985, Non freezing cold injury: the pathogenesis. *J. Roy. Nav. Med. Serv.*, **71**, 3–8.

Gaydos, H.F. and Dusek, E.R., 1958, Effects of localised hand cooling versus total body cooling on manual performance. *J. Appl. Physiol.*, **12**, 377–380.

Golden, F. St C., Hampton, I.F.C., Hervey, G.R. and Knibbs, A.V., 1979, Shivering intensity in humans during immersion in cold water. *J. Physiol.*, **290**, 48P.

Hayes, P.A. and Belyavin, A., 1986, Unpublished data.

Hayes, P.A., Smith, F.W. and Sowood, P.J., 1985, Distribution and quantification of human body fat using nuclear magnetic resonance (NMR) imaging: comparison with a skinfold caliper method. *J. Physiol.*, **369**, 160P.

Hayes, P.A., Sowood, P.J. and Cracknell, R., 1985b, Reactions to cold water immersion with and without waves. *IAM Report No. 645* (Farnborough: RAF IAM).

Hayward, M.G. and Keatinge W.R., 1979, Progressive symptomless hypothermia in water: possible cause of diving accidents. *Brit. Med. J.*, **1**, 1182.

Hayward, M.G. and Keatinge W.R., 1981, Roles of subcutaneous fat and thermoregulatory reflexes in determining ability to stabilise body temperature in water. *J. Physiol.*, **320**, 229–251.

Hayward, J.S., Eckerson, J.D. and Collis, M.L., 1977, Thermoregulatory heat production in man: prediction equation based on skin and core temperatures. *J. Appl. Physiol.: Respirat. Environ. Exercise Physiol.*, **42**, 377−384.

IMO, 1974, *International Convention For the Safety of Life at Sea* (SOLAS), 1974, Chap III (revised) and Assembly Resolution A521 (13) entitled: Recommendations on Testing of Life-saving Appliances (International Maritime Organisation).

Jacobs, I., Romet, T., Frim J. and Hynes A., 1984, Effects of endurance fitness responses to cold water immersion. *Aviat. Space Environ. Med.*, **55**, 715−720.

Kaufman, W.C., 1982, Cold-weather clothing for comfort or heat conservation. *Physician and Sports*, **10**(2), 71–75.

Keatinge, W.R., 1957, The effect of general chilling on the vasodilation response to cold. *J. Physiol.*, **139**, 497−507.

Keatinge, W.R., Coleshaw, S.R.K., Millard, C.E. and Axelsson, S., 1986, Exceptional case of survival in cold water. *Brit. Med. J.*, **292**, 171−172.

Keys, A. and Brozek, J., 1953, Body fat in adult man. *Physiol. Rev.*, **33**, 245−325.

Larsson, B., Svardsudd, K., Welin, L., Wilhelmson, L., Bhorntorp, P. and Tibbin, G., 1984, Abdominal adipose tissue distribution, obesity, and risk of cardiovascular disease and death: 13 year follow up of participants in the study of men born in 1913. *Brit. Med. J.*, **228**, 1401−1404.

Light, I.N. and Dingwall, R.H.N., 1985, *Basic Anthropometry of 419 Offshore Workers, 1984* (Aberdeen: Robert Gordon Institute of Technology, Offshore Survival Centre).

Nunneley, S.A., Wissler, E.H. and Allan, J.R., 1985, Immersion cooling effect of clothing and skinfold thickness. *Aviat. Space Environ. Med.*, **56**, 1177−1182.

Padbury, E.H., 1984, A practical evaluation of the thermal problems of man in the diving environment. Ph.D. Thesis, University of London.

Provings, K.A. and Norton, R., 1960, Tactile discrimination and skin temperature. *J. Appl. Physiol.*, **15**, 155−160.

Rennie, D.W., Corvino, B.G., Howell, B.J., Song, S.J., Kang, B.S. and Kong, S.K., 1962, Physical insulation of Korean diving women. *J. Appl. Physiol.*, **17**, 961−966.

Richardson, G., 1985, A review on human thermoregulation and its simulation. *IAM Report No. R 443* (Farnborough: RAF IAM).

Royal College of Physicians of London, 1966, *Report of the Committee on Accidental Hypothermia*.

Scholander, P.F., Hammel, H.T., Andersen, K.L. and Loyning, Y., 1958, Metabolic acclimation to cold in man. *J. Appl. Physiol.*, **12**, 1−8.

Skreslet, S. and Aarefjord, F., 1968, Acclimatisation to cold in man induced by frequent scuba diving in cold water. *J. Appl. Physiol.*, **24**, 177−181.

Sloan, R.E.G. and Keatinge, W.R., 1973, Cooling rates of young people swimming in cold water. *J. Appl. Physiol.*, **15**, 371−375.

Sowood, P.J., Cohen, J.B. and Hayes, P.A., 1986, Unpublished data.

Stitt, J.T., Hardy, J.D. and Stolwijk, J.A.J., 1974, PGE fever: its effect on thermoregulation at different low ambient temperatures. *Amer. J. Physiol.*, **227**, 622–629.

Strong, L.H., Gee, G.K. and Goldman, R.F., 1985, Metabolic and vasomotor insulative responses occurring on immersion in cold water. *J. Appl. Physiol.*, **58**, 964–977.

Van Someren, R.N.M., Coleshaw, S.R.K., Mincer, P.J. and Keatinge, W.R., 1982, Restoration of thermoregulatory response to body cooling by cooling hands and feet. *J. Appl. Physiol.: Respirat. Environ. Exercise Physiol.*, **58**, 1228–1233.

Vaughan, W.S., 1975, Diver temperature and performance changes during long-duration, cold water exposure. *Undersea Biomed. Res.*, **2**, 75–88.

Volz, P.A. and Ostrove, S.M., 1984, Evaluation of a portable ultrasonoscope in assessing the body composition of college-age women. *Med. Sci. Sports Exer.*, **16**, 97–102.

Webb, P., 1980, *Physiological Design Goals for Thermal Protection of Divers.* Prepared for NMRDC by Webb Associates under contract N0094-80-C-0193.

Werner, J., 1980, The concept of regulation for human body temperature. *J. Therm. Biol.*, **5**, 75–82.

Wissler, E.H., 1984, Mathematical simulation of human thermal behaviour using whole body models. In *Heat Transfer in Medicine and Biology*, edited by A. Stitzer and R.C. Eberhart (New York: Plenum), Chapter 4.6.

Wyndham, C.H., Williams, C.G. and Loots, 1968, Reactions to cold. *J. Appl. Physiol.*, **24**, 282–287.

# 13. Constant Wear Thermal Protection Garments For Helicopter Personnel

## Igor B. Mekjavic and Patrick J. Sullivan

## 1. Introduction

Helicopters maintain an instrumental role in the expanding offshore industry. They are used for shuttling materials, supplies and personnel between the mainland and offshore installations, and are also used for rescue operations. In the North Atlantic, helicopter crews and passengers are faced with the hazard of drowning and hypothermia in the event of an accidental ditching (Allan, 1983). For this reason, it has become imperative to supply all personnel on shuttle flights with thermal protection garments offering adequate buoyancy. A recent review of the Canadian Forces aircrew experience in the past 20 years has revealed that in 92% of all cases involving ditching of aircraft in water, the crew had less than 1 minute to respond and in 78% of the cases, less than 15 seconds (Brooks and Rowe, 1984). Such evidence precludes the use of well–insulated garments (insulated buoyant immersion suits, commonly known as survival suits), which are used on board ships and oil rigs and donned only in cases of emergency. On the contrary, it recommends the utilization of garments which are worn continuously throughout the flight. However, in view of mounting evidence (reviewed in Table 13.1) demonstrating the hazardously high levels of air temperature that may be anticipated in aircraft cockpits as a result of solar radiation (the 'greenhouse' effect) and inadequate air conditioning or ventilation, it becomes apparent that pilots' performance may be hindered by the thermal stress imposed upon them, in particular if they are required to wear well–insulated suits in such conditions. Sem-Jacobsen (1971) emphasizes that most aircraft accidents are a result of material or human breakdown, and that 'pilot error' is most often due to physical or mental overstressing of the pilots, and not due to reckless performance.

Ideally, helicopter crew suits should offer adequate insulation and buoyancy in the event of accidental cold water immersion, but should not induce heat strain under normal operations. This suggests that garments need to allow proper ventilation of the suit microenvironment, thus enabling adequate sensible and insensible heat loss. In contrast to thermal protective garments for other offshore occupations, which have a significant buoyancy feature, the buoyancy factor for helicopter crew suits is very critical. Helicopter crews should be able to overcome the inherent buoyancy of the suit in the event that they have to escape from the flooded cabin of a ditched helicopter, thus excessive buoyancy may be detrimental, as it may trap the wearer in the cockpit or cabin.

The development of new fabrics has aided designers in attaining optimal insulation/ventilation ratios, by incorporating a variety of wet-suit (water enters the microenvironment) and dry-suit (a layer of air is trapped in the microenvironment) designs. There now remains a lack of agreement with regards to the preferred design of such constant wear thermal protective garments. Recently, Brooks (1986) suggested that wet-suit designs are not acceptable for North Atlantic operations, and that garments for aircraft personnel need to be totally waterproof. In contrast, Allan *et al.* (1986) argue that there are certain advantages to the wet-suits worn externally over working clothing. For shuttle purposes, it may not be necessary to equip all crew members and passengers with waterproof suits, especially since in most cases there are numerous aircraft and boats available for immediate rescue, and the total time of immersion may be brief. Indeed, Brooks and Rowe (1984) reported that in all cases investigated over a 20-year period, survivors were rescued in under three hours and in 16 of the 34 cases in under 15 minutes.

Despite the increasing focus on the problem of optimal attire for aircrew operating in the North Atlantic and their passengers, the issue of acceptability of wet suits and suitability of dry suits remains unresolved, partly due to the lack of experimental evidence comparing the two types of garment designs under simulated survival and normal operational situations. This chapter reviews several investigations conducted in our laboratory evaluating the performance characteristics of several commercially available helicopter crew suits designed on the wet-suit and dry-suit principles, in an attempt to address these concerns.

## 2. Methods and results

The tests reviewed below were designed to evaluate the pathways of thermal exchange both under simulated operational and accident conditions. Evaluations of the garments were conducted using human subjects and were approved by the Simon Fraser University Ethics Review Committee.

*Table 13.1.*   Review of recent studies evaluating environments in different aircraft (Gaul and Mekjavic, 1987).

| Authors | Aircraft | Location | Time of year | Conditions | Ambient temperature | Cockpit conditions | Subjects' thermal status |
|---|---|---|---|---|---|---|---|
| Bollinger & Carwell (1975) | RF-4C | S. Carolina, USA | Summer Winter | Low-level flights Low level flights | 25·9-31·6°C 13·3-18·8°C | 30-51°C 25-45°C | N/A |
| Brown & Williams (1982) | N/A Lab. work | England | N/A | Lab. investigation: head cooling and thermal comfort | N/A | Lab. temp., 40°C | $T_{ac} = 36·5-38·3°C$[a] $T_{oc} = 36·6-38·1°C$[b] |
| Epstein et al. (1980) | N/A Lab. work | Israel | N/A | Psychomotor deterioration during exposure to heat | N/A | Lab. temps, 24°C 37°C 50°C | $T_{rc} = 37·0-38·5°C$[c] |
| Gibson et al. (1980) | Flight simulator | England | N/A | In liquid conditioned suits | N/A | N/A | $T_{ac} = 37·9-38·5°C$ |
| Gribetz et al. (1980) | Snow Pony Bell 47 Bell 206 | Israel | Sep. | Measurements pre- and post-flight | 12-22°C | 22-36°C | $T_{rc} = 37·5°C$ |
| Harrison & Higenbottam (1977) | Buccaneer | England | Summer | Ground standby | 30-38°C | 10-20°C higher than ambient | N/A |
| Harrison et al. (1978) | Harrier Buccaneer Gazelle Scout | FR Germany | Summer | Ground operations | N/A | 35-40°C | N/A |
| Higenbottam et al. (1977) | Wessex 5 | N. Europe | Feb. | In flight 0·5-2 h | −3-+5°C | 5-20°C | N/A |
| Livingstone et al. (1977) | Musketeer Kiowa | Manitoba, Canada | Summer | 1. Closed aircraft (plane and helicopter) | 25·5°C (78°F) | 35-39°C (95-102°F) | $T_{rc} = 37·6-37·8°C$ |
| | | | | 2. Closed with subject | 26°C (79°F) | 36-42°C (97-107°F) | $T_{rc} = 37·2°C$ |
| | | | | 3. Helicopter doors off with subject | 26°C (79°F) | 39·5°C (103°F) | $T_{rc} = 37·3-37·6°C$ |

| Reference | Aircraft | Location | Month | Condition | | | Temperature |
|---|---|---|---|---|---|---|---|
| | | | | 4. Flight in helicopter with and without doors | 26–29°C (79–84°F) | 21–24°C (70–75°F) | $T_{rc}$ = 36·3–36·6°C |
| | | | | 5. Flight in plane | 28°C (82·5°F) | 25–34°C (77–93°F) | $T_{rc}$ = 37·9–38·1°C |
| Nunneley et al. (1978) | Simulated condition of F-4 | Texas, USA | N/A | Lab. investigation of heat stress effects on performance | N/A | Lab. conditions, 35°C | $T_{rc}$ = 37·0–37·5°C; $T_{sk}$ = 32–36°C[d] |
| Nunneley & James (1977) | F-111A | S.W. USA | Summer | Low altitude, high speed | 35–40°C | 12–33°C | $T_{sk}$ = 29–39°C |
| Nunneley & Maldonado (1983) | N/A | Texas, USA | N/A | Lab. investigation of head and torso cooling | N/A | Lab. conditions, 35°C | $T_{rc}$ = 37·0–37·8°C; $T_{sk}$ = 25–39°C |
| Nunneley & Myhre (1976) | F-15 | California, USA | Sep. | On ground: in sun / in shade | 30–34°C | 34·7–51·9°C | $T_{rc}$ = 36·9–37·2°C; $T_{sk}$ = 33–35·8°C |
| Nunneley et al. (1982) | N/A | Texas, USA | N/A | Lab. investigation of head cooling equipment | 30°C | Suit conditions, 43°C, 30°C | $T_{oe}$ = 37·3–38·6°C; $T_{sk}$ = 32·7–38·9°C |
| Nunneley & Flick (1981) | A-10 | Arizona, USA | Jul. | Low-level flight | 26–42°C | 4–8° higher than ambient | $T_{ac}$ = 37·0–37·8°C; $T_{rc}$ |
| Nunneley & Stribley (1979) | HPA[e] | S. USA[f] | Summer | Ground operations, low-level and high altitude flights | N/A | 32–38°C | N/A |
| Nunneley et al. (1981) | F-4E | N. Florida, USA | Apr.–Jun. | Ground operation, low-level flight | 19–33°C | 20–50°C | $T_{ac}$ = 36·2–37·5°C |
| Stribley & Nunneley (1978) | F-4 A-10 F-111A | Texas, USA | Summer | Ground operations, take off and in-flight | N/A | 12–45°C | N/A |

[a] $T_{ac}$ = Core body temperature using an auditory canal probe.
[b] $T_{oe}$ = Oesophageal temperature.
[c] $T_{rc}$ = Core body temperature using a rectal probe.
[d] $T_{sk}$ = Skin temperature.
[e] HPA = High performance aircraft.
[f] Data from 30 active bases of USAF Air Training Command.

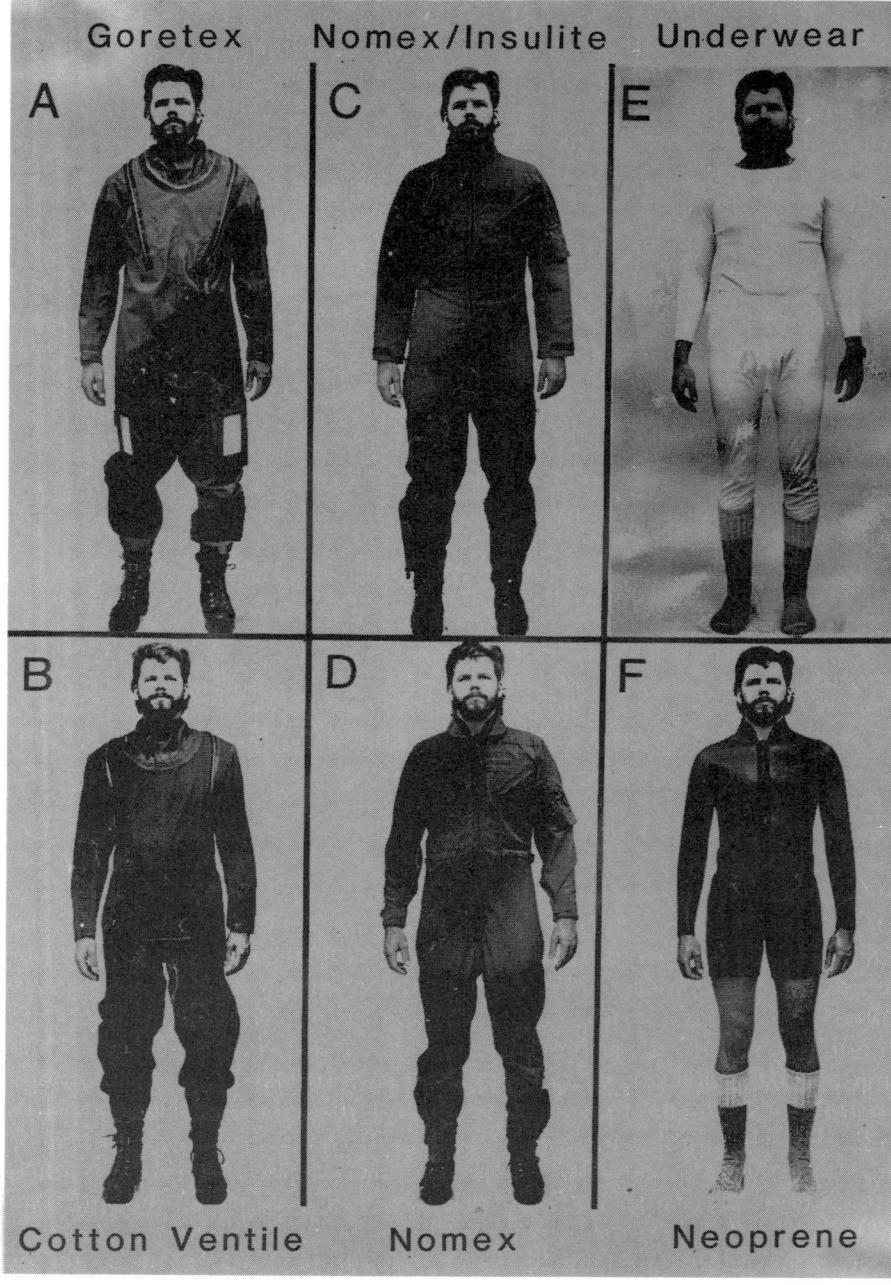

*Figure 13.1.    Constant wear thermal protection garments evaluated in the present study. Suits A and B are dry-suits, thus preventing the ingress of water during water immersion. Suits C and D are defined as wet-suits as they do not prevent water entering the suit microenvironment.*

## Suits

The suits investigated, shown in Figure 13.1, are currently being used by helicopter crews in many countries and represent the dry-suit (suits A and B) and wet-suit (suits C and D) principles. Although suits A and B had identical designs, the former was constructed of a polytetrafluoroethylene film (PTFE), manufactured by Goretex, which when stretched and annealed at high temperatures becomes permeable (Lomax, 1985), and the latter of high density woven long fibre Egyptian cotton, known as Cotton Ventile. Goretex material reportedly has a high water vapour permeability as a result of its 9 billion pores/in$^2$ (about 1·4 billion pores/cm$^2$), and is impermeable to liquid water. Cotton Ventile material allows water vapour transmission through the interfibre spacing, but once the fibres become wetted, they expand and reduce the interfibre space, preventing liquid water penetration. Both suits A and B incorporated thin rubber neck and wrist seals and a 'yoke' style waterproof zipper. Thin rubber boots were also attached to the trouser leg. Suit C incorporated a Nomex outer shell with an Insulite lining (tradename Mac 10), while suit D was a coverall constructed of Nomex material. All suits were manufactured by Mustang Industries Inc. (British Columbia, Canada).

During the performance evaluation of these suits under simulated environmental conditions, subjects wore long cotton underwear and cotton socks under suits A, B and C (assembly E in Figure 13.1). Suit D was worn in combination with a short neoprene wet-suit (F). Subjects wore heavy boots during both series of exposures, and also wore leather gloves during the cold water immersion. Subjects were fitted with the appropriate suits, according to the measurements suggested by the manufacturer (Mustang Industries Inc.).

## Ventilation index

Thermal comfort of personnel wearing constant wear thermal garments in warm conditions is determined in part by the permeability and insulative characteristics of the fabric. Increased barriers to conductive, convective and evaporative heat loss, as posed by increased insulation and decreased water vapour permeability, will enhance the stress of the warm environment on the pilot. Adequate airflows through the garment are necessary, to ensure proper heat exchange. Birnbaum and Crockford (1978; see also the review by Crockford in this volume) have proposed a unique method for evaluating the ability of air exchange between the external environment and the garment microenvironment, and have termed this the Ventilation Index of the garment. The method entails the determination of the volume of air trapped in the microenvironment of the garment ($V_\mu$, litres) and the rate at which air can be

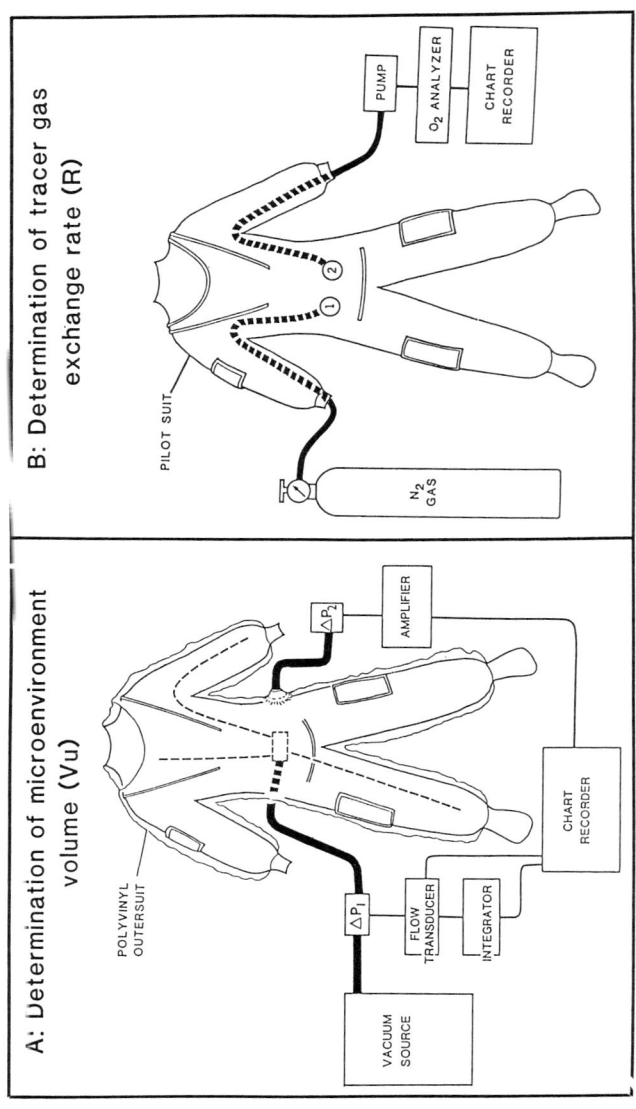

*Figure 13.2.  Experimental arrangement for determining the ventilation index of garments. (A): Microenvironment volume. Air is evacuated from the internal environment of a polyvinyl suit surrounding the subject, wearing the test garment. The volume of air evacuated from the internal environment, as the differential pressure between the ambient and internal environment decreased from −9mmH$_2$O to −300mmH$_2$O, was termed the environmental volume (Ve). The procedure was repeated with subjects clad only in shorts. The environmental volume determined from the shorts only procedure was subtracted from the volume obtained with the garment, and the difference defined as the microenvironment volume (Sullivan et al., 1987a). (B) Rate of air exchange. The microenvironment is flushed with pure nitrogen (N$_2$) through a series of perforated tubes running the length of each limb and the front and back of the torso. Simultaneously, through a similar network of tubes positioned under the test garment, the air is extracted and passed through an oxygen analyser. The rate of air exchange (R, min$^{-1}$) is derived, by determining the rate of rise of the O$_2$ fraction in the microenvironment, once the N$_2$ flushing is terminated (Sullivan et al., 1987b).*

exchanged between the external environment and garment microenvironment ($R$, min$^{-1}$). The rate of air exchange will include the exchange occurring through the fabric and garment vents and openings. The fabric characteristics alone are therefore not necessarily adequate indicators of the ability of air exchange between the suit microenvironment and the external environment.

There are some inherent problems with the original method proposed by Crockford and his co-workers, which have been addressed by Lotens and Havenith, in this volume, and Sullivan *et al.* (1987a). Nevertheless, continuous improvement of the original concept may eventually develop an international standard for the determination of the Ventilation Index of garments.

The method used to obtain the Ventilation Index of the helicopter crew suits in the present study is based on the original tracer gas method, but incorporates some changes in the determination of $V\mu$ (Sullivan *et al.*, 1987b). The diffusion of a tracer gas through the suit material and vents was monitored as illustrated in Figure 13.2b, which enabled the quantification of the rate of air exchange, indicating the air permeability of the suit. The air exchange rate ($R$) was adjusted for the volume of air trapped in the suit microenvironment ($V_\mu$), to obtain the Ventilation Index ($Q = V_\mu \times R$ l min$^{-1}$) of a particular garment assembly. The microenvironment volume ($V_\mu$) was determined by encapsulating the subject wearing the test garment with a polyvinyl outer suit (Figure 13.2a). The air within the polyvinyl suit was evacuated with a vacuum source until the pressure inside the suit fell to 300 mm $H_2O$ below the ambient pressure. The volume of air evacuated from the time at which internal pressure became lower than the external pressure, to the time this differential pressure reached $-300$ mm $H_2O$, was termed the microenvironment volume. Sullivan *et al.* (1987a) adjusted this volume to account for the compliance of the outer suit by placing subjects in the polyvinyl suit in shorts and repeating the procedure. Under this condition $V_\mu = 0$, and any empirically determined value will be due to the compliance of the fabric. Thus, to obtain the true $V_\mu$, the volume obtained with shorts was subtracted from the volume obtained while wearing the suit assembly.

The results presented in Table 13.2 indicate the ability of the suits to exchange the microenvironment air with ambient air—the highest ventilation indices were observed for the Nomex and Cotton Ventile suits, with the Nomex/Insulite and Goretex suits demonstrating the least ventile characteristics.

## Fabric properties

Numerous tests have now been accepted as standards for the quantification of the physical characteristics of a given fabric. In situations

*Table 13.2.    Ventilation indices (Q, l min$^{-1}$) of four helicopter crew suits: A, Goretex; B, Cotton Ventile; C, Nomex/Insulite; and D, Nomex coverall.*

| Subject | A | B | C | D |
|---|---|---|---|---|
| 1 | 0·03 | 0·84 | 0·06 | 1·64 |
| 2 | 0·02 | 0·95 | 0·08 | 1·47 |
| 3 | 0·04 | 1·04 | 0·09 | 2·06 |
| 4 | 0·03 | 0·63 | 0·06 | 1·44 |
| 5 | 0·02 | 1·11 | 0·05 | 1·06 |

where a protective garment is being developed, it is essential to analyse the appropriate physical characteristics of the fabrics under consideration before actually constructing the garment (Umbach, this volume; Gilling *et al.*, 1972). In studies such as the present one, where an evaluation is being made of several garments and design concepts, the analysis of both the garments as a whole and fabrics alone enables conclusions to be drawn regarding the suitability of the fabric and design. For the purpose of the present investigation, therefore, the most important properties were that of air permeability, water vapour transmission and thermal resistance.

*Air permeability*

The assessment of ventilation index described above incorporates the rate of air exchanged through the garment fabric and via the vents and openings of the suit. The contribution of each pathway of air exchange may be quantified by also determining separately the air permeability of only the fabric.

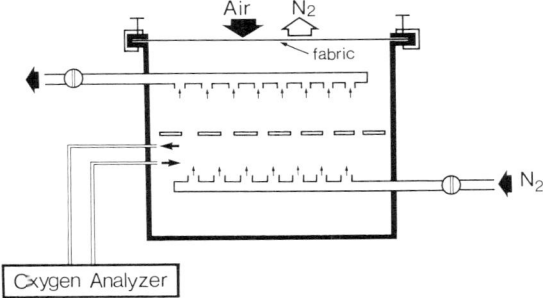

*Figure 13.3    The rate of air exchange was determined for the fabric alone utilzing the same principle as for assessing R for garments. The fabric was clamped to a plexiglass vessel (diameter = 128 mm, height = 72 mm) and the internal environment of the vessel flushed with pure $N_2$. The air within the vessel was continuously monitored for $O_2$ concentration. The rate of air exchange was determined from the termination of $N_2$ flushing, at which point the outlet value was also closed.*

A method was designed to allow the determination of fabric air permeability using oxygen as a tracer gas, as shown in Figure 13.3. A 6-inch (15 cm) diameter fabric sample was clamped on a plexiglass container, which was then flushed with pure $N_2$. The $O_2$ fraction ($FO_2$) within the environment of the container was continuously sampled, and the rate of rise of $FO_2$ in the container determined (Figure 13.4) once the $N_2$ flushing was terminated (both inlet and outlet valves in Figure 13.3 were closed). For the dry suits A and B, the ventilation index of the fabric, as determined by the method outlined above, was similar to the indices of ventilation observed for the entire suits (Table 13.2). Thus for dry suit designs, an analysis of only the fabric using the device illustrated in Figure 13.3 enables a very close approximation of Q (suit ventilation index).

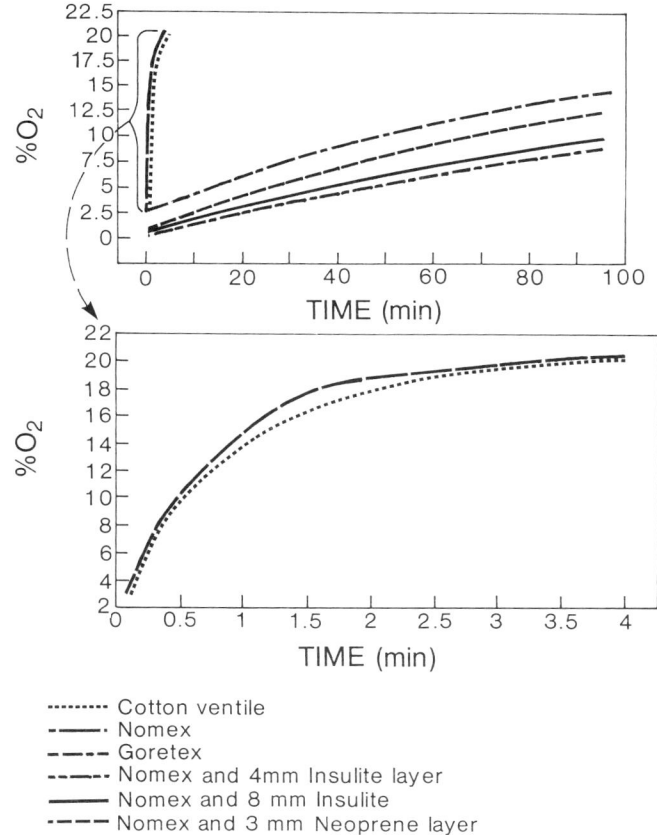

........ Cotton ventile
——· Nomex
———- Goretex
—·——·· Nomex and 4mm Insulite layer
———— Nomex and 8 mm Insulite
·———— Nomex and 3 mm Neoprene layer

*Figure 13.4.    The result of the air exchange test for the fabrics utilized in the construction of the helicopter crew suits. The time constant of the rate of rise of oxygen concentration in the plexiglass container depicted in Figure 13.3 was used as an index of air exchange.*

*Table 13.3   Thermal resistance of the fabric layers incorporated in suits A, B, C and D.*

| Suit | Fabric layers | Thermal resistance $(m^{-2} K^{-2} W^{-1})$ |
|------|---------------|---------------------------------------------|
| A | Goretex | 2·0 |
| B | Cotton Ventile | 0·96 |
| C | Nomex shell with liner: | |
| | ¼" Insulite | 19·2 |
| | ⅛" Insulite | 9·6 |
| D | Nomex shell with 3 mm neoprene wet suit | 7·3 |

*Insulation*

The thermal resistance of the fabrics incorporated in suits A, B, C and D was determined using a modified version of BS4745 (see Gilling *et al.*, 1972, for details). By monitoring the temperature across the fabric layers, the thermal resistance of the various fabrics were determined, as presented in Table 13.3.

## Clothing area factor

For the precise determination of thermal exchanges between the surface of the skin and the external environment, it is important to determine the effective surface area available for heat exchange. The ratio of the surface area of the clothed body to the surface area of the nude body is the clothing area factor ($f_{cl}$), and depends on garment design, fabric thickness and stiffness (McCullough and Jones, 1983). In the past, the increase in the surface area due to the clothing has been predicted from the insulative value (an indication of thickness) of the fabric (Fanger, 1972; Seppanen *et al.*, 1972). However, with high-tech fabrics and designs offering a range of $V_\mu$, this may lead to errors.

By assuming that a clothed individual may be represented by a cylindrical model (Figure 13.5), it is possible to predict the clothing

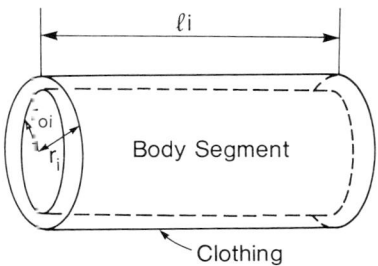

*Figure 13.5.   Cylindrical model of body with corresponding clothing component (Kakitsuba et al., 1987).*

surface area, knowing the volume of the body component and the volume of the air component between the clothing layer and the body component. Assuming that the surface area of the head is 7% of the body surface area, then 93% is covered by the protective garments and the ratio of the radiation area of the clothed body ($Ar_c$) to that of the unclothed body ($Ar_b$) is defined by the factor $f_{cl}$, and may be predicted as:

$$f_{cl} = [1 + 0.93\,(V_\mu/V_b)]^{0.5}$$

where $V_\mu$ = microenvironment volume (litres); and $V_b$ = body volume determined by the water displacement method (litres).

Using the photographic method suggested by Horikoshi and Kobayashi (1982) for determining radiation surface area ($Ar$), the radiation surface area was determined for subjects clad either in shorts or wearing the four different helicopter pilot suits. Subjects were suspended in a seated position in the centre of a $1.8 \times 0.72 \times 0.72$ m frame constructed of steel tubing. Photographs were taken with a 180° OP fish-eye camera lens from 90 equidistant points on the six sides of the frame, each point separated by 0.36 m. The correlation between the photographically determined values of $f_{cl}$ and those predicted by the equation is shown in Figure 13.6.

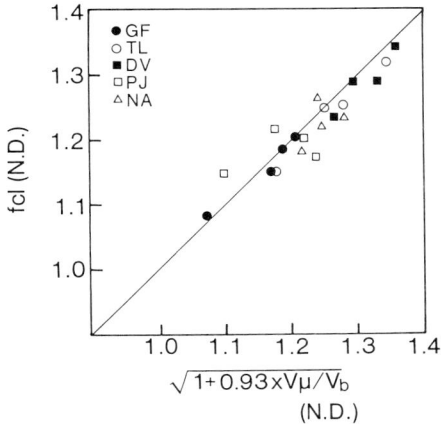

*Figure 13.6. Predicted values of $f_{cl}$ compared with values obtained photographically. Symbols represent five subjects (GF, TL, DV, PJ, NA) wearing the four crew suits depicted in Figure 13.1 (Kakitsuba et al., 1987).*

## Cold water immersion

The stress of cold water immersion on man may in extreme cases prove to be fatal. A number of extensive investigations into the aetiology of these fatalities have revealed several mechanisms, which may be responsible:

*The immediate onset of immersion.* The 'gasp' response at onset of immersion has been implicated as the major contributor to cold water immersion drownings (Keatinge and Evans, 1961; Martin *et al.*, 1978) in cases where the duration of the immersion was of such short duration that hypothermia was not likely to have ensued.

*Prolonged immersion in cold water.* Progressive cooling of internal body regions will lead to symptoms associated with the pathophysiological state of hypothermia (Mekjavic and Bligh, 1987) and ultimately death.

*The rewarming period.* Upon rescue or emergence from the cold water, the core regions of the body continue to cool. In addition, exiting the water will eliminate the hydrostatic pressure, which during immersion aids venous return and maintains the stability of mean arterial blood pressure. The combination of decreasing myocardial temperature and sudden elimination of the hydrostatic pressure during the post-immersion phase may instigate ventricular fibrillation (Golden, 1983).

The benefits of the helicopter crew suits were analysed in terms of the two phases of immersion. Subjects were immersed in 15°C water for four hours or until their rectal temperature (15 cm) decreased to 35°C. Subjects were aware that they could withdraw from the experiment at any time and that the immersion would be terminated at their request.

*Onset of immersion*

It has been established by several investigators that the sudden stimulation of peripheral cold receptors will result in several-fold increases in pulmonary ventilation (Keatinge and Evans, 1961; Martin and Cooper, 1978; Hayward and Eckerson, 1984). It appears that certain areas of the body surface may be more thermosensitive and thus contribute more to the overall respiratory response.

Using the method of mouth occlusion pressure developed by Milic–Emili *et al.* (1981), the respiratory drive was determined by monitoring the mouth pressure while subjects inspired against an occluded airway for 100 ms. The pressure developed at 100 ms of inspiration ($P_{0.1}$) is indicative of the central inspiratory activity and thus respiratory drive. As illustrated in Figure 13.7 the highest values of immersion $P_{0.1}$ were obtained when subjects wore only shorts. The sudden increase in respiratory drive during immersion was retarded for conditions when subjects wore the wet-suits (C and D in Figure 13.1) and almost eliminated when subjects wore the dry-suits (A and B in

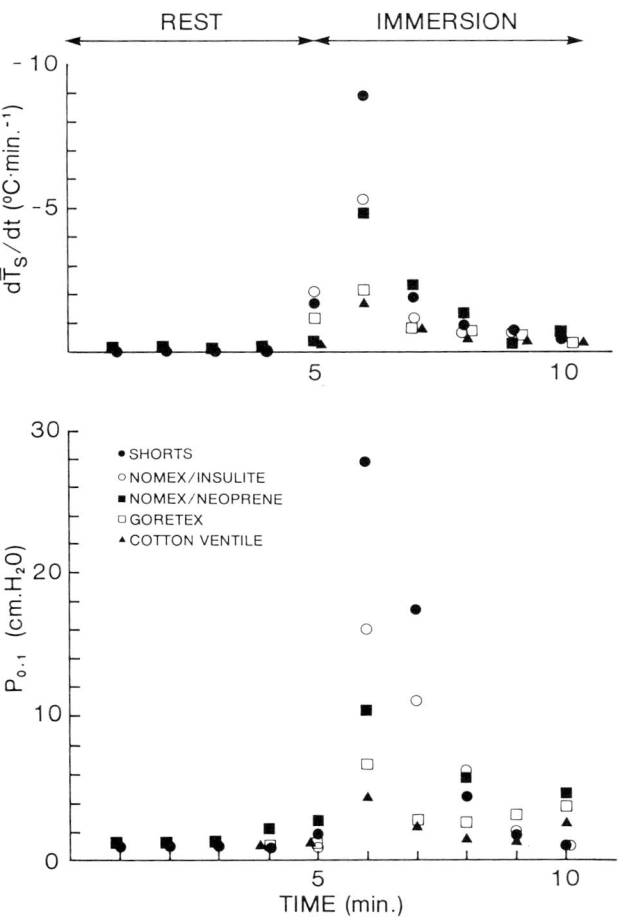

*Figure 13.7.* (a) *The rate of change of mean skin temperature* (dT_S/dt) *and* (b) *mouth occlusion pressure* (P_{0-1}) *for subject DC. The response is indicative of the responses of all subjects. The peak* dT_S/dt *coincides with the peak* P_{0-1} *observed at the onset of immersion (Mekjavic et al., 1987).*

Figure 13.1). In practical terms, this would suggest that subjects would be more likely to suffer from breathlessness and may ingest water due to the high ventilatory rate during the initial minute of immersion wearing the wet-suits. Decreasing the magnitude of the peripheral cold stimulus will consequently decrease the respiratory drive and the feeling of breathlessness. Such amelioration of the immersion respiratory response is observed for the conditions when subjects wore dry-suits. It is also interesting to note that the gain of the response is subject-dependent, indicating the different thermoresponsiveness of the subjects.

It may be worthwhile to consider insulating regions noted to contribute significantly to the immersion respiratory response, thus reducing
the risk of breathlessness and water ingestion, leading to drowning on
sudden cold water immersion.

### Prolonged immersion

As evident from Figure 13.8, the core cooling rates observed were
much greater for the dry suits (A, Goretex and B, Cotton Ventile) than
for the wet suits (C, Nomex/Insulite and D, Nomex with a short
neoprene suit). Three of the five subjects were able to maintain their
rectal temperature well above 35°C while immersed in 15°C water for
four hours while wearing the Nomex suit over a short neoprene suit.
The immersion times were similar for the Nomex/Insulite suit, but
were dramatically reduced for both dry-suits.

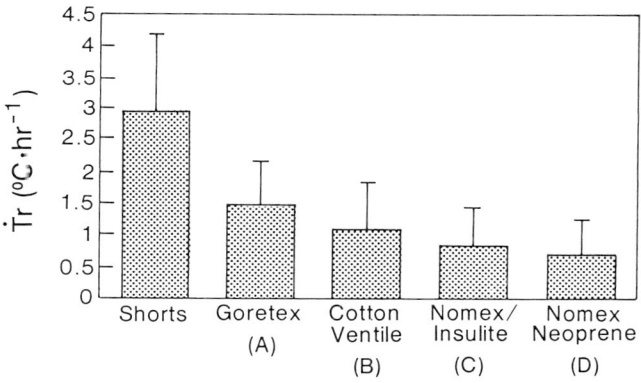

*Figure 13.8.   Rectal temperature cooling rates (mean±se) for five subjects immersed in 15°C
water wearing either only shorts or one of the four crew suits depicted in Figure 13.1, with the
appropriate combination of undergarments.*

The enhanced thermal protection offered by the wet-suit design
reflects the significantly greater thermal resistance offered by the wet-
suit fabric layers compared to the fabric layers of the dry-suit designs.
In addition, the microenvironment volumes of the wet-suits were less
than those observed for the dry-suits, indicating a better fit and a
smaller clothing surface area.

## Hot air environment

An attempt was made to evaluate the test garments under simulated
operational conditions. On a warm summer's day, the cockpit temperature may start at 20°C at the onset of a shuttle flight. As has been

reported earlier (Table 13.1), the temperature may then proceed to rise to alarmingly high levels during regular shuttle flights with an average of three hour duration (Gaul and Mekjavic, 1987). The laboratory tests, designed to simulate such flights, were conducted in an environmental chamber at Simon Fraser University. The ambient temperature was set at 20°C at the onset of the experiment. The temperature was increased linearly to 40°C over one hour and remained at that level for an additional two hours.

To standardize the mean radiant temperature, subjects sat on a chair situated in the centre of a $1.8 \times 0.72 \times 0.72$ m metal frame, elevated 0.5 m from the ground, with all six sides covered by a black cloth internally and a white cloth externally. During the three hours of the experiment, both physiological responses and psychomotor performance were monitored.

## Physiological responses

During the warm air exposures, rectal temperature (15 cm) and mean skin temperature were monitored continuously.

As shown in Figure 13.9(a), a significant elevation in rectal temperature was observed in trials when subjects wore suit C, incorporating a Nomex shell with an Insulite layer. The rectal temperature increase over the 3-hour hot air exposure was similar for the remaining conditions (shorts only, suits A: Goretex, and B: Cotton Ventile and Nomex/Neoprene). There was no significant difference in the increase in mean skin temperature between the different conditions (Figure 13.9(b)).

## Psychomotor performance

Subjects performed a series of three motor tests three times during the hot air exposure: immediately prior to the onset of the experiment ($T_a = 20°C$), at minute 90 ($T_a = 40°C$) and at 150 min ($T_a = 40°C$). Each experimental series required 30 minutes for completion and consisted of the following tests:

1. Simple reaction time. Subjects were required to press a key in response to a light stimulus.
2. Choice reaction time. Subjects were required to press the appropriate key (choice of four) in response to a light stimulus.
3. Tracking. Subjects controlled the direction of a cursor displayed on a video monitor, with two keys of a keyboard. The subjects were required to maintain the cursor within boundary lines displayed on the

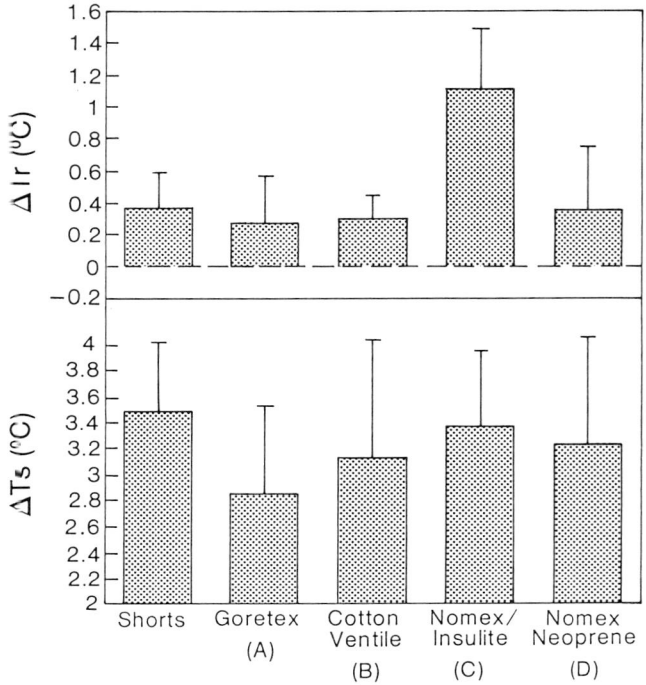

Figure 13.9.    The change in (a) rectal ($\Delta T_r$) and (b) mean skin ($\Delta T_s$) temperature (mean±se of five subjects) for subjects exposed to a hot air environment while wearing either shorts or one of the crew suits depicted in Figure 13.1, with the appropriate undergarments. Ambient temperature was elevated from 20°C to 40°C linearly in 90 minutes. Thereafter the temperature remained stable at 40°C. The total duration of the exposure was three hours.

video screen. The speed of the cursor was varied to a maximum of 132 mm s$^{-1}$.

Analysis of the results of the psychomotor performance tests revealed no significant difference between the results of the pre-exposure tests and the results of the tests conducted at minutes 90 and 150 of the exposure, for all types of garments (Sullivan et al., 1987c).

## 3. Discussion

The results of the various tests conducted on the four constant wear thermal protective garments—Goretex, Cotton Ventile, Nomex shell with insulite lining, and Nomex shell with short neoprene wet-suit—for helicopter personnel suggest that there is no clear preference of design. Contrary to the suggestions of Brooks (1986), the wet-suits offered adequate insulation in cold water and enabled unhindered performance in the hot air environment. However, the experiments demonstrated that the goodness of fit of the wet-suits is imperative for optimal performance under simulated accident conditions.

The Goretex and Cotton Ventile dry-suits indeed prevented the ingress of water into the microenvironment during the water immersion, but the fabric offered minimal thermal resistance. As the fit of the suit is not as important in the dry-suit concept as it is for the wet-suits, it may be expected that the trapped air volume within the microenvironment will act as added insulation. Unfortunately, as a result of the hydrostatic pressure acting upon the immersed regions, all air is displaced into the upper portion of the suits. Usually, the latex neck seal was overexpanded as a result of the increased air pressure in the microenvironment, and some escaped through the neck seal. Such entrapment of large quantities of air may be beneficial during cold water immersion, but may hinder escape from an inverted helicopter in water. Personnel in the cabin of such a helicopter may have to release the air from the microenvironment, via the wrist and neck seals, in order to be able to swim out of the cabin. As illustrated by Pasche and Gordon (1982) and emphasized by Brooks and Potter (1986), occupants of such a helicopter may have to overcome inherent buoyancy to escape from a flooded compartment. The figure of 137 N, recommended to the Canadian General Standards Board as the maximum inherent buoyancy for helicopter crew suits, does not take into account the buoyancy that may exist as a result of the trapped air. Perhaps a figure for a maximal microenvironment volume could be included in this recommendation.

Analogous to the varying degrees of cooling, which will exist with varying degrees of leakage of water into the suit microenvironment (Allan *et al.*, 1985; Steinman *et al.*, 1985; Steinman and Kubilis, 1987), different volumes of air trapped in dry-suits will offer different magnitudes of insulation. The magnitude of the volume of air trapped by the dry-suits was not controlled in the present study. On the contrary, subjects were allowed to exhaust the air via the neck latex seal, once immersed in the water, in order to simulate the worst possible situation. Further work is needed to investigate the benefits of trapped air in dry-suits, during cold water immersion.

Results of the hot air exposure study reveal that, with the exception of the Nomex/Insulite suit, subjects wearing suits A, B and C exhibited similar elevations in core and skin temperature as when clad only in shorts, in similar ambient conditions. Suit C created an almost impermeable water vapour barrier over approximately 90% of the body surface (assuming that the surface area of the uncovered head and hands amounts to 7% and 3% respectively), whereas suits A and B had some degree of permeability. Although the short neoprene suit worn under

suit D created a similar impermeable barrier on the torso, the unhindered evaporation offered from the extremities and the head prevented significant elevations in rectal and skin temperature.

The ability of the subjects to perform psychomotor tasks at the same
level of competence both at the onset and at the end of the 3-hour
exposure to a hot environment indicates that the effect of heat load on
performance was not noticeable. The magnitude of the heat load
imposed on the subjects under the simulated cockpit conditions has
been suggested to induce decrements in both reaction time and tracking. As the results reported in the present study do not indicate any
such decrement, it may suggest that the tests used were not sufficiently
sensitive to detect any changes in psychomotor performance.

A further difficulty arises when attempting to compare such results
of studies with subjects wearing clothing. Not without a certain
amount of criticism, the Effective Temperature (ET) of the ambient air
has been the most common index of heat stress. However, for clothed
subjects, the temperature and relative humidity of the microenvironment of the clothing may be a more accurate index of heat stress
imposed on the subject.

Similar problems arise when reporting the vapour permeability of
fabrics. Ross (1987) lists a total of 10 methods available for evaluating
the transmission of water vapour across a fabric layer. Methods range
from the simple determination of change in mass of a dish filled with
water and covered by a piece of the test fabric, to the more complex
monitoring of the amount of infra-red radiation absorbed by water
vapour, penetrating through the fabric. In addition, each method incorporates a specific combination of air temperatures and relative humidi-
ties on either side of the fabric. Since the driving force, namely the
$\Delta PH_2O$ across the fabric, differs among all methods, this gives rise to
results which are markedly different not only in terms of magnitude,
but also relative to one another (Keighley, 1985). Although many tests
exist for evaluating the vapour permeability, the results are not directly
comparable. Figure 13.10 illustrates the difference in the results of two
tests for water vapour transmission for a range of Goretex and Cotton
Ventile fabric samples. Both methods involve the determination of the
amount of water evaporated from a dish covered with a fabric test
sample. One method (hatched bars) utilizes water at 35°C, while the
second method (solid bars) uses 20°C water. Assuming that ambient air
temperature and relative humidity are identical in both tests, the evaluation using 35°C water in the dish will establish a greater $PH_2O$
gradient across the fabric, thus enhancing the transmission of water
vapour through the fabric.

It would also be preferable to use a method for evaluating water
vapour permeability that would be applicable on fabric samples and

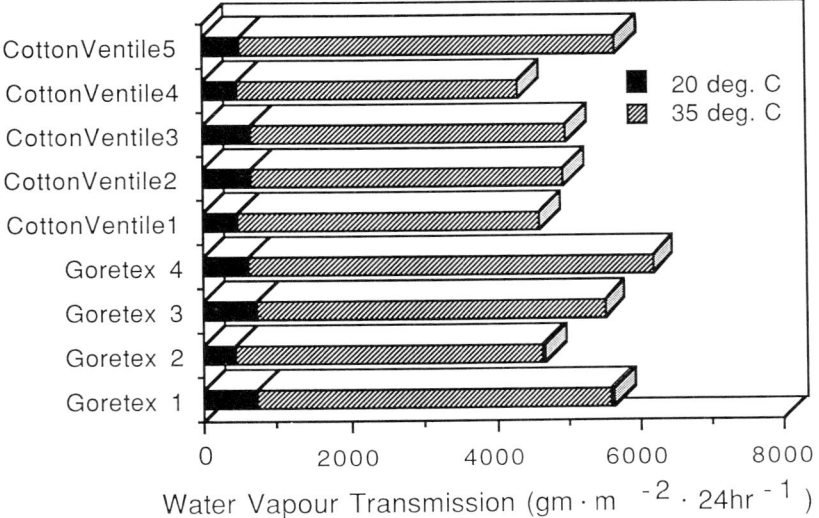

*Figure 13.10.   Evaluation of water vapour transmission of five Cotton Ventile samples (1: L28 Cerol; 2: L34 Cerol; 3: L34 Perlit; 4: L24 Perlit; 5: L19 Cerol) and four Goretex samples (1: 2 layer sample 1; 2: 3 layer 4 oz; 3: 2 layer sample 2; 4: 3 layer 124) by determining the loss of weight of a dish filled with water and covered with the test fabric. Solid bars indicate values obtained using a water temperature of 20°C, and the hatched bars indicate values obtained using a water temperature of 35°C (adapated from Keighley, 1985).*

that could also be used on garments being worn. This would enable the determination of the proportion of water vapour transmitted to the external environment through the garment openings, as opposed to transmitted directly through the garment fabric. Such an approach has been incorporated in the assessment of the rate of air exchange ($R$, $min^{-1}$) in this study. The same method was used to estimate $R$ for both the total garment assembly and a sample of the fabric. Unfortunately, the rate of air exchange cannot always be utilized as an indication of the water vapour permeability in modern multi-layer fabrics. In addition, the assessment of ventilation index is conducted in thermoneutral environments, with a dry garment. It is very likely that the air permeability and water vapour transmission characteristics of a garment may deteriorate as its water content increases, a result of either rain or profuse sweating. Of some concern to pilots surveyed (Gaul and Mekjavic, 1987) is also the elimination of the water impermeability trait of garments such as Goretex when exposed to hydrocarbons and salt solutions. Since helicopter pilots working in coastal regions are constantly exposed to salt spray and may often sustain a drop of oil or gasoline on their garments, their suits' potential efficacy as thermal protection during cold water immersion may progressively decrease.

## Acknowledgements

The authors are indebted to Dr Naoshi Kakitsuba for his help and contribution in many aspects of our studies. We would also like to thank C. Gaul, K. Mittleman, W. Burke, H. Michna and Dr D. Goodman, for without their help these studies would not have been completed with such success and expedience. P. J. Sullivan is recipient of the GREAT award from the Science Council of British Columbia.

This work was supported by a grant from the Science Council of British Columbia.

## References

Allan, J.R. 1983, Survival after helicopter ditching: A technique for policy-makers. *Int. J. Aviat.*, **1**, 191–196.

Allan, J.R., Higenbottam, C. and Redman, P.J., 1985, The effect of leakage on the insulation provided by immersion–protective clothing. *Aviat. Space Environ. Med.*, **56**, 1107–1109.

Allan, J.R., Elliott, D.H. and Hayes, P.A., 1986, The thermal performance of partial coverage and wet suits. *Aviat. Space Environ. Med.*, **57**, 1056–1060.

Birnbaum, R.R. and Crockford, G.W., 1978, Measurement of the clothing ventilation index. *Appl. Ergonomics*, **9**, 194–200.

Bollinger, R.R. and Carwell, G.R., 1975, Biomedical cost of low-land flight in a hot environment. *Aviat. Space Environ. Med.*, **46**, 1221–1226.

Brooks, C.J., 1986, Ship/rig personnel abandonment and helicopter crew/passenger immersion suits: The requirements in the North Atlantic. *Aviat. Space Environ. Med.*, **57**, 276–282.

Brooks, C.J. and Potter, P.L., 1986, The establishment of 137 N as the Canadian General Standards Board maximum acceptable inherent buoyancy limit for passenger helicopter immersion suits. *Ann. Physiol. Anthropol.*, **5**, 152.

Brooks, C.J. and Rowe, K.W., 1984, Water survival: 20 years Canadian Forces aircrew experience. *Aviat. Space Environ. Med.*, **55**, 41–51.

Brown, G.A. and Williams, G.M., 1982, The effect of head cooling on deep body temperature and thermal comfort in man. *Aviat. Space Environ. Med.*, **53**, 583–586.

Epstein, Y., Keren, G., Moisseiev, J., Gasko, D. and Yachin, S., 1980, Psychomotor deterioration during exposure to heat. *Aviat. Space Environ. Med.*, **51**, 607–610.

Fanger, P.O., 1972, *Thermal Comfort* (New York: McGraw-Hill).

Gaul, C.A. and Mekjavic, I.B., 1987, Helicopter pilot suits for offshore application: A survey of thermal comfort and ergonomic design. *Appl. Ergonomics*, **18.2**, 153–158.

Gibson, T.M., Allan, J.R., Lawson, C.J. and Green, R.G., 1980, Effect of induced cyclic changes of deep body temperature on performance in a flight simulator. *Aviat. Space Environ. Med.*, **51**, 356–360.

Gilling, D.R., Cooper, D.S. and Dickins, T.L., 1972, *The Transfer of Heat and Moisture Through Clothing Fabrics*, Army Personnel Research Establishment, Farnborough, U.K., Report No. 44/72(R).

Golden, F.S., 1983, Rewarming. In *The Nature and Treatment of Hypothermia*, edited by R.S. Pozos and L.E. Wittmers (Minneapolis, MN: University of Minnesota Press), pp. 194–208.

Gribetz, B., Richter, E.D., Krasna, M. and Gordon, M., 1980, Heat stress exposure of aerial spray pilots. *Aviat. Space Environ. Med.*, **51**, 56–60.

Harrison, M.H. and Higenbottam, C., 1977, Heat stress in an aircraft cockpit during ground standby. *Aviat. Space Environ. Med.*, **48**, 519–523.

Harrison, M.H., Higenbottam, C. and Rigby, R.A., 1978, Relationships between ambient, cockpit and pilot temperatures during routine air operations. *Aviat. Space Environ. Med.*, **49**, 5–13.

Hayward, J.S. and Eckerson, J.D., 1984, The physiological responses and survival time prediction for humans in ice water. *Aviat. Space Environ. Med.*, **55**, 206–212.

Higenbottam, C., Marcus, P. and Waddell, J., 1977, Thermal data from helicopters operating in a sub-arctic environment. *Aviat. Space Environ. Med.*, **48**, 640–644.

Horikoshi, T. and Kobayashi, Y., 1982, Configuration factors between the human body and rectangular planes and the effective radiation area of the human body. *Trans. Architect. Inst. Japan*, **22**, 92–100.

Kakitsuba, N., Michna, H. and Mekjavic, I.B., 1987, Clothing surface area as related to body volume and clothing microenvironment volume. *Aviat. Space Environ. Med.*, **58**, 411–416.

Keatinge, W.R. and Evans, M., 1961, The respiratory and cardiovascular response to immersion in cold water. *Quart. J. Physiol.*, **46**, 83–94.

Keighley, J.H., 1985, Breathable fabrics and comfort in clothing. *J. Coated Fabrics*, **15**, 89–104.

Livingstone, S.D., Bell, D.G. and Kuehn, L.A., 1977, *Measurement of Heat Stress: Kiowa and Muskateer Cockpits*, Defence and Civil Institute of Environmental Medicine, Technical Report.

Lomax, G.R., 1985, Coated fabrics: Part 1 – Lightweight breathable fabrics. *J. Coated Fabrics*, **15**, 115–126.

Martin, S. and Cooper, K.E., 1978, The relationship of deep and surface skin temperatures to the ventilatory responses elicited during cold water immersion. *Can. J. Physiol. Pharmacol.*, **56**, 999–1004.

Martin, S., Diewold, D.J. and Cooper, K.E., 1978, The effect of clothing on the initial ventilatory responses during cold water immersion. *Can. J. Physiol. Pharmacol.*, **58**, 886–888.

McCullough, E.A. and Jones, B.W., 1983, *Measuring and Estimating the Clothing Area Factor*, Institute of Environmental Research, Kansas State University, Technical Report No. 83–02.

Mekjavic, I.B. and Bligh, J., 1987, The pathophysiology of hypothermia. *Int. Rev. Ergonomics,* **1**, 201–218.

Mekjavic, I.B., La Prairie, A., Burke, W. and Lindborg, B., 1987, Respiratory drive during sudden cold water immersion. *Resp. Physiol.,* in press.

Milic-Emili, J., Whitelaw, W.A. and Grassino, A.E., 1981, Measurement and testing of respiratory drive. In *Regulation of Breathing,* Part II, edited by T.F. Hornbeim (New York: Marcel Dekker).

Nunneley, S.A. and Flick, C.F., 1981, Heat stress in the A-10 cockpit: Flights over desert. *Aviat. Space Environ. Med.,* **52**, 513–516.

Nunneley, S.A. and James, G.R., 1977, Crew skin temperatures measured in flight. *Aviat. Space Environ. Med.,* **48**, 44–47.

Nunneley, S.A. and Maldonado, R.J., 1983, Head and/or torso cooling during simulated cockpit heat stress. *Aviat. Space Environ. Med.,* **54**, 496–499.

Nunneley, S.A. and Myhre, L.G., 1976, Physiological effects of solar heat load in a fighter cockpit. *Aviat. Space Environ. Med.,* **47**, 969–973.

Nunneley, S.A. and Stribley, R.F., 1979, Fighter index of thermal stress (FITS): Guidance for hot-weather aircraft operations. *Aviat. Space Environ. Med.,* **50**, 639–642.

Nunneley, S.A., Dowd, P.J., Myhre, L.G. and Stribley, R.F., 1978, Physiological and psychological effects of heat stress simulating cockpit conditions. *Aviat. Space Environ. Med.,* **49**, 763–767.

Nunneley, S.A., Stribley, R.F. and Allan, J.R., 1981, Heat stress in front and rear cockpits of F-4 aircraft. *Aviat. Space Environ. Med.,* **52**, 287–290.

Nunneley, S.A., Reader, D.C. and Maldonado, R.J., 1982, Head temperature effects on physiology, comfort and performance during hyperthermia. *Aviat. Space Environ. Med.,* **53**, 623–628.

Påsche, A. and Gordon, S., 1982, *Test of Survival Suits for Offshore Helicopter Transportation,* Norwegian Underwater Technology Center, Report No. 43–82.

Ross, B.A., 1987, Breathable coatings—the how and why. *J. Coated Fabrics,* submitted.

Sen-Jacobsen, C.W., 1971, Physiological aspects of aircraft accident investigation. *Aerosp. Med.,* **42**, 199–204.

Seppanen, D., McNall, P.E., Munson, D.M. and Sprague, C.H., 1972, Thermal insulating values for typical indoor clothing ensembles. *ASHRAE Trans.,* **78**, 120–130.

Sprague, C.H. and Munson, D.M., 1974, A composite ensemble method for estimating thermal insulating values of clothing. *ASHRAE Trans.,* **80**, 120–129.

Stribley, R.F. and Nunneley, S.A., 1978, Physiological requirements for design of environmental control systems: Control of heat stress in high performance aircraft. American Society of Mechanical Engineers, Pamphlet 78–ENAS–22, 1–8.

Sullivan, P.J., Mekjavic, I.B. and Kakitsuba, N., 1987a, Determination of clothing microenvironment volume. *Ergonomics,* **30**(7), 1043–1052.

Sullivan, P.J., Mekjavic, I.B. and Kakitsuba, N., 1987b, Ventilation index of helicopter pilot suits. *Ergonomics,* **30**(7), 1053–1061.

Sullivan, P.J., Gaul, C.A., Goodman, D. and Mekjavic, I.B., 1987c, Heat stress and psychomotor performance in simulated helicopter cockpit conditions, in preparation.

Steinman, A.M. and Kubilis, D.S., 1987, *Survival at Sea: The Effects of Protective Clothing and Survivor Location on Core and Skin Temperature*, US Coast Guard Report, in preparation.

Steinman, A.M., Nemiroff, M.J., Hayward, J.S. and Kubilis, P.S., 1985, *A Comparison of the Protection against Hyperthermia Provided by Coast Guard Anti-exposure Clothing in Calm versus Rough Seas*, US Coast Guard Report No. CG–D–17–85.

# PART IV
# THERMOREGULATORY MODELLING

# 14. A Review of Human Thermal Models

**Eugene H. Wissler**

## 1. Introduction

Physiology is essentially an empirical science not heavily dependent on mathematical models for expressing quantitative relationships between quantities under various conditions. Since graphs and an occasional equation or two generally suffice to describe their research results, many physiologists deny that they use mathematical models, even though these simple relationships are, in fact, mathematical models. However, that term is usually reserved for more complex sets of equations developed to answer questions stemming from a need to understand human behaviour under particular circumstances. Such situations may involve survival, in which case it is impossible to study the entire process in the laboratory, or they may simply involve a sufficiently broad range of conditions that it is impractical to study all possible combinations experimentally, and interpolation and extrapolation become necessary to deal with conditions that have not been studied directly.

The use of such models is illustrated by the application to survival during accidental immersion in cold water. Of the many possible questions that arise in connection with this problem, very few can be answered without using a mathematical model. One question that could be answered directly is, 'does a particular anti-exposure system provide sufficient protection to maintain one's rectal temperature above a critical level (typically 36°C) for a specified period of immersion in water of given temperature?' If one equates the expected survival time to the time required to cool to some safe temperature, such as 36°C, very conservative estimates of survival time can be obtained without using a model. However, if the expected survival time is defined in terms of some other criterion involving extrapolation of experimental cooling data beyond the range of accessible values, then that procedure defines a mathematical model, although it may not be a very good one.

## 2. Applications of human physiological models

Here we shall mention several situations in which transport of heat and mass is important in man. Emphasis is placed on thermal response in one atmosphere air, although pressure and $PO_2$ are also important environmental factors, and a general model should include inert gas exchange and response to hypoxia. The rationale for employing a model and requirements placed on the model by each application will be discussed. In general, these requirements are stated in terms of: (*a*) the amount of detail required to describe adequately the temperature field within the body; (*b*) the complexity of thermoregulatory responses involved; and (*c*) the kind of garments and boundary conditions.

The most straightforward application of human thermal modelling is simulation of work in a hot environment with uniform environmental conditions, which occurs in many situations. High rates of perfusion throughout the body and the absence of highly nonuniform surface conditions allow one to obtain useful results with relatively simple models. Different work–rest cycles can be defined depending on specific circumstances. Various garments can also be used, ranging from light-weight summer clothing to heavy protective garments of the kind worn by firemen, foundry workers, and military personnel. In the case of prolonged exercise which causes heavy garments to become wet with sweat, their properties may change appreciably during the exposure, thereby complicating matters significantly.

A somewhat more demanding application is the simulation of aircrew performance in fighter aircraft. Modern high performance aircraft require such a high level of proficiency from aircrew members that even moderate decrements owing to physiological stress can have serious consequences. Pilots may be required to wear an anti-gravity suit, a positive-pressure bladder and an anti-exposure garment for flights over cold water or a CBR overgarment when there is danger of exposure to chemical or biological agents. Since all of these items impair thermoregulation during exposure to heat, it can be anticipated that personal cooling in the form of air or liquid cooled vests will be provided for aircrew members within a few years.

Although this is a rather complex system to model, previous efforts to simulate pilot performance in high performance aircraft have been reasonably successful, probably because the level of exercise and intensity of heat stress are not great, and large temperature gradients are not produced within the body. Even though surface conditions may vary markedly owing to nonuniform solar load and airflow and asymmetrical boundary conditions imposed by the contour seat, anti-gravity suit and positive-pressure bladder, enhanced skin blood flow tends to reduce the effect of nonuniform boundary conditions. Therefore, it is possible to obtain a reasonable description of gross response by defining mean values of garment thermal resistance and vapour transport properties over relatively large areas of body surface and by specifying mean environmental temperatures and radiant loads. However, that procedure introduces an element of uncertainty into the analysis that could be avoided by using a more detailed model. When an air or liquid cooled vest is used, additional questions arise about the effect of local cooling on thermoregulatory responses, which for the most part have been studied only when relatively uniform conditions exist over the surface of the body.

Accidental immersion in cold water was mentioned above as an important application of physiological models. These applications generally involve a prediction of survival time under various conditions. Because large internal temperature gradients are generated during cold exposure, this application is especially challenging, even when boundary conditions are relatively uniform, as illustrated in Figure 14.1. If, in addition, boundary conditions are highly nonuniform, as illustrated in Figure 14.2, where an individual is floating only partially immersed, significant questions arise about the validity of computed results. Van Someren *et al.* (1983) have shown that thermoregulatory

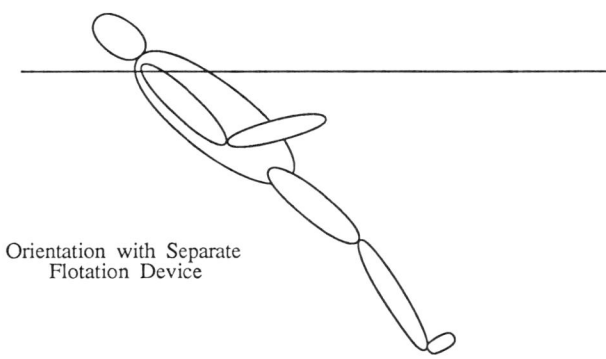

Orientation with Separate
Flotation Device

*Figure 14.1.   Representation of an individual supported by a life preserver that only maintains the head above water.*

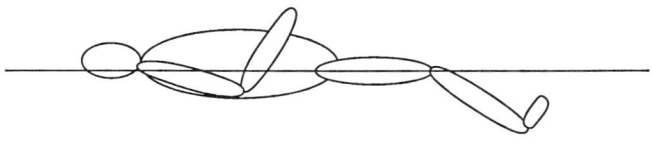

Orientation with Integral Flotation

*Figure 14.2.   Representation of an individual who is wearing an anti-exposure garment with integral flotation.*

responses to cold are strongly influenced by hand and foot temperatures, which calls attention to the fact that little is known about the integrated response to highly nonuniform skin temperatures. This is reinforced by observations reported from several laboratories (e.g., see Påsche and Ilmarinen, 1986) that rectal temperature can actually increase after several hours of partial immersion in cold water, even when the subject eventually becomes severely hypothermic. It appears that the special characteristics a model must possess to deal adequately with this problem are sufficiently detailed in the computed temperature field, as are the control equations for vasoconstriction and shivering which make adequate allowance for the distinctly nonuniform skin temperature generated in this case.

A different application is provided by hypothermia, which is sometimes employed during open-heart surgery to correct congenital defects in infants. Induction of hypothermia can be accomplished either by surface or extracorporeal cooling. In either case, the objective is to maintain cerebral temperatures at a level low enough to avoid tissue damage during cardiac arrest. It is possible that significant differences exist in the degree to which that objective is accomplished by the two methods of cooling. Mathematical modelling of heat transfer within the head offers the possibility of clarifying the principal factors involved in these procedures. Since the structure of the head is quite nonuniform, any model used for this purpose must provide good spatial resolution of the temperature field. It is also necessary to know how the metabolic and blood perfusion rates vary with position and time.

Modelling of thermoregulation in severely burned individuals is also an important medical application in which several aspects of thermoregulation are significantly altered, including the metabolic and perfusion rates, evaporative heat loss and $CO_2$ loss. Even though there is considerable variability between patients with nominally similar burns, the general features of physiological changes accompanying severe burns have been identified and quantified to a limited extent. Since this is a complicated process involving several related systems, a mathematical model should be of considerable value in developing an internally consistent description of thermoregulation in burn patients. One possible application of such a model would be to define a range of acceptable environmental conditions for burn wards and operating rooms used for skin grafting burn patients.

## 3. Brief description of representative human thermal models

A number of models have been developed during the past 40 years to describe human thermal response to various conditions. Although the scientists and engineers who developed those models employed a variety of approaches, all of them can be characterized in terms of the following factors: (*a*) the amount of detail provided about the temperature field within the body; (*b*) the way thermoregulatory responses are handled; and (*c*) the treatment of garments and boundary conditions. The way in which those factors are handled in each model determines its capability for simulating various conditions within the rather broad range of possible human experiences.

Some models attempt to provide a fairly detailed description of the temperature field within the body, while others define only central and mean skin temperatures. As noted previously, the latter approach is more likely to produce acceptable results under conditions of heat stress when temperature is relatively uniform throughout the body. On the other hand, it cannot be expected to deal adequately with exposure to cold when large temperature gradients develop within the body, which renders inadequate a two- or three-temperature description of the entire field. In addition, temperature fields that are not axially symmetric, as assumed by most models, can be generated either by non-uniform physical properties and metabolic rates within an element, or by spatially varying environmental conditions imposed over the surface. When such gradients are large, the level of detail provided about the temperature field becomes an important characteristic of models.

The amount of detail incorporated into a model is generally limited by the fortitude of the modeller and the speed and amount of memory provided by the computer on which the model is run. Basic decisions that must be made early in the development of a new model are: What coordinate system will be used? Will symmetry be assumed? and What level of detail (how many temperatures) will be computed? The answers to these questions are important, as noted above, but once the decisions have been made, constructing a numerical solution for the heat conduction equation is a fairly straightforward process, which need not concern us greatly as long as it is done correctly.

Developing an adequate description of relevant physiological phenomena is much more difficult than computing temperatures. Three processes are involved in thermoregulation: cardiovascular responses, sweating and shivering. The first is by far the least well understood, in part because local perfusion rates are almost impossible to measure, while rates of sweat production and shivering are relatively easy to measure. In addition, the cardiovascular system transports metabolic reactants and products, as well as heat, with the consequence that cardiovascular responses to thermal stress during exercise differ from those during rest. A general limitation that applies to all three phenomena is that, except for a relatively small number of cases, they have been studied only under steady-state conditions with relatively uniform surface conditions, and, therefore, our understanding of transient responses and the effect of regional variations in skin temperature are rather limited.

Cardiovascular responses are influenced by the requirement for oxygen delivery to metabolically active tissue, as well as by thermoregulatory requirements. Since the capacity of the circulatory system is influenced by fitness, posture, and central blood volume, cardiovascular responses to thermal stress are rather complex and imperfectly understood at the present time. It is not surprising that they have been incorporated into various mathematical models in different ways, and therefore this forms an important distinction between models. Close coupling between metabolic and thermal requirements suggests that it would be advantageous to incorporate material balances for oxygen, carbon dioxide and lactate into models, but that adds a new dimension to the problem and is not done very often.

Sweating and shivering are essentially pure thermal responses whose intensities are determined by the magnitude of afferent signals generated by central and peripheral thermoreceptors. Those receptors are located principally in the hypothalamus and skin, although there is evidence to suggest that receptors may also be located at other deep sites. The general features of sweating and shivering are reasonably well established for steady-state, 'normal' conditions, which include fairly uniform skin temperature during exposure to heat or immersion in water, and a physiological variation of skin temperature during exposure to cold air. It is well known, for example, that elevating the central temperature by 1°C produces an increase in sweat rate $10-20$ times larger than the increase owing to the same elevation of mean skin temperature, although the central drive is also modulated locally depending on the local skin temperature.

Shivering, on the other hand, is driven largely by reduced skin temperature, with the hypothalamic temperature playing a secondary role. The dominant nature of skin temperature is illustrated by the fact that shivering can be suppressed completely by warming the skin, even when the central temperature is below 36°C. It seems to be quite clear that rapidly decreasing skin temperature also stimulates shivering; for example, immersion in cold water or exposure to cold air is generally accompanied by a burst of shivering that peaks 2 or 3 minutes after the onset of rapid cooling and lasts for approximately 5 minutes. Sufficiently intense central cooling produced, for example, by breathing cold hyperbaric gas is also capable of stimulating shivering, presumably in response to a rapidly decreasing hypothalamic temperature, even when skin temperature remains in the comfort zone.

Clothing plays a very important role in determining human thermal response and must be adequately described by a useful mathematical model. The thermal properties of garments are characterized primarily by two factors; one is the resistance to sensible heat transfer and the other is resistance to water transport through the garment. It has been customary to employ a heated manikin to measure those factors, which are defined in terms of the clo value for sensible heat transfer and the ratio $i_m$/clo for vapour transport. While the properties of clothing are deceptively easy to measure in that way, there is reason to suspect that information obtained from such routine measurements may not adequately characterize the garment. A few manikins provide segmental values for the two resistances, although it is more common to measure only mean values for the entire garment. Although that may be acceptable in certain cases, there are others in which regional variations may be of considerable importance. For example, a dry, anti-immersion suit with built-in flotation clearly provides less thermal insulation on posterior surfaces than on anterior surfaces, and a mean value may not provide a good characterization of the garment. Another limitation of most manikins is that they do not introduce enough water into the garment to simulate profuse sweating over a prolonged period of time. In the author's opinion, more attention needs to be given to characterizing the thermal properties of garments. It is necessary to define measurement procedures that provide the data necessary to characterize a given garment, and to standardize those procedures so that results obtained in different laboratories will be reasonably consistent with each other.

To summarize, mathematical models currently available for analysing human responses to environmental stress and exercise are likely to be limited by one or more of the following factors:

1. Insufficient spatial resolution to describe important features of the temperature field within the body.

2. Inadequate description of one or more aspects of thermal physiology.

3. Inadequate description of the clothing worn by the subject.

In many cases, these factors are interrelated. For example, when dealing with a subject floating partly immersed in cold water, it is probably necessary to use a model that allows circumferential variation in temperature if heat storage and local skin temperatures are to be computed with good accuracy. Successful computation of temperatures also requires that thermal properties of the garment be known on various regions of the body. However, even when a very detailed temperature model is used with well-defined garment properties, good results are not guaranteed, because not enough is currently known about thermoregulation when the skin temperature is very nonuniform, as it is likely to be in this case.

# 4. Current mathematical models

Since many of the human thermal models developed throughout the years were only used by the original developer to analyse a few situations and are no longer in current use, they have not had much influence on subsequently developed models. A representative set of models that are either in current use, or are significant for other reasons, is presented in Table 14.1. Since these models were chosen on the basis of the author's personal bias and knowledge, and not as the result of a thorough review of the literature, omission of a model from that tabulation does not imply lack of significance. However, the tabulation does provide a reasonable indication of the kind of models currently available.

# 5. Assessment of current modelling capabilities

As the previous discussion indicates, our ability to simulate various human responses to exercise and thermal stress depends on the situation. Here, representative conditions will be evaluated in terms of the probability of generating useful results using currently available mathematical models. This section should not be taken too seriously because it contains many subjective evaluations that are simply a reflection of the author's personal opinion and experience. While it is hoped that this evaluation has some value, a more careful evaluation should be carried out by a group composed of individuals who have the experience and knowledge necessary to make valid judgements about the requirements that useful models should possess.

The meaning of 'useful results' depends on particular requirements. For one thing, it is necessary to decide whether computed and measured results will be compared for individuals or for a representative group of subjects. In many cases, comparisons are made using data for individuals because the number of identical runs conducted with different subjects is too small to define a true group response. However, it is generally conceded that models are most useful when used to describe mean behaviour. It should be noted that, quite often, individual differences from the mean can be attributed to factors that are well known to physiologists, but are simply overlooked in defining the experimental protocol. For example, although cold response is strongly affected by subcutaneous fat thickness, cold studies continue to be performed without determining and reporting some measure of subcutaneous fat thickness. A great deal of the unexplained variation between individual responses would probably disappear if proper allowance were only made for that factor.

*Table 14.1.  Current mathematical models.*

| Author | Description | Computer |
|---|---|---|
| Givoni and Goldman | Empirical<br>Computes $T_{re}$ and HR during exercise in the heat. Allows for subject wt and SA; $T_{db}$, $T_{wb}$ and $V$; $H_{met}(t)$; Acc; Clo and $i_m$/Clo.<br>Well validated for hot environments; not applicable to cold exposure.<br>(1971, 1972, 1973a, b) | Pocket calculator. A highly portable version has been developed at ARIEM (Natick). |
| Stolwijk | Physiological<br>Computes $T_c$, $T_m$, $T_f$ and $T_{sk}$ for the head, trunk and extremities; $T_l$; and Sw, Sh, and BF$_l$. Used in designing the Apollo life support system.<br>Allows for subject wt and SF; $T_{db}$, $T_{wb}$ and $V$; $H_{met}(t)$; Clo and $i_m$/Clo.<br>Well validated for hot environments; not applicable to cold exposure.<br>(1966) | Personal computer |
| Montgomery | Physiological<br>An adaptation of Stolwijk's model in which additional temperatures are computed in each element to improve the description of response to cold exposure.<br>Validated for one set of cold immersions while wearing a wet suit.<br>(1974) | Personal computer |
| Gordon | Physiological<br>Computes 154 temperatures in 14 elements; otherwise, this model is rather similar to Stolwijk's, with the exception that Sh depends on $q_{sk}$.<br>(1976) | Minicomputer |
| Werner | Physiological<br>Computes several hundred thousand temperatures. Currently steady-state only. For more detail, refer to his chapter in this volume.<br>(1986) | Supercomputer |
| Wissler | Physiological<br>Computes 225 temperatures in 15 elements plus $O_2$, $CO_2$ and lactate concentrations. Allows for subject wt and SF; $T_{db}$, $T_{wb}$ and $V_t$; $H_{met}(t)$; Clo and $i_m$/Clo; air and liquid cooled or heated garments.<br>Well validated for hot and cold environments, both 1 ATA and hyperbaric.<br>(1985) | Mainframe |

Notation:
Clo = clothing insulation.
$H_{met}(t)$ = time variable metabolic rate.
HR = heart rate.
$i_m$ = clothing permeability for water vapour.
$q_{sk}$ = thermal flux through skin.
Sh = rate of shivering metabolism.

SA = body surface area.
SF = mean skinfold thickness.
Sw = sweat rate.
$T_{db}$ and $T_{wb}$ = dry and wet bulb termperatures.
$t$ = time.
$V$ = environmental fluid velocity.

When agreement between computed and measured results is unsatisfactory, the entire responsibility for lack of agreement should not necessarily be attributed to the model, because there is also a need to define more precisely procedures for measuring experimental values. For example, it is necessary to decide which central temperature will be measured because the oesophageal, rectal and tympanic temperatures may differ by as much as 1°C. Furthermore, there are several different ways to determine mean skin temperature depending on the number and location of measurement sites, and the weights assigned to each temperature. It should also be noted that a time-dependent sweat rate is only rarely measured; at best, one has information about nude and clothed weight loss during an experiment. And, as noted previously, determining the properties of garments is not a simple task either. Therefore, experimental and computed results must both be evaluated carefully before a decision is made about the validity of the model.

Having made those comments, it will be proposed that a model can be said to yield 'useful results' if computed and measured values agree within ±0·5°C for central temperature, within ±2°C for mean skin temperature, and within ±10% for metabolic and sweat rates. The difficulty of producing a useful simulation for various conditions will be defined using a scale of 1 to 5, with higher numbers denoting increased difficulty, as follows.

| Difficulty | Probability of obtaining a useful result |
|:---:|:---:|
| 1 | Excellent (90%) |
| 2 | Good (70%) |
| 3 | Fair (50%) |
| 4 | Poor (30%) |
| 5 | Unlikely (10%) |

Since extensive experimental study and computation would be required to determine exact probabilities for even a single case, which clearly has not been done, probabilities presented in the following section were assigned on the basis of the inherent difficulty of the simulation, the extent to which models have been validated for either the specific or closely related conditions, and the author's perception of the importance of factors that are not properly accounted for in currently available models. If nothing else, a high degree of difficulty identifies a problem that offers possibilities for additional study. See Table 14.2.

## 6. Future directions

Since computer simulation provides an important tool for dealing with complex problems, including many that involve human performance, it can be anticipated that work on model improvement will continue as long as significant progress can be achieved at reasonable cost. As has been stressed above, improvement is needed in three areas: computation of temperature field, description of physiological phenomena, and description of garments.

The amount of detail that can be incorporated into a model is heavily dependent on the speed and size of the computer available to run the model. When the author developed his first finite-difference model approximately 25 years ago, it ran on a CDC 1604 computer (comparable to an IBM 704), which was a million dollar machine, but today comparable computing power is provided by a $5 000 IBM PC-AT. That first model did not include metabolic balances, which were added as larger and faster computers became available, and now even the full model runs on a minicomputer costing around $20 000.

At the same time that traditionally designed computers have become less expensive, alternative computer architectures have been developed that offer opportunities for improving models. Supercomputers capable of performing several million instructions per second and providing a million words of random access memory are available at many locations. In addition to being very large and fast, these computers perform vector operations, which offers an opportunity for achieving enhanced performance through careful structuring of the computations.

Another architecture offering intriguing possibilities is the highly parallel processor, sometimes referred to as a hypercube computer because it consists of $2^n$ smaller machines each connected to $n$ nearest neighbours; this arrangement is illustrated in Figure 14.3 where each node represents a computer. A model developed for such a computer can be structured so that each major element of the body has its own processor which carries out the calculations necessary to compute new temperatures and metabolic variables given values at a previous time. This would permit one to perform simultaneously the rather complex computations for all of the elements, instead of performing them one element at a time, as is necessary on conventional computers. Another processor would be used to co-ordinate computations carried out in the parallel processors, and direct the transfer of data between machines in preparation for taking the next time step. Even though the individual processors are not as fast as a supercomputer, 15–20 of them running in parallel might give a considerable increase in effective speed. How fruitful this approach will be remains to be seen, but human thermal modelling with its high degree of parallelism seems to provide a good opportunity for evaluating this approach to simulation.

Since computation of two and three-dimensional temperature fields requires more computing power than current conventional machines can provide, developing higher order models has not been practical in the past. Now that larger machines are becoming accessible in many

Table 14.2. *An assessment of the difficulty of simulating human responses to exercise and environmental stress.*

| Application | Difficulty | Comments |
|---|---|---|
| 1. Exercise in a hot environment | | |
| a. Astronauts | 1 | Protective garments isolate astronauts from the environment producing relatively uniform temperatures throughout the body. This system was studied extensively during the Apollo project. |
| b. Foundry workers | 2 | Moderate levels of exercise in heat are relatively easy to |
| c. Firemen | 2 | simulate, but the properties of protective clothing are not well defined. While there should be no surprises, models have not been adequately validated for these cases. |
| d. Military personnel wearing a CBR overgarment | 1·5 | Similar to the previous case, except that garments have been evaluated in use, and models have been validated. |
| e. Personal cooling (air or liquid cooled garments) | 2·5 | Use of personal cooling devices increases difficulty because of uncertainty about the properties of these devices and lack of knowledge about physiological responses to localized cooling. Simulation of liquid cooling devices has been validated, but additional data are required for air cooling. |
| f. Aircrew in fighter aircraft | 3 | The complexity of systems worn by aircrew members in high-performance aircraft and the asymmetric boundary conditions owing to solar load and the seat complicate this problem. Nevertheless, it has been shown that models can yield useful results. |
| g. Occupational exposure to microwave radiation | 5 | Unfortunately, it does not appear to be possible to compute accurate energy absorption rates. |
| 2. Exposure to cold | | |
| a. Accidental immersion in cold water | | |
| i. Fully immersed | 3 | While there is still much to be learned about such exposures, there has also been considerable effort devoted to experimental study and simulation of cold immersion with reasonable success. |
| ii. Floating on the surface | 4 | Large variations in surface conditions and lack of knowledge about thermoregulation when there are large variations in skin temperature hinder modelling efforts. |
| b. Rewarming severely hypothermic subjects | 5 | Very little is known about thermoregulation in severely hypothermic subjects. |
| c. Work under arctic conditions | 4 | More information is needed about the properties of arctic clothing, and, in particular, about moisture accumulation in cold garments. |
| d. Marching through cold water | 4 | Not enough is known about thermoregulation under such highly nonuniform conditions. |

*Table 14.2   Continued.*

| Application | Difficulty | Comments |
|---|---|---|
| e. Complete or partial loss of hot water while diving | 2·5 | While this is a difficult problem, such conditions have been simulated with good accuracy. |
| f. Lost-bell situations | 3 | Assumptions must be made about heat transfer mechanisms in the bell. |
| 3. Medical applications | | |
| a. Thermoregulation in severely burned patients | 4 | Not enough is known about the physiological responses of such individuals. |
| b. Hypothermia during open-heart surgery | 4 | Additional information is needed about thermoregulation in an aesthetized, hypothermic subjects, and about heat loss from an open thorax. |
| c. Hyperthermia for treating cancer | 4 | Cardiovascular responses to intense local heating are not fully understood. |
| d. Clinical use of hyperbaric oxygen | 5 | Basic mechanisms are not well understood. |
| e. Decompression sickness | 5 | The fundamental causes of decompression sickness are still not known. |

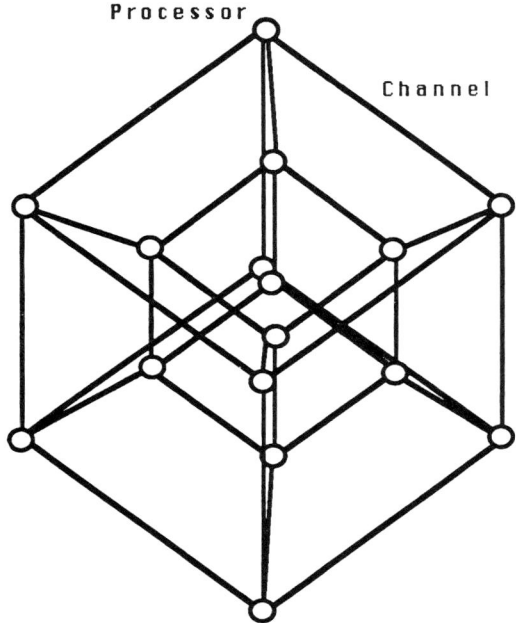

*Figure 14.3.   Schematic representation of a 16-node hypercube computer in which each processor is connected by data channels to its four nearest neighbours.*

locations, it is reasonable to expect that more detailed models will be developed. Indeed, Werner (1986) has already developed a model providing 1 cm resolution throughout the entire body, and it is doubtful that a more detailed model will be developed soon. One can question whether that degree of refinement is really necessary, but there is no easy answer to that question. However, it is worth noting that a number of important, poorly understood phenomena require that level of detail for adequate description, and new models, such as Werner's, should be useful in gaining a fuller understanding of those phenomena. For example, it was mentioned earlier that the three commonly used sites for measuring central temperature yield inconsistent results under most conditions, but reasons for the disagreement are not well understood. Since this is a purely physical problem, more adequate modelling of regions around the measuring sites, including heat transfer from blood in major arteries and veins to tissue, should provide useful information.

Multidimensional models can be constructed in several ways. One is to use a standard co-ordinate system, such as Cartesian, cylindrical, polar or elliptical, and the other is to use a finite element approach. In either case, allowance must be made for nonuniform physical and physiological properties, and variable boundary conditions. The use of finite difference techniques and a regular co-ordinate system may offer some computational simplification compared to finite element methods, while finite element methods handle irregular geometries more effectively. Both methods require computation of approximately 200 temperatures in each major element, which yields 3 000 total temperatures for a 15-element model of man. Hence, a two-dimensional model will be at least an order of magnitude larger than Wissler's current one-dimensional, 15-element model, but still two orders of magnitude smaller than Werner's new 1 cm model. As is usually the case, a trade-off is involved in deciding which approach to employ. Models that use a large number of points to obtain excellent resolution become very costly to run, with the consequence that it is difficult to perform the many computations necessary to evaluate unmeasurable physiological parameters and validate the model, while smaller models provide less spatial resolution, but can be run fairly cheaply.

A new model based on an elliptical co-ordinate system is currently under development at the University of Texas at Austin. The first approximation of a thigh section is shown in Figure 14.4, where different shadings represent bone, active muscle, resting muscle and fat. Isotherms generated after 20 minutes of exercise are shown in Figure 14.5. It is clear that this model is capable of describing the asymmetric temperature profiles that occur as a consequence of nonuniform physical and physiological properties. Subsequent versions of this model will undoubtedly employ a finer grid near the surface to provide better temperature resolution in the region where large

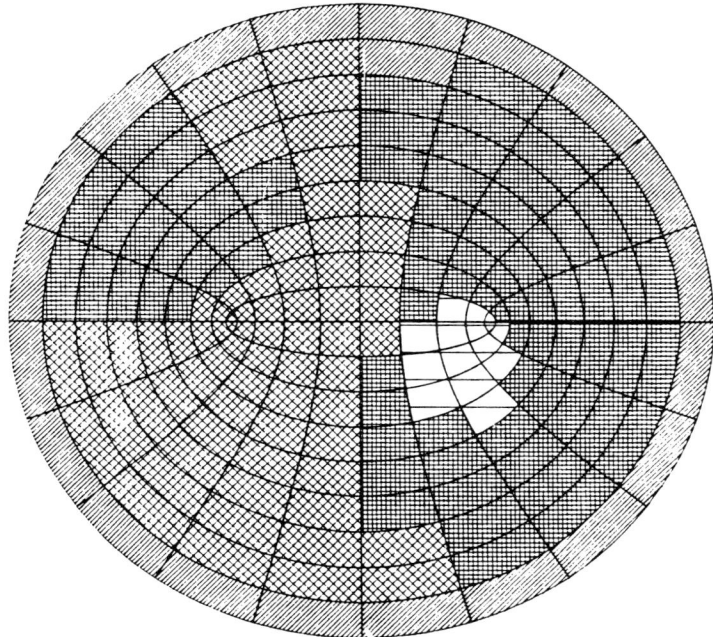

*Figure 14.4. Elliptical co-ordinate system used to represent a human thigh. The white area represents bone; also shown are active and inactive muscle and a layer of subcutaneous fat.*

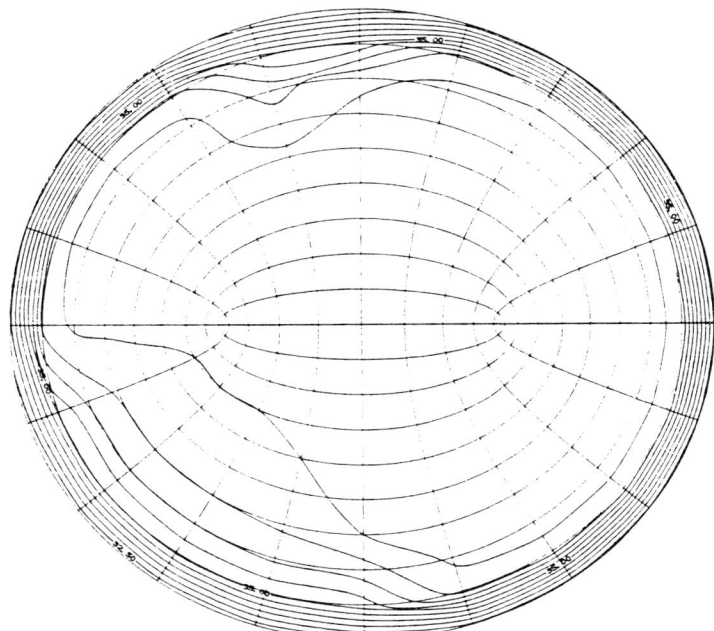

*Figure 14.5. Isotherms computed for the system shown in Figure 14.4 after 20 minutes of moderate exercise.*

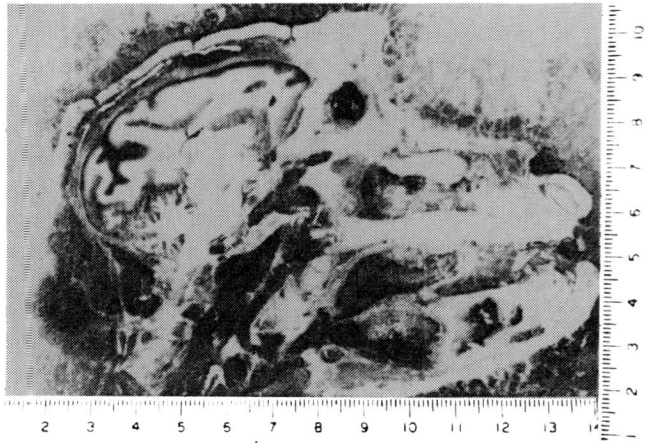

*Figure 14.6.   Sagittal section through the monkey head, 1 cm off midline (Olsen et al., 1984).*

gradients occur. Using a fine grid near the surface will also allow more representative description of the subcutaneous fat distribution, which Hayes mentions in Chapter 12 of this volume.

The kind of resolution that finite element methods can provide is illustrated by a study completed recently at the University of Texas Health Science Center in Dallas by Olsen *et al.* (1984). This experimental and computational study was designed to evaluate surface and extracorporeal cooling as alternative modalities for inducing hypothermia in infants prior to open heart surgery. The adult male macaque

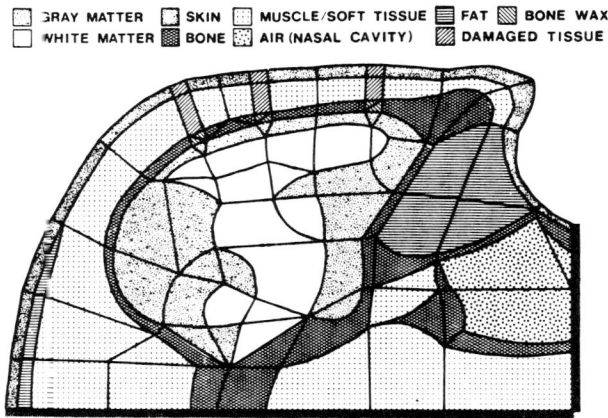

*Figure 14.7.   Finite element representation of the monkey head shown in Figure 14.6. Note that the three-dimensional model is formed by rotating this section about the heavy base line (Olsen et al., 1984).*

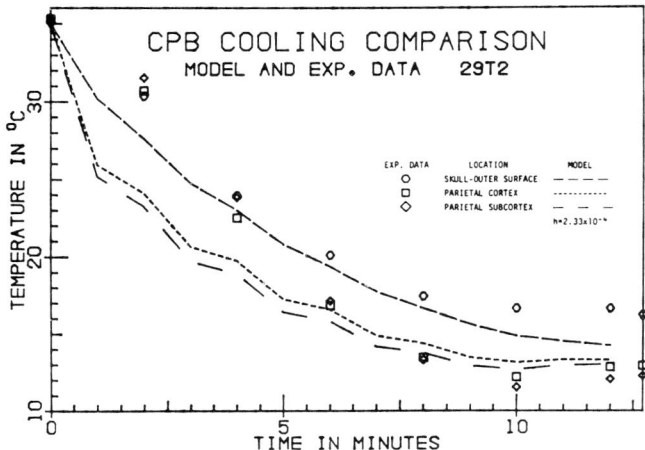

*Figure 14.8.    A comparison of computed temperatures with values measured at three sites during cardiopulmonary bypass (Olsen et al., 1984).*

monkey was used as an experimental model in this study. Figure 14.6 shows a sagittal section through the monkey head 1 cm off midline, while the finite element representation of the physical system is shown in Figure 14.7. It should be noted that the three-dimensional representation is a cylinder formed by rotating the cross-section shown in Figure 14.7 around the horizontal axis at the base of the figure. While such a representation is completely inappropriate for describing the anatomy of the lower head, it should provide a reasonably good description of the cranium and brain.

Starting at time = 0, arterial blood was cooled in a stepwise manner during the next 12 minutes while the animal was supported by cardiopulmonary bypass. Computed temperatures at the three sites designated as damaged tissue in Figure 14.7 are shown together with measured temperatures in Figure 14.8. It appears that computed temperatures decrease somewhat more rapidly than measured temperatures, although the final values reached at the end of the cooling period were in good agreement, as one would expect. Computer isotherms at the end of the cooling period are shown in Figure 14.9, where it can be seen that this method is indeed capable of providing rather good spatial resolution.

## 7. Conclusions

Mathematical modelling of physiological phenomena is becoming an increasingly useful tool for predicting human performance under various conditions, and it is essential that those who use models develop an

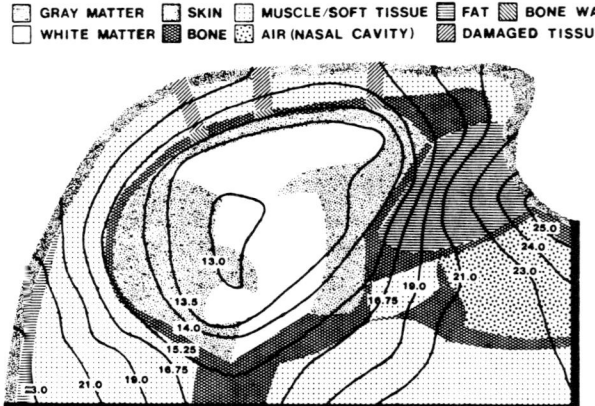

GRAY MATTER  SKIN  MUSCLE/SOFT TISSUE  FAT  BONE WAX
WHITE MATTER  BONE  AIR (NASAL CAVITY)  DAMAGED TISSUE

**CPB COOLING SIMULATION: $T_{art}$=13.0°C, t=12 min**

*Figure 14.9. Computer isotherms at the end of cardiopulmonary bypass (Olsen et al., 1984).*

appreciation for their limitations. It would probably be worthwhile to carry out a thorough assessment of the capability of models currently in use by analysing a large number of situations, but that would be a challenging undertaking. Even though a large-scale evaluation is not likely to take place soon, opportunities for evaluating models occur whenever they are used to simulate new situations, and much can be learned from careful, critical comparison of computed and measured results.

It is important to recognize that new experimental results are essential for continued model development. Important relationships, such as those involved in thermoregulation, cannot be deduced from first principles; they can only be derived from penetrating new experimental studies. Classical physiologists who work in the laboratory and theoreticians of various kinds who work on models have a common objective of gaining a more perfect understanding of the human organism. Each group has an important contribution to make in achieving that objective, although they view the problem from different perspectives.

It is hoped that this chapter provides a realistic evaluation of the current state of model development, and encourages new efforts to obtain the knowledge necessary to achieve a significant improvement in our capability for predicting human behaviour in the areas discussed.

## Acknowledgements

It is a pleasure to acknowledge financial support provided for our work by the USAF School of Aerospace Medicine at Brooks AFB, Texas.

# References

Givoni, B. and Goldman, R.F., 1971, Predicting metabolic energy cost. *J. Appl. Physiol.*, **30**, 429−433.

Givoni, B. and Goldman, R.F., 1972, Predicting rectal temperature response to work, environment, and clothing. *J. Appl. Physiol.*, **32**, 812−822.

Givoni, B. and Goldman, R.F., 1973a, Predicting heart rate response to work, environment, and clothing. *J. Appl. Physiol.*, **34**, 201−204.

Givoni, B. and Goldman, R.F., 1973b, Predicting effects of heat acclimatization on heart rate and rectal temperature. *J. Appl. Physiol.*, **35**, 875−979.

Gordon, R.G., Roemer, R.B. and Horvath, S.M., 1976, A mathematical model of the human temperature regulatory system—transient cold exposure response. *IEEE Trans. Biomed. Eng.*, **23**, 434−444.

Hayes, P.A., 1986, The physiological basis for the development of immersion protective clothing. *Ann. Physiol. Anthrop.*, **5**, 154−155.

Montgomery, L.D., 1974, A model of heat transfer in immersed man. *Ann. Biomed. Eng.*, **2**, 19−46.

Olsen, R.W., Hayes, L.J., Wissler, E.H., Nikaidoh, H. and Eberhart, R.C., 1984, Influence of extracerebral temperature on cerebral temperature distributions during induction of deep hypothermia with subsequent circulatory arrest. *ASME Technical Paper 84-WA/HT-62*, Presented at the 1984 ASME Winter Annual Meeting in New Orleans.

Påsche, A. and Ilmarinen, R., 1986, Temperature parameters for manned survival suit evaluation. *Ann. Physiol. Anthrop.*, **5**, 155−156.

Stolwijk, J.A.J. and Hardy, J.D., 1966, Temperature regulation in man—a theoretical study. *Pflugers Arch.*, **291**, 129−162.

Stolwijk, J.A.J. and Hardy, J.D., 1977, Control of body temperature. In *Handbook of Physiology—Reaction to Environmental Agents*, edited by D.H.K. Lee (Bethesda, MD: American Physiological Society), pp. 45–67.

Van Someren, R.N.M., Coleshaw, S.K.R., Mincer, P.J. and Keatinge, W.R., 1983, Restoration of thermoregulatory response to body cooling by cooling hands and feet. *J. Appl. Physiol. Respirat. Environ. Exercise Physiol.*, **53**, 1228−1233.

Werner, J. and Buse, M., 1986, Three-dimensional simulation of cold and warm defence in man. *Ann. Physiol. Anthrop.*, **5**, 173.

Wissler, E.H., 1985, Mathematical simulation of human thermal behavior using whole body models. In *Heat Transfer in Biology and Medicine*, edited by A. Shitzer and R.C. Eberhart (New York: Plenum).

# 15. Three-dimensional Simulation of Cold and Warm Defence in Man

**Jurgen Werner and Monika Buse**

## 1. Introduction

Warm and cold defence in man operates within a system having distributed variables and parameters. This means that:

1. All variables (e.g., temperature, heat flow, etc.) should be regarded as functions of time and of three-dimensional local co-ordinates.
2. All parameters (e.g., density, conductivity, etc.) should be considered as locally distributed parameters.
3. Geometry and anatomy of the body should be adequately represented.

Of the variety of computer models for the open-loop passive thermal system (see Chapter 14) the proposals of Wissler (1961, 1964, 1970, 1985) come nearest to these requirements using primarily multi-element cylinder models. Werner (1975, 1977, 1984) endeavoured to take into account the principle of distributed variables and parameters for the closed thermoregulatory control loop. However, the use of a comprehensive data bank of the true geometry and anatomy of the body (Kelterbaum *et al.*, 1977) could not be put into practice until super-computers became available within the last few years.

## 2. Mathematical simulation

The data bank was developed on the basis of photogrammetric analysis of anatomical models. Figure 15.1 for example shows the contour lines of the trunk after elimination of all organs, Figure 15.2(a) gives an example of a sectional map obtained for one contour line and Figure 15.2(b) presents the corresponding computer representation of the same

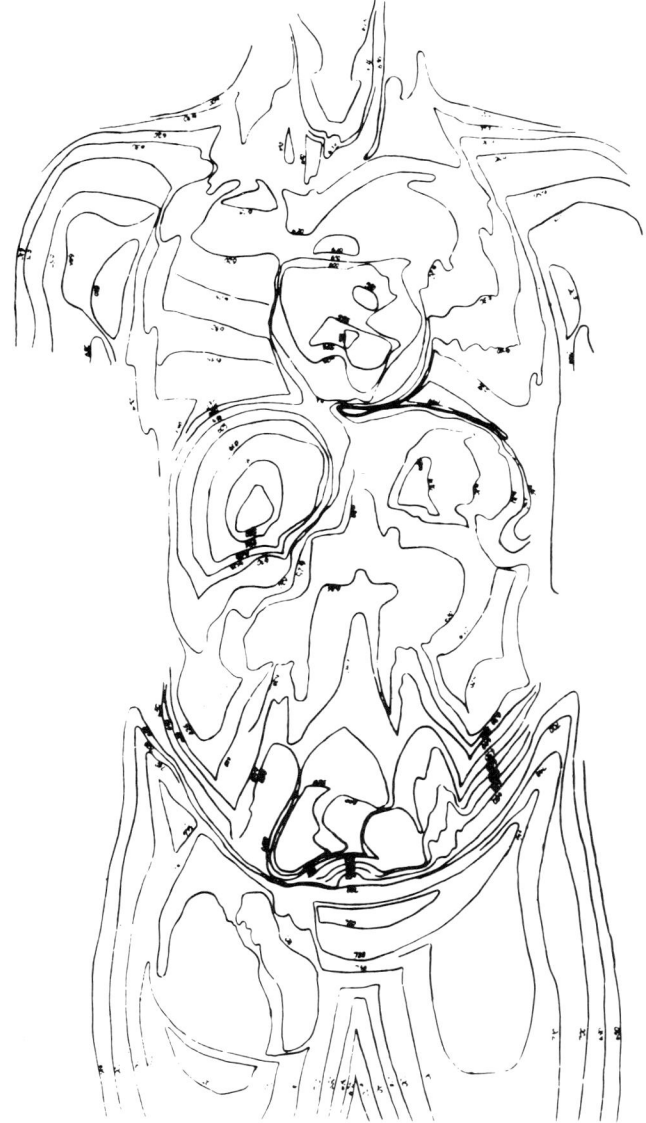

*Figure 15.1.   Photogrammetric analysis: contour lines of the trunk after the elimination of all organs.*

section. Sixty-three types of tissue are differentiated. The local grid is 1 cm for the trunk and 0·5 cm for the other parts of the body. This results in a digital representation of the body with about 400 000 points associated with the physical parameters of the tissue.

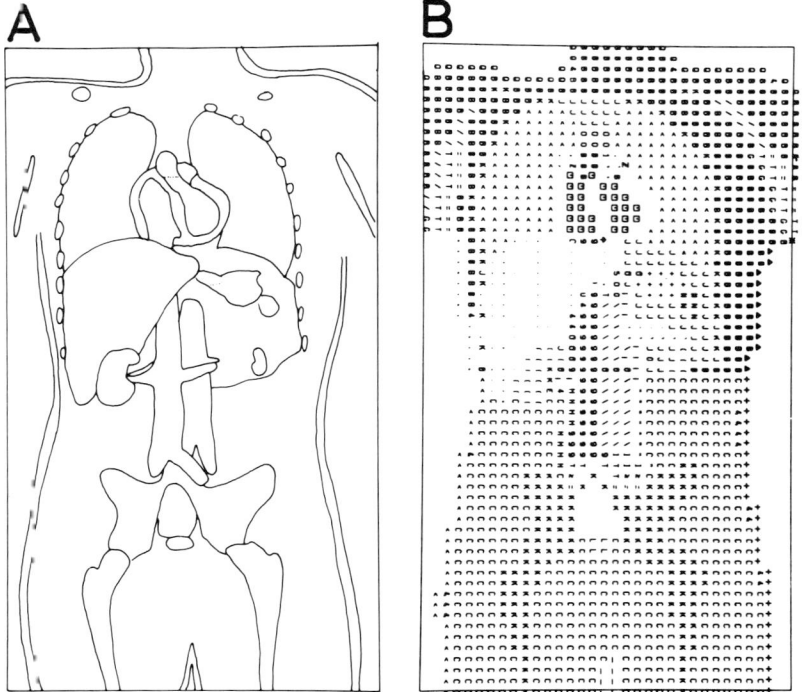

*Figure 15.2. A example of a section map for one contour line: (a) analogue; (b) computer presentation.*

The simulation is principally based on the following partial differential equations for the passive system.

Heat transport via heat conduction and capillary blood flow:

$$\varrho\ (\xi)\ c\ (\xi)\ \frac{\delta T\ (\xi,t)}{\delta t} = \text{div}\ [\lambda\ (\xi)\ \text{grad}\ (T\ (\xi,t))] + M\ (\xi,\ T) - \dot{Q}\ (\xi,\ T) \tag{1}$$

where $\varrho$ = density; $\xi$ = three-dimensional local co-ordinate; $c$ = specific heat; $T$ = temperature; $t$ = time co-ordinate; $\lambda$ = conductivity index; $M$ = specific metabolic heat production; and $Q$ = specific heat transport via capillaries.

Heat transport via blood flow in large vessels (radius $r_B > 2$ mm):

$$\frac{r_B}{2}\ \varrho_B\ c_B\ \left[\frac{\delta T_B\ (\xi,t)}{\delta t} + v\ (\xi)\ \frac{\delta T_B\ (\xi,t)}{\delta s}\right] = h_B\ [T\ (\xi,t) - T_B\ (\xi,t)] \tag{2}$$

where $v$ = flow velocity; $s$ = direction vector of flow; and $h_B$ = heat transfer coefficient tissue/blood.

Heat loss via skin:

$$-\lambda \, (\xi) \, \frac{\delta T \, (\xi,t)}{\delta n} = \varepsilon \sigma \, (T^4 - T^4_{ra}) + a_c \, (T - T_A) + a_e \, [p_v \, (T) - p_v \, (T_A)] \tag{3}$$

where $n$ = normal vector; $\varepsilon$ = emissivity index; $T$ = Boltzman constant; ra = radiation; $a_c$ = heat transfer coefficient for conduction + convection; $A$ = air; $a_e$ = heat transfer coefficient for evaporation; and $p_v$ = water vapour pressure.

Heat loss via respiration:

$$\dot{Q}_{res} = \dot{V} \, (p_A \cdot C_A \, (T_{res} - T_A) + \Delta h_v \cdot \Delta p''_{H2O})$$

$$\Delta p''_{H2O} = p_{sl} \cdot p''_{H2O} \, (T_l) - p_{sA} \cdot p''_{H2O} \, (T_A) \tag{4}$$

where res = respiratory; l = lung; $V$ = tidal volume; $\Delta h$ = evaporative enthalpy; $p_s$ = saturated water vapour pressure; and $p''$ = water vapour contents.

For the receptors, the integrative controllers and the effectors the following principal equations are used:

Receptors:

$$f_a \, (\zeta,t) + \tau_R \, (\zeta) \, \frac{\delta f_a \, (\zeta,t)}{\delta t} = K_R \, (\zeta) \left\{ T \, (\zeta,t) + \tau_D \, (\zeta) \, \frac{\delta T \, (\zeta,t)}{\delta t} \right\} \tag{5}$$

where $f_a$ = afferent frequencies; $\zeta$ = three-dimensional local co-ordinate of receptor site; $\tau_R$, $\tau_D$ = time constants; and $K_R$ = gain.

Controller:

$$f_{ei} \, (\eta,t) + \tau_c \, (\eta) \, \frac{\delta f_{ei} \, (\eta,t)}{\delta t} = \int c_i \, (\zeta,\eta) \cdot f_a \, (\zeta,t) \, d\zeta \tag{6}$$

where $f_e$ = efferent frequencies; $\eta$ = three dimensional local co-ordinate of effector site; $\tau_c$ = time constant; c = coupling matrix between receptor and effector site; and $i$ = 1, 2, 3.

Effectors:

$$Y_i \left(\zeta, t\right) + \tau_e \left(\eta\right) \frac{\delta \gamma_i \left(\eta, t\right)}{\delta t} = K_{Y_i} \left(\eta\right) \cdot f_e \left(\eta, t\right) \tag{7}$$

where $Y$ = effector; $\tau_e$ = time constant; and $K_i$ = gain.

The importance of taking into account the geometry and inhomogeneity of the human body has been confirmed in an earlier study (Buse and Werner, 1985a). Therefore we endeavoured to represent the passive system's equations $(1)-(4)$ and parameters as exactly as possible in order to be able to test how far the controller equations $(5)-(7)$ could be simplified for obtaining an adequate description of the system's operation.

The simulation has been implemented on the vector computer Cyber 205 by an implicit alternating direction method according to Douglas and Rachford (see Buse and Werner, 1985b). The solution needs a core capacity of 3·5 million words (64 bit).

## 3. Results

### Topography of temperatures

The simulation has delivered a realistic picture of the topography of temperatures under neutral conditions (Figure 15.3). Compatibility of reality and simulation was tested by experiments and by comparison of values given in the literature. It was achieved solely on the basis of physical considerations and physiological data. An adjustment of parameters of the passive system was not necessary, so that the computation should be equally suited to delivering temperature profiles which cannot be obtained experimentally.

Figure 15.4 shows temperature profiles of transverse sections of the head and trunk in the indifferent status ($T_A = 30°C$). In the head, temperature is relatively low and constant. A decrease towards skin temperature starts only in the skull. This is due to the high metabolic and blood flow rate of the brain. The profiles of the extremities are more convex, due to the lower rate of blood flow, but equally, rather homogeneous and smooth. In the trunk, the transverse profile depends very much on the longitudinal co-ordinate. Figure 15.4(b) shows one example of a profile at the level of the heart and the transition to the arms. It is noticeable that there is a slight temperature decrease in the region of the spinal cord (c) with high metabolic and blood flow rate in the spinal cord and lower blood flow rate in the adjacent tissues. The profile also shows the influence of the higher metabolic rate of the heart even in the adjacent tissue (d).

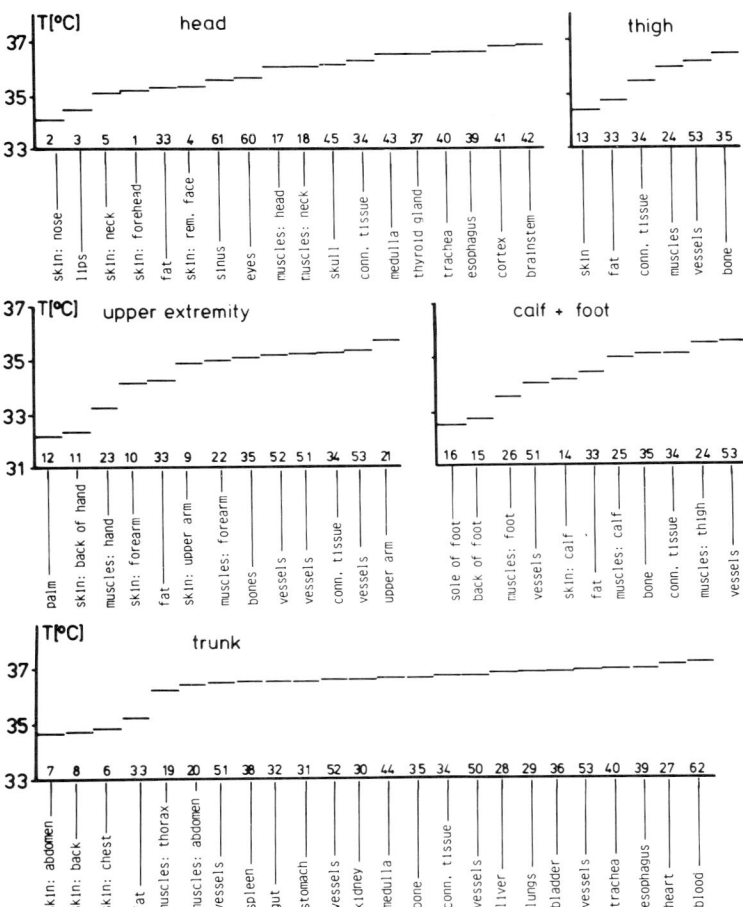

*Figure 15.3.    The topography of temperatures in the indifferent status (T$_A$ = 30°C). Numbers on the x axis denote the numbers of the tissue code, 1–63.*

## Control strategy

An unsolved problem is the control strategy of the system, that is the question whether the inhomogeneous pattern of effector distribution is maintained in the cold and warmth or whether active modification and control is distributed.

Therefore the lumped parameter approach delivering the best results was first determined. As controller input the following equation seems to be a reasonable simplification.

$$Y = 0·8 \; \bar{T}_{core} + 0·1 \; \bar{T}_{skin} + 0·1 \; \bar{T}_{muscle} - 37 \qquad (8)$$

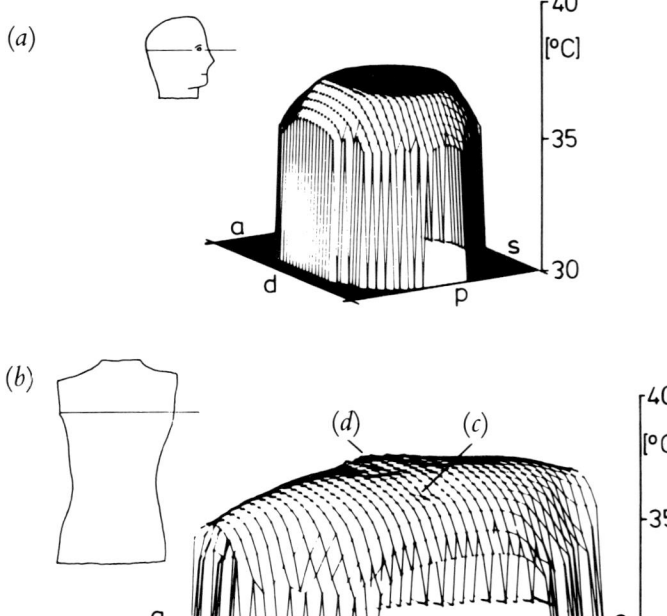

*Figure 15.4.    Examples of transverse profiles, p = posterior, a = anterior, d = dexter, right, s = sinister, left. The slices are seen (site seen inset) from candal to cranial. $T_A = 30°C$. (a) Head; (b) trunk; for (c) and (d) see text.*

A correct global metabolic heat production $M$ in $W\ m^{-3}$ may be computed by:

$$M = -5000\ Y + 648 \qquad (9)$$

However, maintaining the pattern of distribution of heat production yields too high amounts in the arms and the legs which do not contribute much to the stabilization of central trunk and head temperatures. This may only be achieved by the concentration of heat production in the proximal areas. Thus, for cold defence a locally distributed control of heat production in the skeletal muscles must be required. The minimal requirements of a distributed control strategy are the following:

$$
\begin{aligned}
M &= -5200\ Y + 648 \text{ for head and trunk} \\
M &= -\ 200\ Y + 648 \text{ for arms and legs} \\
M &= \qquad\qquad 648 \text{ for hands and feet}
\end{aligned}
\qquad (10)
$$

A similar but minor relevance of the distributed control concept has been confirmed with regard to blood flow rate. However, the influence of distributed minimal rates is more relevant than locally differing gains. When compared to the following quasi-optimal lumped control for blood flow $Q$ (in units of m$^3$ blood m$^{-3}$ tissue g$^{-3}$ s$^{-1}$)

$$Q = 0{\cdot}01\ Y + 0{\cdot}0001 \tag{11}$$

the following simplest distributed strategy delivers more realistic results:

$$
\begin{aligned}
Q &= 0{\cdot}01\ Y + 0{\cdot}0003 \text{ for forehead, nose, mouth}\\
Q &= 0{\cdot}01\ Y + 0{\cdot}0001 \text{ for remainder of head, upper arm, thigh}\\
Q &= 0{\cdot}008\ Y + 0{\cdot}0001 \text{ for trunk}\\
Q &= 0{\cdot}012\ Y + 0{\cdot}0001 \text{ for forearm, calf}\\
Q &= 0{\cdot}012\ Y + 0{\cdot}0003 \text{ for hand, foot}
\end{aligned}
\tag{12}
$$

On the other hand, the influence of distributed controller structures is not so obvious in warm defence with regard to evaporative heat loss, as there are very small temperature differences in the body after stronger warm load. The lumped approach for the sweating rate $S$ in g$^{-2}$ s$^{-1}$

$$S = 0{\cdot}1\ Y + Y_{min} \tag{13}$$

($Y_{min}$ different for 16 areas) yields results which are very near to the following quasi-optimal distributed approach:

$$
\begin{aligned}
S &= 0{\cdot}12\ Y + Y_{min} \text{ for forehead, nose, mouth}\\
S &= 0{\cdot}08\ Y + Y_{min} \text{ for remaining skin}
\end{aligned}
\tag{14}
$$

$Y_{min}$ is different for 16 areas; from 177 (back of hand) to1 766 (palm). Therefore for sweating rate it cannot definitely be decided whether the distributed control is present or not.

## Temperature profiles in the cold and in the warm

Using the distributed approaches outlined above longitudinal profiles for head, arm, leg and trunk were computed for various ambient temperatures as shown in Figure 15.5. Central temperatures vary from 35°C to 37·5°C in the head and trunk. The central profiles in the arms and legs decrease enormously in the cold. At $T_A$=35°C and 40°C, central temperatures of the extremities are nearly constant and very near to central temperatures of the trunk and head. The strong temperature decrease in the area of the knee is due to the relatively high percentage of bone.

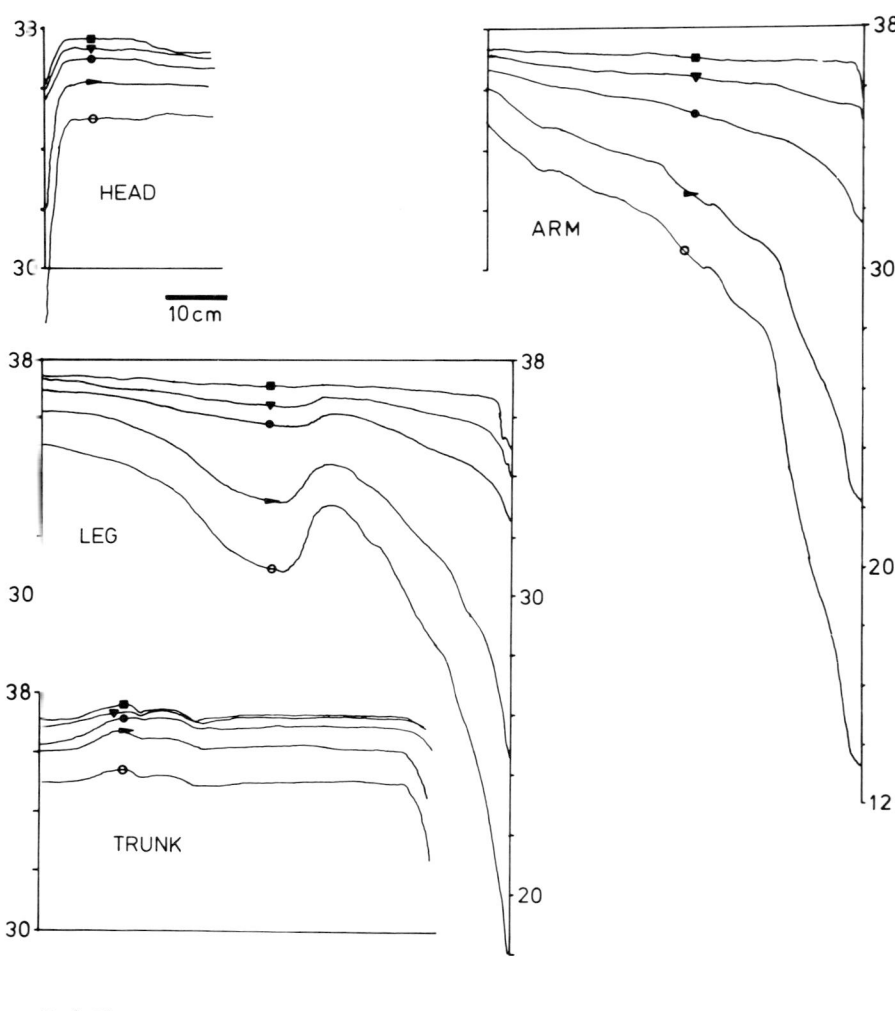

*Figure 15.5. Longitudinal profiles for various environmental temperatures, $T_A$ (y axis).*

Figure 15.6 demonstrates isotherms of the trunk at the level of the collar bone, at environmental temperatures of A: 10°C, B: 20°C, C: 30°C, D: 40°C. A and B demonstrate the reduction of the limited central warm area, whereas at $T_A=40$°C, temperatures lower than 36°C are present only in the area of the shoulders.

The simulation of dynamic effects is possible, but fails at present on account of the small working storage (0·5 million words) of the Cyber 205 version at the Ruhr University, and inadequate changing frequency

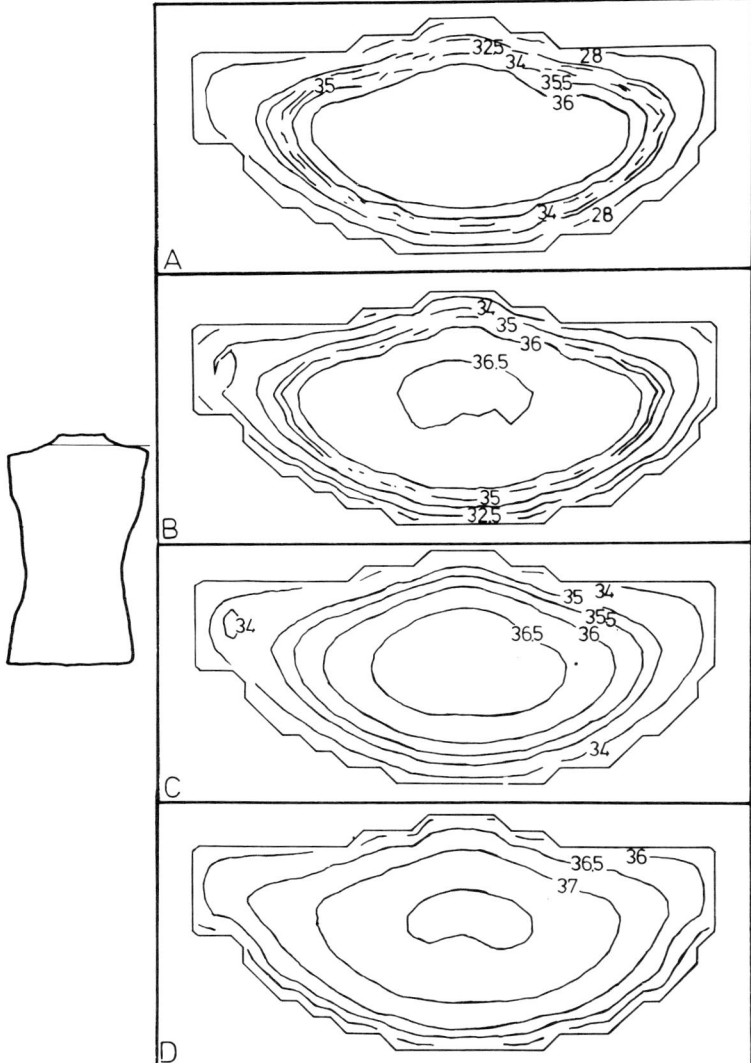

*Figure 15.6.* Isotherms in the trunk at the level of the collar bone (see left) for $T_A = 10°C$ (A), $20°C$ (B), $30°C$ (C) and $40°C$ (D).

of segments makes the computation inaccurate. Nevertheless, first tests show that the program is adequate and that the time course of temperature change in various segments under different ambient conditions will be computed correctly as soon as the described memory expansion becomes available.

## Acknowledgement

This work was supported by the Deutsche Forschungsgemeinschaft, grant no. SFB 114.

## References

Euse, M. and Werner, J., 1985a, Heat balance of the human body: influence of variations of locally distributed parameters. *J. Theor. Biol.*, **114**, 34—51.

Euse, M. and Werner, J., 1985b, ADI-Verfahren zur Losung von Warmelei-tungsproblemen in allgemeinen dreidimensionalen Korpern. *PARS-Mitteilungen*, 75—84.

Kelterbaum, J., Werner, J. and Schon, H., 1977, Makroskopische Topographie des menschlichen Korpers: Gewinnung der Rohdaten und deren EDV-gerechte Aufarbeitung in einer Datenbank. *EDV Med. Biol.*, **4**, 123—128.

Werner, J., 1975, Zur Temperaturregelung des menschlichen Korpers. Ein mathematisches Modell mit verteilten Parametern und ortsabhangigen Variablen. *Biol. Cyb.*, **17**, 53—63.

Werner, J., 1977, Mathematical treatment of structure and function of the human thermoregulatory system. *Biol. Cyb.*, **25**, 93—101.

Werner, J., 1984, *Regelung der menschlichen Korpertemperatur* (Berlin: De Gruyter).

Wissler, E.H., 1961, Steady state temperature distribution in man. *J. Appl. Physiol.*, **16**, 734—740.

Wissler, E.H., 1964, A mathematical model of the human thermal system. *Bull. Math. Biophys.*, **26**, 147—166.

Wissler, E.H., 1970, The use of finite difference techniques in simulating the human thermal system. In *Physiological and Behavioral Temperature Regulation*, edited by J.D. Hardy, A.P. Gagge and J.A.J. Stolwijk (Springfield, IL: C.C. Thomas), pp. 367—388.

Wissler, E.H., 1985, Mathematical simulation of human thermal behavior using whole body models. In *Heat Transfer in Medicine and Biology*, edited by A. Shitzer and R.C. Eberhart (New York: Plenum), pp. 325—374.

# PART V
# THE HYPERBARIC
# ENVIRONMENT

# 16. Physiological Limitations of Human Performance in Hyperbaric Environments

## Peter B. Bennett

## 1. Introduction

One of the prime human facets which has enabled us to progress from the simple cave dwellers to space explorers, is our quest to extend the limits of our immediate environment and to use increasingly complex technology to help us achieve this end. In this regard, even climbing the high mountain tops of this planet soon showed us that the body is itself a major constraint to our aspirations. Millions of years of evolution decree that the human body continues to live only within very narrow bounds of oxygen concentration, temperature and ambient pressure.

Thus in reaching out for an ultimate goal, the Moon, a mass of physiological and medical data had first to be acquired on the affects of such factors as hypoxia, weightlessness, lowered pressure, $g$ forces, temperature, etc. These facts of aerospace physiology and medicine were then coupled with superb engineering technology to control or counteract, where possible, these factors and to provide the most efficient manner of getting the astronaut to his work site on the Moon. It was then necessary to provide correctly adapted tools and equipment to enable him to work at his optimum, with maximum safety.

Diving technology is following the same evolutionary path but still has a long way to go. Any visit to an offshore oil rig would show that divers, even when diving to a mere 150 m, are, by comparison with their cousins the astronauts, equipped with the crudest of tools and life support systems.

The past 20 years has seen an amazingly rapid extension of diving depth. In the first decades of this century British Royal Navy divers were diving with difficulty to only about 60 m and were limited by intoxication and loss of consciousness which we now know is due to nitrogen narcosis (Bennett 1966, 1982b). Yet it was not until the mid-1930s in the USA that Captain A. Behnke, MC, USN, showed us that the nitrogen constituent of air was responsible and that substitution of

299

helium for the nitrogen would prevent, as Cousteau called it, "rapture of the deep" (Behnke *et al.*, 1935; Behnke and Yarbrough, 1939). An additional difficulty, unknown until 150 years ago, was that divers ascending too rapidly contracted severe pains in the joints which were called 'the bends' and were first described by the French physiologist, Paul Bert, in 1878, as being due to gas bubbles in the blood and tissues of the body.

Progress continued slowly, and it was not until 1962 that a dive to 300 m was achieved. This was by Hannes Keller, who was not a diver but a Swiss mathematician. His record-setting dives were the spur for the next 20 years of aggressive research. This resulted in a marked increase in our knowledge of deep diving physiology and medicine, the constraints of the human body to such diving, and what can be done to mitigate these constraints. The research has extended our knowledge of limitations to deep diving and revealed conditions, such as the high pressure nervous syndrome, which appear when diving deeper than 200 m (Bennett, 1982a) and aseptic bone necrosis (McCallum and Harrison, 1982). Saturation diving technology has also greatly improved the comfort and safety of men diving for extended periods (Workman *et al.*, 1962; Bond, 1966).

The author has been privileged to have worked in diving physiology and medicine since 1953 and so has been actively involved with this rapid growth of knowledge which culminated in the record research dives at Duke Medical Center (Bennett *et al.*, 1981, 1982) to 2 250 feet (about 700 m) for 1 day and for nearly 2 weeks at 1 969 feet (about 600 m). This means that since 1962, when only 10% of the ocean floor was accessible to man at 1 000 feet, now some 20% may be available. The area of untapped resources beneath the oceans equals the area of the Moon, therefore its importance should not be underestimated. Increasing awareness of the vast potential of this future resource is shown by the rapid growth of the offshore oil and gas drilling industry and the growth of other sub-sea-dependent industries such as minerals, food and new pharmaceuticals from the oceans. These eventually may prove to be of greater significance to us than outer space where all the planets are far more hostile and difficult to reach.

We need, however, to proceed carefully and on an informed basis and never forget that the underwater environment is very alien and hostile. Every phase of entry, stay and return from such high-pressure conditions continues to have physiological and medical dangers which must be carefully considered if the work or research is to be conducted safely. Dangers occur in three main areas: the basic life support systems; the medical factors resulting from saturation diving which may require longer decompression than required in a return from the Moon; and most important of all the physiological stresses imposed by breathing special gases at high pressures and the pressure effect itself.

## 2. Life support

A number of factors are vital to the life support of the diver (i.e., basic factors necessary for maintenance of the body within safe physiological limits). These include: effective communications, correct humidity and temperatures, control of breathing gases, fire hazards, shower and toilet facilities, adequate nutrition and relaxation activities.

Many of these involve the interaction of engineer and physiologist to produce an effective solution to a problem. For example, when breathing oxygen-helium there is a severe distortion of speech due to changes in the principles of resonance which affect the final frequency response of the complex resonating system within the throat. Electronic engineers have produced voice processors to correct voice distortion but recent dives have shown that at depths over 300 m these can be ineffective and especially bad for special gas mixtures such as Trimix (i.e. helium-nitrogen-oxygen). Considerable care is thus required in deep diving research to ensure back-up methods of communication by signals, writing pads, etc. until one is sure communications are effective. Further research is also required using advanced electronic technology to provide more effective voice processors.

The correct humidity and temperature are also vital for health and safety, especially in saturation diving. Usually the humidity is held at 60−70% for comfort and to reduce the risk of skin and ear infections. These infections grow easily unless care is taken with chamber and body cleanliness and appropriate preventive techniques are used, as we will discuss later (Farmer, 1982). Some expensive long duration saturation dives have had to be aborted due to acute and very painful ear infections. As will also be discussed later, it is necessary to maintain cabin temperatures of approximately 29−32°C for safety and comfort in oxygen-helium deep diving (Webb, 1982). Variations of as little as 0·5°C can result in a significant perception of cooling or heating by the divers.

Breathing gases must be very accurately controlled, both with regard to the vital oxygen supply and any attendant undesirable carbon dioxide accumulation. For example, in a dive to 1 312 feet (about 400 m), the oxygen percentage would need to be maintained at only 1·227% to be comparable to sea level conditions and the carbon dioxide should be no more than 122 ppm (0·000122%). These low levels, which become even lower at greater depths, place considerable strain on the effectiveness of modern monitoring and measuring apparatus. Any small toxic gas or material inadvertently taken into the chamber may cause serious toxicity to the divers at high pressures. Chamber gases therefore need regular monitoring and analysis by sensitive instruments at the limits of present technology.

Nutrition is an often ignored factor in diving comfort and safety but there are many reports of significant weight loss during long saturation dives. This may be from not eating due to the nausea of high pressure nervous syndrome, or increased metabolism from the cold effect of helium, or the decreased activity by living in a saturation pressure chamber with limited mobility and exercise. Good, hot, nutritious meals are vital and are often the one real enjoyment a diver may have over a prolonged period of isolation from society during an extended saturation dive. Recent work suggests that weight loss may be due to fluid loss only.

In order to maintain psychological stability and to prevent the ever-present risk of boredom, attention must be paid to relaxation activities with games, books, tape recorders, radio, and projection of films or TV through a port, etc.

A major safety factor is fire suppression methods and procedures. Although in deep diving there is little risk of internal fire due to low oxygen partial pressure, during the shallower parts of a decompression profile, oxygen concentration rises to a combustible range and all combustibles must be removed from the chamber.

## 3. Medical stresses

### Diver selection

It is necessary for the divers to have intensive and regular medical examination. These medical studies should include lung and long bone X-rays for aseptic bone necrosis (McCallum, 1982), pulmonary function tests, electroencephalography (EEG), evoked EEG potential measurement and both neuropsychiatric and clinical psychological studies to establish psychiatric stability of the diver. Such tests are important as a baseline of data against which to compare problems occurring after a decompression incident or some other developing effect on a diver's health. The psychiatrist needs to understand the problems of isolation and possible overcrowding that can occur in saturation diving and also the stresses on the families of divers who may be isolated for long periods in a dangerous situation.

One needs to know too if the divers are especially sensitive to pressure, nitrogen narcosis and oxygen toxicity or if they are predisposed to decompression illness. The latter information can only be achieved by a test dive. However, diver selection remains a difficult task. It is often difficult to assess a diver's reactions to a deep dive until he is exposed.

### Infections

A moist humid environment allows rapid growth of infections of the ear and skin, principally due to pseudomonas and common fungal infections. Prophylactic measures are available, especially for the ears (Edmonds *et al.*, 1981; Farmer, 1982) and the divers must be trained in absolute cleanliness and in keeping the skin dry. Regular ear prophy- laxis and reporting to the physician the mildest of early symptoms is vital. Prophylaxis consists of putting a solution of 2% acetic acid in each ear for 5 minutes. Once ear pain occurs it may be intransigent to treatment and a diver may have to be brought to the surface, as the pain may be excruciating.

## Common medical and surgical illness during saturation diving

Illness such as common colds, stomach upsets, etc. can be treated easily with suitable drugs from outside the saturation chamber. It should be remembered, however, that there is very little information on how drugs interact with pressure, and care must be exercised in their use. Surgery, if required at pressure, could be a major problem. Anaesthesia is usually best by regional methods or steroidal anaesthetics, as most general anaesthetics are contraindicated. One would also need a surgical team trained to function at pressure if an emergency should arise.

### Aseptic bone necrosis

The risk of aseptic bone necrosis (ABN) increases significantly in saturation diving (McCallum, 1982). Little is really known of the cause of ABN or whether it is related to decompression sickness or not. It is likely that the mere presence of so-called 'silent' bubbles in the blood stream arising during any decompression from whatever depth affects the medullary blood supply to the bone. Problems occurring in the shaft are not as serious as at an articular surface such as the hip joint, which can result in serious disability.

### Post-diving neurological deficits

There are many documented cases of neurological dysfunction with loss of memory and personality changes resulting from central nervous system decompression sickness. An indication that 'near miss' de- compression accidents may result in neurological deficits during decompression from compressed air dives has been reported (Vaernes and Eidsvik, 1982).

Despite much speculation to the contrary there is no evidence of any similar damage occurring from deep oxygen-helium or Trimix dives with their very slow decompression schedules (Shields *et al.*, 1983). Nevertheless, considerable care is required and annual clinical psycho- logical testing is advised for all personnel regularly diving under these conditions.

# 4. *Physiological stresses*

Most life support factors may be provided by good engineering, good equipment and knowledge of the limits of pressure exposure. The actual physiological stresses of diving, however, are less amenable to solution and require careful and continual attention to compression profiles, gas mixtures and decompression methods on the basis of accrued knowledge from past research.

These stresses include: oxygen toxicity, nitrogen narcosis, respiratory failure, temperature, pressure itself—high pressure nervous syndrome, and decompression safety, as illustrated in Figure 16.1.

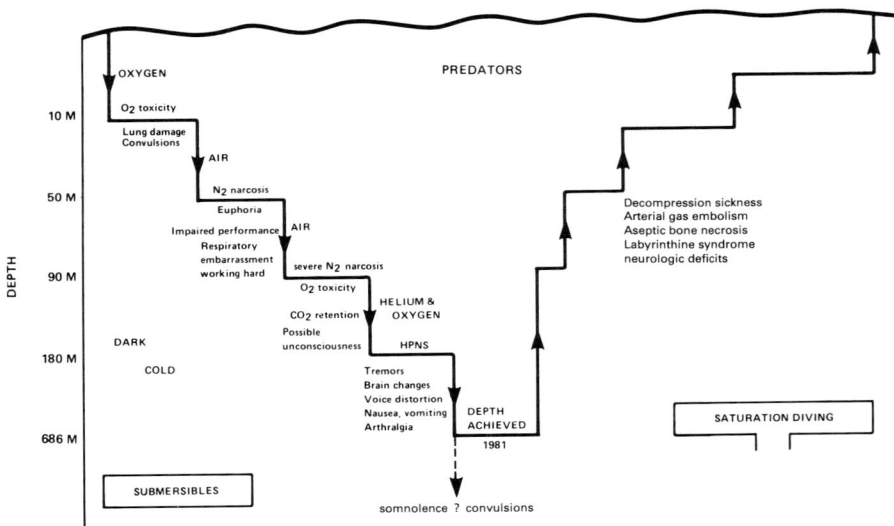

*Figure 16.1. Physiological and other limitations to diving.*

## Oxygen toxicity

Oxygen toxicity occurs when oxygen partial pressure is in excess of 0·5 ata. Its severity increases progressively with elevation of the inspired oxygen pressure and with the duration of exposure. In fact, under the correct combination of oxygen pressure and exposure conditions any living cell will become functionally compromised and eventually die. Two areas of oxygen toxicity are of major importance, namely those affecting the pulmonary and central nervous systems (Bean, 1945; Clark and Lambertsen, 1971; Clarke, 1982).

Pulmonary toxicity is manifest by destruction of capillary and alveolar epithelium, alveolar cell hyperplasia, oedema, haemorrhage, arterial

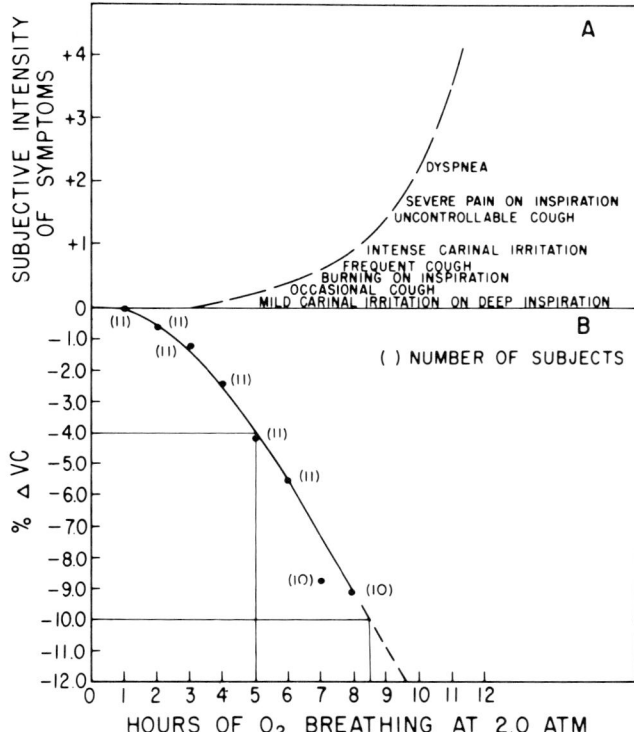

*Figure 16.2.    The rate of decrease in vital capacity and increasing severity of symptoms during continuous oxygen breathing at 2·0 ATA (from Clark and Lambertsen, 1971).*

thickening and hyalinization, fibrin formation, atelectasis and consolidation. This results in the paradox of an individual becoming hypoxaemic and dying while living in a high oxygen environment.

Clarke and Lambertsen (1971) have attempted to correlate the subjective signs and symptoms of oxygen toxicity in humans, with a fall in vital capacity (Figure 16.2). Very often however, the first signs of pulmonary toxicity are substernal pain and throat irritation. If allowed to continue, the symptoms become severe with coughing and dyspnoea.

As with most of the physiological limitations of hyperbaric exposure there is a wide variation in human susceptibility. One of the most useful methods for delaying toxicity is to institute air breaks such as 25 minutes of oxygen breathing followed by 5 minutes of air, repeated. This type of profile is especially useful in the treatment of decompression accidents with recompression and utilization of oxygen partial pressures as high as 2–4 ata. Care is then required to provide oxygen sufficient to resolve decompression sickness but insufficient to produce pulmonary toxicity.

Central nervous system oxygen toxicity at pressures in excess of 2 ata is principally manifest by tonic, clonic convulsions similar to those observed in epilepsy (Figure 16.3). If oxygen pressure is maintained at a 2 ata level then progressive neuronal destruction will occur resulting in permanent paralysis or death (Donald, 1947; Lambertsen, 1965).

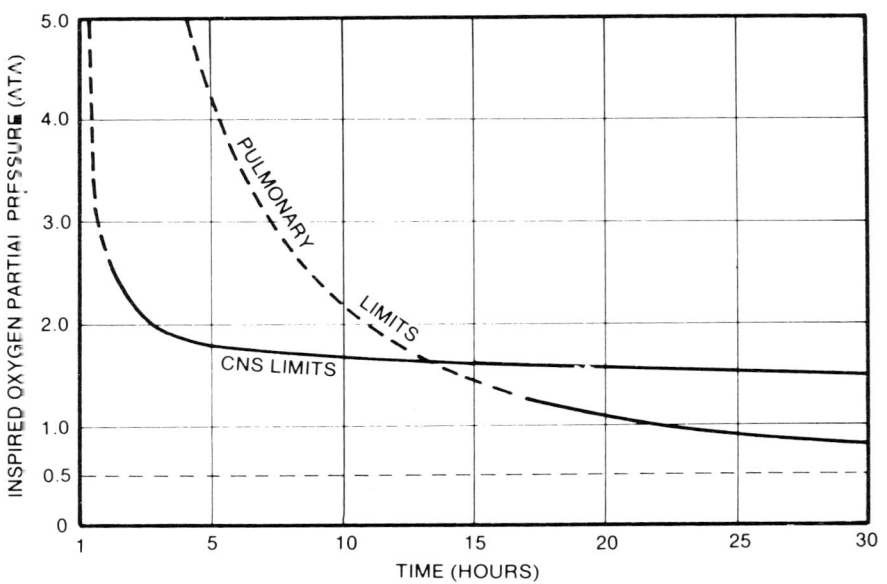

*Figure 16.3.   Predicted human pulmonary and CNS tolerance to high pressure oxygen (from Edmonds et al., 1976).*

The causes of the toxicity are due to oxygen at increased pressure acting as a widespread metabolic and biochemical poison. Recent interest has turned to the production of free radicals such as the superoxide anion and the role of superoxide dismutase, an enzyme which catalytically scavenges this free radical. Administration of superoxide dismutase and catalase in micelles to rats has produced a profound effect in reducing toxicity (Yusa *et al.*, 1984).

Nevertheless, oxygen toxicity remains a major limitation to performance in hyperbaric environments and the widespread nature of the cellular damage produced appears to make the solution of the problem extremely difficult.

## Nitrogen narcosis

The nitrogen component of compressed air is as much a limitation on human hyperbaric performance as is excess oxygen, but for different reasons. The physiologically inert nitrogen—under pressure—is responsible for the signs and symptoms of intoxication and narcosis (Behnke *et al.*, 1935; Bennet, 1966, 1982b). The effect is felt first at depths of 15–30 m. At 30 m divers experience feelings of stimulation, excitement, euphoria, a slowing of mental activity, impairment of memory and concentration. Intellectual functions are usually impaired more than psychomotor. As the depth increases narcosis becomes more severe until at depths deeper than 100 m stupefaction or loss of consciousness results. The effects are similar to alcoholic intoxication, hypoxia and the early stages of gaseous anaesthesia. Indeed at sufficient pressure nitrogen will certainly cause anaesthesia.

Many factors potentiate the narcotic effect, such as alcohol (Jones *et al.*, 1979; Fowler *et al.* 1985), fatigue and hard work (Adolfson, 1965), apprehension and anxiety (Davis *et al.*, 1972). An increase in exogenous or endogenous carbon dioxide (Hesser *et al.*, 1971, 1978) also produces similar potentiation. An amnesia of events occurring at increased pressure while narcotic is common. On return to lower pressures of less than 15 m the narcosis disappears.

The mechanism of the narcosis is not understood but it is related to the mechanisms of general anaesthesia. Evidence suggests that the synapse is the site of action (Bennett, 1966), and that failure to propagate the potential across the synaptic gap is likely to be due to impairment of the mechanisms for the release of the various neurotransmitters in the polysynaptic systems of the brain, such as the reticular activating system of the brain stem and cortex (Bennett, 1966).

Prevention of narcosis has been achieved as in oxygen toxicity by control of the cause, rather than from any detailed understanding of the mechanism involved. Thus divers breathing air are advised not to dive deeper than 40 m and to breathe helium–oxygen instead of air.

Many of the physical constants of gases correlate with narcotic potency, in particular, lipid solubility (Table 16.1). Examination of the inert gases shows that helium and neon are less narcotic than nitrogen—the former markedly so—whereas argon, krypton and xenon are more narcotic. Thus helium has been the gas of choice for deep diving in order to prevent signs and symptoms of narcosis. The lower density of helium also helps to prevent respiratory problems.

## Respiratory embarrassment

A factor of considerable importance, especially to the diver working in water, is the effect of the increased density on respiration (Lanphier and

*Table 16.1.　Correlation of narcotic potency of the inert gases, hydrogen, oxygen and carbon dioxide with lipid solubility and other physical characteristics.*

| Gas | Molecular weight | Solubility lipid | Temperature (°C) | Oil–water solubility ratio | Relative narcotic potency |
|---|---|---|---|---|---|
| | | | | | (least narcotic) |
| He | 4 | 0·015 | 37 | 1·7 | 4·26 |
| Ne | 20 | 0·019 | 37·6 | 2·07 | 3·58 |
| H$_2$ | 2 | 0·036 | 37 | 2·1 | 1·83 |
| N$_2$ | 28 | 0·067 | 37 | 5·2 | 1 |
| A | 40 | 0·14 | 37 | 5·3 | 0·43 |
| Kr | 83·7 | 0·43 | 37 | 9·6 | 0·14 |
| Xe | 131·3 | 1·7 | | 20·0 | 0·039 |
| | | | | | (most narcotic) |
| | | | 40 | | |
| O$_2$ | 32 | 0·11 | 40 | 5·0 | |
| CO$_2$ | 44 | 1·34 | | 1·6 | |

Camporesi, 1982). The effect is central to the design of underwater breathing apparatus (Morrison and Reimers, 1982). There are only about two diving apparatus that will permit divers to breathe effectively at 500 m and one of these is still in the development stage.

Apart from the more obvious effect of a decreased efficiency of ventilation while breathing compressed air of increasing gas density, dyspnoea or breathlessness has been frequently reported, especially during recent deep oxygen–helium research dives. Dwyer *et al.* (1977) at 1 400 feet (about 450 m) and Spaur *et al.* (1977) at 1 609 feet (about 500 m) reported that divers working in the water rapidly became exhausted at moderate work rates and experienced severe breathlessness. Thalmann and Piantadosi (1981) noted similar results at 1 815 feet (about 600 m) with profound dyspnoea, limiting underwater exercise to only 30−40% of surface values. During the Atlantis deep Trimix dives at Duke Medical Center (Moon *et al.*, 1980; Salzano *et al.*, 1981), it was evident that dyspnoea was more prevalent when breathing heliox than when breathing Trimix. A submerged diver whose ventilation is hampered by a breathing apparatus will be in a life-threatening situation if he suffers acute dyspnoea. This is a vital area of further research.

An increase in tissue carbon dioxide tensions, produced by the increased effort of breathing a denser gas, synergistically potentiates nitrogen narcosis, oxygen toxicity and decompression sickness. The advent of hydrogen–helium–oxygen mixtures for diving may be a considerable help but the precise role dyspnoea plays in diving awaits elucidation.

# Temperature

A typical temperature of sea water in northern latitudes may be $0-4°C$ so that protection from cold by special clothing, wet suits, dry suits and hot water suits is vital. In deep saturation dives, where divers are remote from surface help and where they may be diving several times daily, the effects of hypothermia may become critical (Webb, 1982). To prevent such hypothermia the water temperature would have to be 33°C.

At depths greater than 200 m the loss of heat through breathing oxygen–helium becomes greater than may be produced metabolically and it is not sufficient to rely solely on heating the body because heliox has six times the thermal conductivity of air. Hoke *et al.* (1975) exposed divers to warm heliox at 820 feet (about 250 m) and gave them cold gas via the breathing apparatus. Violent shivering ensued after 15–20 minutes. Piantadosi *et al.* (1981) warmed divers at depths to 1 804 feet (about 600 m) breathing oxygen–helium at 14°C with similar results. A major problem arising from breathing cold gas is the stimulation of the secretion of large amounts of fluids and mucous in the respiratory tract, which causes choking and gagging and may foul the mouthpiece. Thus the breathing gas must be heated.

Comfort temperatures narrow greatly with increasing pressure from 1·5 to 50 ata. Thus at 2·2 ata divers may tolerate temperature variations of $±2°C$ but at 30 ata this is narrowed to 0·5°C and comfort temperatures may be as high as 32°C or even higher at greater depths. Kuehn and Zumrick (1978) took warm divers at 30 ata, 37 ata and 43 ata and put them in a precooled chamber wearing only swimming trunks to test their voluntary limits. At 15°C it was too cold for pressures above 21 ata, at which tolerance was 1 h. At 30 ata, 20°C tolerance was 1·1 h, at 25 ata tolerance was 2·5 h and at 43 ata it was 1·25 h. Clearly heat loss is rapid and dangerous in diving.

Engineering solutions are available, albeit by the crudest and sometimes most dangerous methods such as hot water supplied to the suit and breathing apparatus, rebreather circuits, etc., and improved design and methods are required. One should never become complacent about the dangers of hypothermia as its effects may be insidious. In slow cooling there is little warning before body temperature falls below 35°C and symptoms appear such as mental confusion, lethargy, poor speech articulation, hallucinations, decreased sensation and impaired motor function.

Again the solution to this problem has been to try to avoid it by prevention. Present technology (heated suits and similar) leaves much to be desired and it is time diver heating technology matched that developed for the astronaut.

## 5. *The high pressure nervous syndrome*

In the 1930s substitution of nitrogen gas by helium eliminated signs and symptoms of nitrogen narcosis and seemed to have cured the problem. In the mid-1960s however, during dives to only 600 feet (about 200 m), with compressions at 30 m min$^{-1}$, divers showed decrements in performance efficiency of some 20% which at 800 feet (about 250 m) were twice as severe (Bennett, 1965; Bennett and Dossett, 1967). These decrements were accompanied by dizziness, nausea, vomiting, tremours of the hands and fatigue. This was the first observation of a condition which has set definite limitations on the manner in which divers can dive to 1 968 feet (about 650 m) and arrive fit and able to work, and is now called the high pressure nervous syndrome (Bennett, 1982a).

Since 1965 considerable research has been performed to understand and control this syndrome better. It is manifested by tremors, myoclonic jerking, nausea, vomiting, fatigue, somnolence and lapses of consciousness, breathlessness or dyspnoea, poor sleep with nightmares and arthralgias. During this period over 170 men have made over 50 physiological research dives between 100 feet (about 30 m) and 2 250 feet (about 700 m) in the Duke University Diving Centre and strategies have evolved for control of the condition.

These involve the use of a number of factors (Bennett, 1975):

1. Selection of least susceptible divers.
2. Choice of a suitable slow rate of compression.
3. Stages during compression.
4. Use of nitrogen or hydrogen (Trimix).
5. Adaptation.
6. Use of excursion dives.

These experimental dives involved various methods of gas breathing (Bennett, 1980):

1. Helium-oxygen compression.
2. Helium-oxygen compression plus excursions.
3. Helium-nitrogen-oxygen; helium-hydrogen-oxygen (Trimix).
4. Trimix plus excursions.

It is not possible here to review all of these data, also available elsewhere (Bennett, 1982a), nor the important record breaking Trimix studies at the F G Hall Laboratory over the last few years (Bennett *et al.*, 1981, 1982; Bennett and McLeod, 1984) and the Duke/German research since 1983 to 600 m (Bennet *et al.*, 1986). On the basis of considerable data it is concluded that it is possible for most men to dive in a fit and working condition to as deep as 2 250 feet (about 750 m), possibly beyond. This should be done with a slow exponential, never linear, compression rate, with long 14-hour + stages at various depths during compression for the purposes of allowing the body to adapt to the pressure. The use of Trimix, with between 5–8% nitrogen in oxygen-helium, is merited at depths greater than 1 500 feet (about 500 m) to assist control of tremors, nausea and other incapacitating effects of HPNS.

## 6. Decompression

The return to the surface from great depths remains a difficult task if decompression sickness with joint pains, paralysis or death is to be avoided. Few of the deep research dives made to date have escaped one or more of the divers experiencing the joint pain of the bends. Yet the decompression profiles have become longer and longer requiring some 31 days for Atlantis III from 2 250 feet (686 m) and 28 days from 2 132 feet (650 m) with the deep part of the decompression as slow as $0.9$ m h$^{-1}$ or less.

The physiology and biophysics of decompression remain a mystery (Hempleman, 1982; Vann, 1982). Decompression even from common shallow dives using US Navy tables are based on unsubstantiated principles and most tables are merely empirical solutions to the need for safe decompression. Once these decompression table concepts are extrapolated to depths greater than 160 feet (50 m) the incidence of decompression sickness rises sharply. Safe decompression for the present relies on practical experience of the wide range of tables available and the underlying physiology of factors such as gas uptake and elimination, haematologic changes and the additional effects of such factors as cold and exercise at work.

Choices need to be made between exponential or linear decompression profiles. For deep diving, there is a strong inclination to utilize linear rates at constant oxygen partial pressures as it decreases the number of variables and builds a unique body of data pointing to safe, though long, decompression tables.

## 7. Conclusions

In conclusion it is clear that we do not yet know sufficient diving physiology and medicine to be able to state with assurance that divers may dive to even shallow depths without certain risks. These, however, are calculated risks and should not deter us from continuing to explore the capabilities of man under the sea. Using the knowledge available, a well-trained staff and good pressure chamber facilities and diving equipment it is possible to proceed in a responsible and safe manner in hyperbaric environments.

Remote operating vehicles (ROV) and one atmosphere systems which are in increasing use today also are methods to help reduce the physiological and medical risks to divers. As in the space programme, and the quest for new worlds, man is learning how to tame the hyperbaric environment and push himself to ever higher pressures, and so open up the world beneath the oceans. The success of these endeavours will depend on the finance and support for continued research into the effects of hyperbaric environments on human performance.

## References

Adolfson, J., 1965, Deterioration of mental and motor functions in hyperbaric air. *Scand. J. Psychol.*, **6**, 26−31.

Bean, J.W., 1945, Effects of oxygen at high pressure. *Physiol. Rev.*, **25**, 1−147.

Behnke, A.R. and Yarbrough, O.D., 1939, Respiratory resistance, oil−water solubility and mental effects of argon compared with helium and nitrogen. *Amer. J. Physiol.*, **126**, 409−415.

Behnke, A.R., Thomas, R.M. and Motley, E.P., 1935, The psychologic effects from breathing air at 4 atmospheres pressure. *Amer. J. Physiol.*, **11**, 554−558.

Bennett, P.B., 1965, *Psychometric Impairment in Men Breathing Oxygen-Helium at Increased Pressures*, Medical Research Council, R N Personnel Research Committee, Underwater Physiology Sub-Committee, Report No. 251.

Bennett, P.B., 1966, *The Aetiology of Compressed Air Intoxication and Inert Gas Narcosis* (Oxford: Pergamon Press).

Bennett, P.B., 1975, A strategy for future diving. In *The Strategy for Future Diving to Depths Greater than 1000 Feet*, 8th Undersea Medical Society Workshop. Report No. WS-6-15-75 (Bethesda, MD: Undersea Medical Society), pp. 71−76.

Bennett, P.B., 1980, Potential methods to prevent the HPNS in human deep diving. In *Techniques for Diving Deeper than 1500 feet*, 23rd Undersea Medical Society Workshop. Report No. 40 WS (DD) 6-30-80 (Bethesda, MD: Undersea Medical Society), pp. 36−47.

Bennett, P.B., 1981, The United States national diving accident network. *EMT J.*, **5**, 323−327.

Bennett, P.B., 1982a, The high pressure nervous syndrome. In *The Physiology and Medicine of Diving*, 3rd edition, edited by P.B. Bennett and D.H. Elliott (London: Bailliere Tindall), pp. 262−296.

Bennett, P.B., 1982b, Inert gas narcosis. In *The Physiology and Medicine of Diving*, 3rd edition, edited by P.B. Bennett and D.H. Elliott (London: Bailliere Tindall), pp. 239−261.

Bennett, P.B. and Dossett, A.N., 1967, *Undesirable Effects of Oxygen-Helium Breathing at Great Depths*, Medical Research Council, R N Personnel Research Committee, Underwater Physiology Sub-Committee, Report No. 251.

Bennett, P.B. and McLeod, M., 1984, Probing the limits of human deep diving. In *Diving and Life at High Pressures*, edited by W.D.M. Paton, D.H. Elliott and E.B. Smith. *Phil. Trans. R. Soc. Lond.*, **B304**, 105–117.

Bennett, P.B., Coggin, R. and Roby, J., 1981, Control of HPNS in humans during rapid compression with trimix to 650 m (2132 ft). *Undersea Biomed. Res.*, **8**, 85–100.

Bennett, P.B., Coggin R. and McLeod, M., 1982, Effect of compression rate on use of trimix to ameliorate HPNS in man to 686 m. *Undersea Biomed. Res.*, **9**, 335–351.

Bennett, P.B., Schafstall, H., Schnegelsberg, W. and Vann, R., 1986, An analysis of fourteen successful Trimix 5 deep saturation dives between 150 m–600 m. *Proceedings 9th Symposium on Underwater Physiology*, Kobe, Japan, September.

Bert, P., 1878, *La Pression Barometrique* (Paris: Masson).

Bond, G.F., 1966, Effects of new and artificial environments on human physiology. *Arch. Environ. Hlth*, **12**, 85–90.

Clarke, J.M., 1982, Oxygen toxicity. In *The Physiology and Medicine of Diving*, 3rd edition, edited by P.B. Bennett and D.H. Elliott (London: Bailliere Tindall), pp. 200–238.

Clarke, J.M. and Lambertsen, C.J., 1971, Pulmonary oxygen toxicity: a review. *Pharm. Rev.*, **23**, 37–133.

Davis, F.M., Osborne, J.P., Baddeley, A.D. and Graham, I.M.F., 1972, Diver performance: nitrogen narcosis and anxiety. *Aerospace Med.*, **43**, 1079–1082.

Davis, J.C. and Elliott, D.H., 1982, Treatment of decompression disorders. In *The Physiology and Medicine of Diving*, 3rd edition, edited by P.B. Bennett and D.H. Elliott (London: Bailliere Tindall), pp. 473–487.

Davis, J.C. (Chairman), 1979, *Treatment of Serious Decompression Sickness and Arterial Gas Embolism*. 20th Undersea Medical Society Symposium at Duke University, UMS Publication No. 34 WS (SDS) 11-30-79 (Bethesda, MD: Undersea Medical Society).

Donald, K.W., 1947, Oxygen poisoning in man, I and II. *Brit. Med. J.*, **1**, 667–672, 712–717.

Dwyer, J., Saltzman, H.A. and O'Brian, R., 1977, Maximal physical work capacity of man at 43·4 ATA. *Undersea Biomed. Res.*, **4**, 359–372.

Edmonds, C., Lowry, C. and Pennefather, J., 1981, *Diving and Subaquatic Medicine* (Melbourne: Diving Medical Center/California: Best Publishing).

Elliott, D.H. and Kindwall, E.P., 1982, Manifestations of the decompression disorders. In *The Physiology and Medicine of Diving*, 3rd edition, edited by P.B. Bennett and D.H. Elliott (London: Bailliere Tindall), pp. 461–472.

Farmer, J.C., 1982, Otologic and paranasal sinus problems in diving. In *The Physiology and Medicine of Diving*, 3rd edition, edited by P.B. Bennett and D.H. Elliott (London: Bailliere Tindall), pp. 505–536.

Fowler, B., Ackles, K.N. and Parlier, G., 1985, Effects of inert gas narcosis on behavior—a critical review. *Undersea Biomed. Res.*, **12**, 369–402.

Hempleman, H.B., 1982, History of evolution of decompression procedures. In *The Physiology and Medicine of Diving*, 3rd edition, edited by P.B. Bennett and D.H. Elliott (London: Bailliere Tindall), pp. 319–351.

Hesser, C.M., Adolfson, J. and Fagraeus, L., 1971, Role of $CO_2$ in compressed air narcosis. *Aerospace Med.*, **42**, 163−168.

Hesser, C.M., Fagraeus, L. and Adolfson, J., 1978, Roles of nitrogen, oxygen and carbon dioxide in compressed air narcosis. *Undersea Biomed. Res.*, **5**, 381−400.

Hoke, B., Jackson, D.L., Alexander, J.M. and Flynn, E.T., 1975, Respiratory heat loss and pulmonary function during cold gas breathing at high pressures. In *Underwater Physiology, Vth Symposium on Underwater Physiology*, edited by C.J. Lambertsen (Bethesda, MD: Federation of American Societies for Experimental Biology), pp. 725−740.

Jones, A.W., Jennings, R.D., Adolfson, J. and Hesser, C.M., 1979, Combined effects of ethanol and hyperbaric air on body sway and heart rate in man. *Undersea Biomed. Res.*, **6**, 15−25.

Kuehn, L.A. and Zumrick, J., 1978, Thermal measurements on divers in hyperbaric helium-oxygen environments. *Undersea Biomed. Res.*, **5**, 213−231.

Lambertsen, C.J., 1965, Effects of oxygen at high partial pressure. In *Handbook of Physiology*, Section 3: Respiration, edited by W.O. Fenn and H. Rahn (Washington, DC: American Physiological Society), Vol. 2, pp. 1027−1046.

Lanphier, E.H. and Camporesi, E.M., 1982, Respiration and exercise. In *The Physiology and Medicine of Diving*, 3rd edition, edited by P.B. Bennett and D.H. Elliott (London: Bailliere Tindall), pp. 99−156.

McCallum, R.I. and Harrison, J.A.B., 1982, Dysbaric osteonecrosis: Aseptic necrosis of bone. In *The Physiology and Medicine of Diving*, 3rd edition, edited by P.B. Bennett and D.H. Elliott (London: Bailliere Tindall), pp. 488−506.

Miller, J.N., Fagraeus, L., Bennett, P.B., Elliott, D.H., Shields, T.G. and Grimstad, J., 1978, Nitrogen-oxygen saturation therapy in serious cases of compressed-air decompression sickness. *Lancet*, **1**, 169−171.

Moon, R.E., Salzano, J.V., Camporesi, E.M. and Stolp, B.W., 1980, Exercise induced dyspnea at 46·7 and 65·6 ATA. *Physiologist*, **23**, 86.

Morrison, J.B. and Reimers, S.D., 1982, Design principles of underwater breathing apparatus. *The Physiology and Medicine of Diving*, 3rd edition, edited by P.B. Bennett and D.H. Elliott (London: Bailliere Tindall), pp. 55−98.

Piantadosi, C.A., Thalmann, E.D. and Spaur, W.H., 1981, Metabolic response to respiratory heat loss induced core cooling. *J. Appl. Physiol.: Resp. Environ. Exercise Physiol.*, **50**, 829−834.

Salzano, J.V., Stolp, B.W., Moon, R.E. and Camporesi, E.M., 1981, Exercise at 47 and 66 ATA. In *Underwater Physiology VII*, Proceedings of the 7th Symposium on Underwater Physiology, edited by A.J. Bachrach and M.M. Matzen (Bethesda, MD: Undersea Medical Society), pp. 181−196.

Shields, T.G., Minsas, B., Elliott, D.H. and McCallum, R.I., 1983, *Long Term Neurological Consequences of Deep Diving* (Stavanger, Norway: A.S. Verbum).

Spaur, W.H., Raymond, L.W., Knott, M.M., Crothers, J.C., Braithwaite, W.R., Thalmann, E.D. and Uddin, D.F., 1977, Dyspnea in divers at 49·5

ATA. Mechanical not chemical in origin. *Undersea Biomed. Res.*, **4**, 183–198.

Thalmann, E.D. and Piantadosi, C., 1981, Submerged exercise at pressure up to 55·55 ATA. *Undersea Biomed. Res.*, **8**, Suppl., 23.

Vaernes, R.J. and Eidsvik, S., 1982, Central nervous dysfunctions after near miss accidents in diving. *Aviat. Space Environ. Med.*, **53**, 803–807.

Vann, R.D., 1982, Decompression theory and applications. In *The Physiology and Medicine of Diving*, 3rd edition, edited by P.B. Bennett and D.H. Elliott (London: Bailliere Tindall), pp. 352–382.

Webb, P., 1982, Thermal problems. In *The Physiology and Medicine of Diving*, 3rd edition, edited by P.B. Bennett and D.H. Elliott (London: Bailliere Tindall), pp. 297–318.

Workman, R.D., Bond, G.F. and Mazzone, W.F., 1962, Prolonged exposure of animals to pressurized normal and synthetic atmospheres. *US Naval Medical Research Laboratory, New London, Research Report*, **26**(5), 1–19.

Yusa, T., Crapo, J.D. and Freeman, B.A., 1984, Liposome mediated augmentation of brain S.O.D. and catalase inhibits CNS $O_2$ toxicity. *J. Appl. Physiol.: Respirat. Environ. Exercise Physiol.*, **57**, 1674–1681.

# 17. Modelling Human Exposure to Altered Pressure Environments

## T. R. Hennessy

## 1. Introduction

It is now accepted that, shortly after commencing a reduction in environmental pressure around humans, inert gas begins to form in some tissues, and if the pressure drop is too large or too rapid, a critical excess quantity of undissolved gas will appear in a particular tissue, resulting in decompression sickness. It is therefore necessary to develop a model capable of predicting the uptake and elimination of both dissolved and undissolved inert gas in these tissues throughout an exposure to altered environmental pressure.

### Gas exchange parameters

The basic tissue parameters for dissolved nitrogen are shown in Figure 17.1. For a perfusion limited tissue, the blood perfusion $P$ (ml blood/ml tissue/min) and the tissue/blood solubility partition coefficient $\alpha_p$ are the chief parameters. The characteristic time scale is given by $\alpha_P/P$ and $P$ ranges from 0·2 to 500 minutes.

For a diffusion limited tissue, the characteristic time scale is given by $L^2/D$, where $L$ is the effective intercapillary distance and $D$ is the tissue diffusion coefficient. The time scale ranges from 15 to 167 minutes.

In practice, diffusion limited and perfusion limited tissue adjoin or are merged with each other. Hence for some, possibly all, tissues the perfusion and diffusion time scales will be of a similar order of magnitude and obviously there will be an interaction between the two gas exchange processes in the tissue.

However, the modelling of gas exchange developed along quite different lines. The earliest tissue model of Zuntz (1897) and Boycott et al. (1908) was based only on a perfusion exchange process, as shown in Figure 17.2. A key approximation is the replacement of the unknown venule inert gas partial pressure $p_v$ with $p_t$. This amounts to assuming

316

## PERFUSION $\boxed{\propto p / P}$

$\propto p$ = tissue / blood sol. partition coeff.

  ~1 aqueous tissue

  ~5 lipid tissue, nitrogen

P  = blood perfusion

$\frac{1}{P}$ ~·2 min well perfused tissue

  ~100 min poorly perfused tissue

SLOW TISSUE   $\propto p / P$ ~ 500 min

FAST TISSUE         ~ ·2 min

---

## DIFFUSION $\boxed{L^2 / D}$

L  = diffusion path length

D  = diffusion coefficient

Well perfused tissue   $L \sim 30 \mu m$

Poorly perfused tissue   $L \sim 300 \mu m$

$D \sim 10^{-6} cm^2 / sec = 6 \times 10^3 \mu m^2 / min$

$L^2 / D \sim$   15 min ( Well perfused )

      ~ 167 min ( Poorly perfused )

*Figure 17.1.   Tissue parameters.*

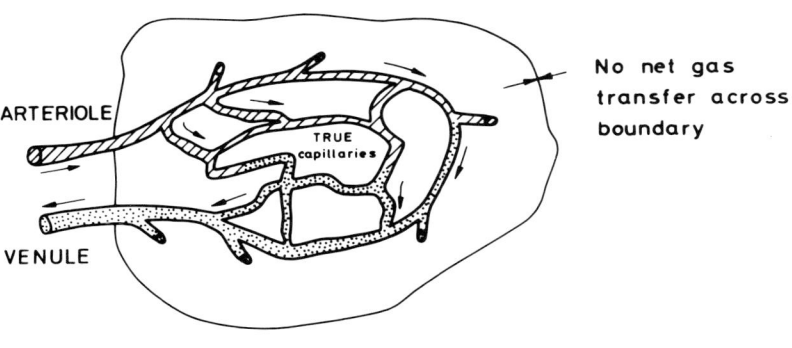

ARTERIOLE

TRUE capillaries

VENULE

No net gas transfer across boundary

$$\frac{dP_t}{dt} = \frac{P}{\propto t / \propto_b} (P_a - P_v) \approx \frac{P}{\propto t / \propto_b} (P_a - P_t)$$

*Figure 17.2.   The Zuntz–Haldane model.*

(*a*)

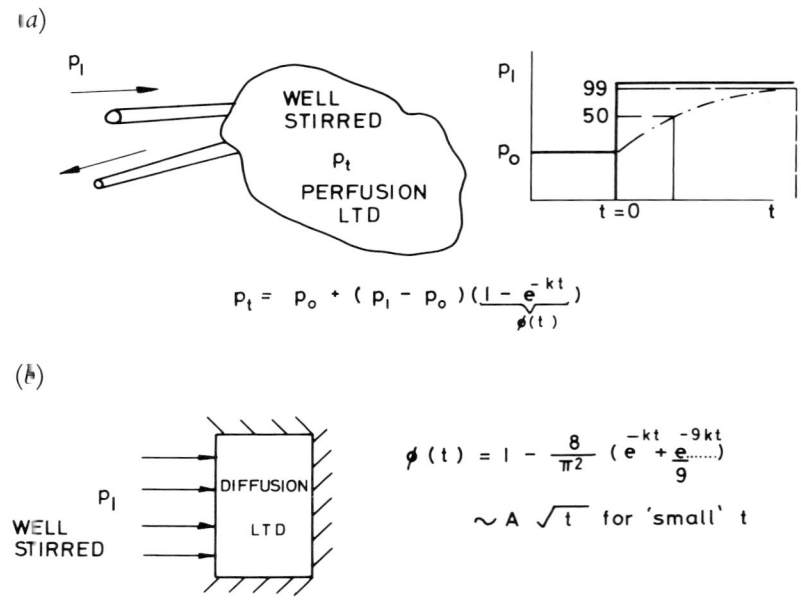

Figure 17.3.    (a) *The Haldane model;* (b) *the Hempleman model.*

that the flow in the venule leaves the tissue in equilibrium with the average partial pressure (equivalent to assuming the tissue to be a well-stirred fluid). This enables the simple gas balance differential equation to be solved, leading to the classical mono-exponential uptake response curve following a step change in arteriole nitrogen partial pressure, as shown in Figure 17.3(*a*).

In 1951 Hempleman (1952) noticed that the nil-decompression curve could be better approximated using a $\sqrt{t}$ rather than a mono-exponential response. This strongly suggested a diffusion rather than a perfusion limited model, and the simplest case is one-dimensional diffusion into a slab (see Figure 17.3(*b*)). The response to a step change in arterial $PN_2$ is multi-exponential which, at sufficiently small times, may be approximated by a simple $\sqrt{t}$ function.

Actually the $\sqrt{t}$ response does not necessarily imply an underlying diffusion process. A wide variety of multi-exponential functions have this property, not only those that satisfy Fick's well-known diffusion equation. This is shown in Figure 17.4 where a mono-exponential (not passing through the origin) closely approximates a $\sqrt{t}$ response over quite long times. The apparent presence of a $\sqrt{t}$ relationship is unfortunately misleading but has had a profound influence on decompression modelling and has recently been resurrected (Thalmann, 1986; Vann, 1986).

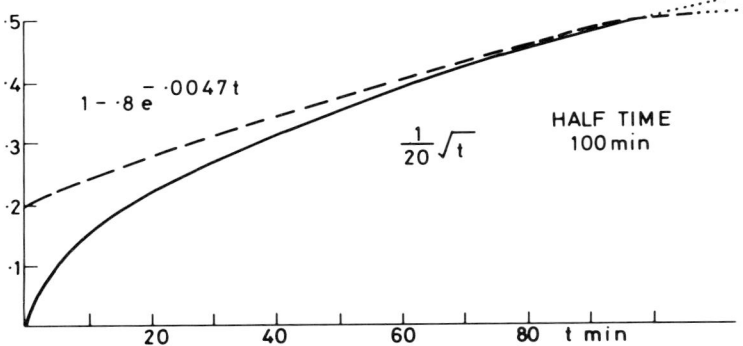

Figure 17.4.   *Mono-exponential vs. square root of time response.*

## 2. Compartmental modelling

Returning to the concept that both perfusion and diffusion contribute to the gaseous uptake response, Figure 17.5 shows a simple attempt to model both intertissue diffusion and perfusion processes. Variations of this model were proposed (from different standpoints) by Perl (1963) and

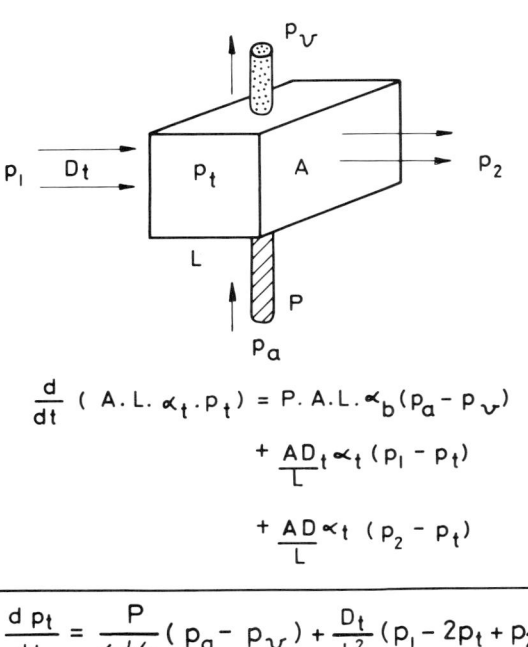

$$\frac{d}{dt}(A.L.\alpha_t.P_t) = P.A.L.\alpha_b(P_a - P_v)$$
$$+ \frac{AD_t}{L}\alpha_t(P_1 - P_t)$$
$$+ \frac{AD}{L}\alpha_t(P_2 - P_t)$$

$$\boxed{\frac{dP_t}{dt} = \frac{P}{\alpha_t/\alpha_b}(P_a - P_v) + \frac{D_t}{L^2}(P_1 - 2P_t + P_2)}$$

Figure 17.5.   *The intertissue perfusion–diffusion interaction.*

1.  SINGLE COMPARTMENT (ZUNTZ-HALDANE)

$$\frac{dp_I}{dt} = k(P_a - P_I) \qquad P_a \xrightarrow{\ k\ } \boxed{P_I} \ \ (P_{01} \text{ initially})$$

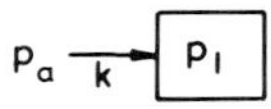

step change (Pa const)

$$P_I = P_{01} + (P_a - P_{01})(1 - e^{-kt})$$

MONO-EXPONENTIAL RESPONSE

2.  DOUBLE COMPARTMENT

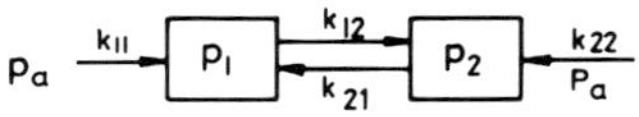

$$P_a \xrightarrow{k_{11}} \boxed{P_I} \ \underset{k_{21}}{\overset{k_{12}}{\rightleftarrows}} \ \boxed{P_2} \xrightarrow{k_{22}} P_a$$

$$(P_{01}, P_{02} \text{ initially})$$

$$\frac{dp_I}{dt} = k_{11}(P_a - P_I) - k_{12}(P_I - P_2)$$

$$\frac{dp_2}{dt} = k_{21}(P_I - P_2) - k_{22}(P_2 - P_a)$$

step change

$$P_I = P_{01} A_{11} e^{-\lambda_1 t} + A_{12} e^{-\lambda_2 t}$$

$$P_2 = P_{02} A_{21} e^{-\lambda_1 t} + A_{22} e^{-\lambda_2 t}$$

DOUBLE EXPONENTIAL RESPONSE

*Figure 17.6.   Compartmental models.*

Hempleman (1963). Obviously the model cannot be solved without further information describing the adjacent structures.

It is possible to partially incorporate these features using a compartmental modelling approach. This offers a convenient, if semi-empirical method of developing a useful computational model which is able to maintain the characteristic gas exchange time-scales. The set of parameters of the model is solved by simply fitting the 'output response' to an experimental data curve. Two popular compartmental models are shown in Figure 17.6.

The advantage is that it is not necessary to have to interpret the biophysical nature of the various parameters in order to use the compartmental model at different environmental pressures, provided that the basic tissue perfusion and diffusion time-scales remain the same. In fact the tissue time-scales will be altered by any change in the inert gas and the compartmental model can only be used to predict a decompression schedule for a single inert gas. Indeed, changes in environmental temperature and metabolic rate will also alter the time-scales and will have a substantial effect on any decompression table developed under a different set of conditions. Such effects are well-known but cannot be incorporated into the above simple compartmental models and so have been usually ignored in early models.

However, using modern numerical optimization methods, it is possible to construct a multi-parameter computational model based on a number of single (or double) compartments. This is able to predict accurately the results of a large set of decompression experiments under different environmental conditions. But as mentioned above, and despite its accuracy, the multi-parameter model cannot be adapted to predict schedules for some other gas mixture or series of mixtures. To be able to do this it is essential that the model is biophysically and physiologically meaningful so that the new perfusion and diffusion time-scales may be predicted. This is a fundamental difficulty in any modelling procedure.

A model with a large number of primarily numerical parameters contradicts the Principle of Parsimony of Parameters or Occam's Razor, where parameters not known to exist should not be inferred unless found to be absolutely necessary. The approach, therefore, should be to carry out a computer search to find the minimum number of parameters necessary to fit the database to achieve some reasonable level of accuracy.

This 'minimal' parameter set will not necessarily be more biophysically meaningful. Nevertheless, it is better to.deal with an economical number of parameters and to gradually increase the number only when forced to do so by new data.

Unfortunately this policy has not been followed in all of the model revisions to date. Various models have been developed which are based on sets of the single-tissue compartments shown in Figure 17.6. Over the years the number of tissue compartments and parameters has been increased in order to fit an ever widening decompression database. The US Standard Air Table was originally generated by 6 such compartments (Des Grange, 1956; Dwyer, 1956). In 1965 Workman introduced a 9-compartment set and NOAA uses 11 compartments (Hamilton *et al.*, 1973; Miller, 1976). Both of these models use a 'safe ascent criterion' (see below) which is a function of depth and compartment. The overall model employs 27 and 33 parameters respectively. Buhlmann (1983) now uses a 16-compartment model with 38 parameters. On the other hand the original DCIEM model (Defence and Civil Institute of Environmental Medicine, Canada), although using a small number of parameters, proved to be unsatisfactory as it was strongly non-linear being based on the rather dubious assumption that gas flow in a porous resistor characterized the gas exchange process in tissue (Weaver *et al.*, 1968).

Remarkably in all these cases no attempt was made to recompute a new minimal set of best-fit compartmental parameters while leaving the simple linear compartmental structure (and hence the number of parameters) the same. Instead the number of compartments and parameters has proliferated to a point where any initial resemblance to the underlying physiology has disappeared.

Indeed it is hard to see how models consisting of 9—16 tissue compartments employing 27—38 parameters is a good modelling strategy for the relatively small set of decompression data and involving a limited number of sites for bends. Unless it can be shown that such a large number of parameters is necessary, such modelling amounts to computational waste.

There is another reason for disregarding the single-tissue compartment with its mono-exponential response. The hallmark of the multi-exponential response is the ability to predict a wide range of tissue half-times (the 50% saturation point) for a given saturation time (the 99% point). This is better put the other way round, namely, the wide range of saturation times that becomes possible for a given half-time. This feature is shown in Figure 17.7.

The chief shortcoming of the mono-exponential response is the fact that the half-time fixes the saturation time for the tissue. Now it is known from saturation decompression data that tissues with very long saturation times must be involved. Thus, if such a tissue is modelled by a single compartment, the half-time is also very large and physiologically unrealistic because it implies an extremely low vascularity, virtually to the point of being anoxic.

A double or triple exponential response is able to provide a realistic half-time but can still have a sufficiently long saturation time, using only 4—6 parameters, as shown in Figure 17.6.

To summarize, without more detailed data on dissolved and undissolved gas in tissue, the aim must be to follow the Principle of Parsimony: use the smallest number of parameters necessary to fit existing

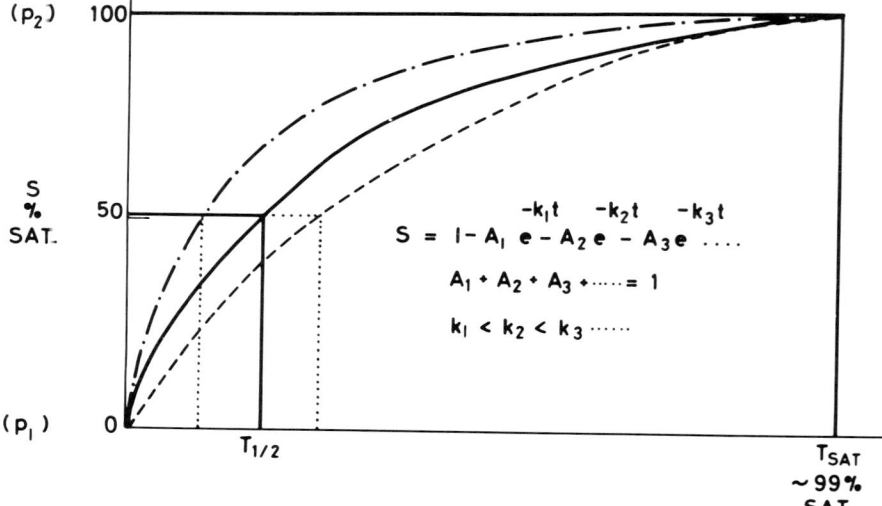

*Figure 17.7.   Half-times and saturation times in a multi-compartment tissue.*

data. Obviously a physiologically realistic structure should be a basic building brick in any modelling because it enables the predicted response for one gas to be converted to another gas without further computation or empirical adjustment.

Thus a model should be refined to a small number of realistic (i.e., reasonably complex) conceptual compartments, each governed by linear equations using a small number of parameters. As new data is obtained best-fit parameters should, in the first instance, be recomputed until a stage is reached where an increase in the number of parameters, or perhaps a new model structure, becomes necessary to improve the goodness of fit.

## 3. New tissue models

As mentioned above, a region of tissue with dimensions of the order of 10−100 mm consists of an aqueous and a lipid component, each with a characteristic distribution of vascularity such that both diffusion and blood perfusion contribute to the exchange of metabolic gases, the so-called intertissue diffusion effect (Perl, 1963; Hempleman, 1963; Hennessy, 1974).

The solubility of a respired gas in fat may be significantly higher than that in water. This means that the relative tissue volume (or inert gas capacity) of each component must appear in any model.

Of course, it may not be possible to identify individual lipid and aqueous tissue zones where either diffusion or perfusion is the dominant gas exchange mechanism. If we assume, however, that the entire tissue response can be approximated by a set of linear differential equations then, by the principle of superposition, we can separate each gas exchange process and lipid and aqueous tissue volumes.

From a macroscopic point of view therefore, the simplest linear model (model A) containing these essential ingredients consists of four compartments—two diffusion-limited and two perfusion-limited water and fat compartments.

To fully represent the overall tissue response, all four compartments must be interconnected to allow intertissue exchange of dissolved gas. Thus each compartment is specified by four parameters and so the response is a quadruple-exponential function. Model A is shown in Figure 17.8 where the upper two compartments are blood perfusion limited and the lower two compartments are diffusion limited.

As shown in Figure 17.8, it is assumed that blood saturated with respired nitrogen at environmental partial pressure enters and leaves the two upper compartments. Blood does not enter the lower compartments because these are diffusion-limited which means that there is no direct connection with blood.

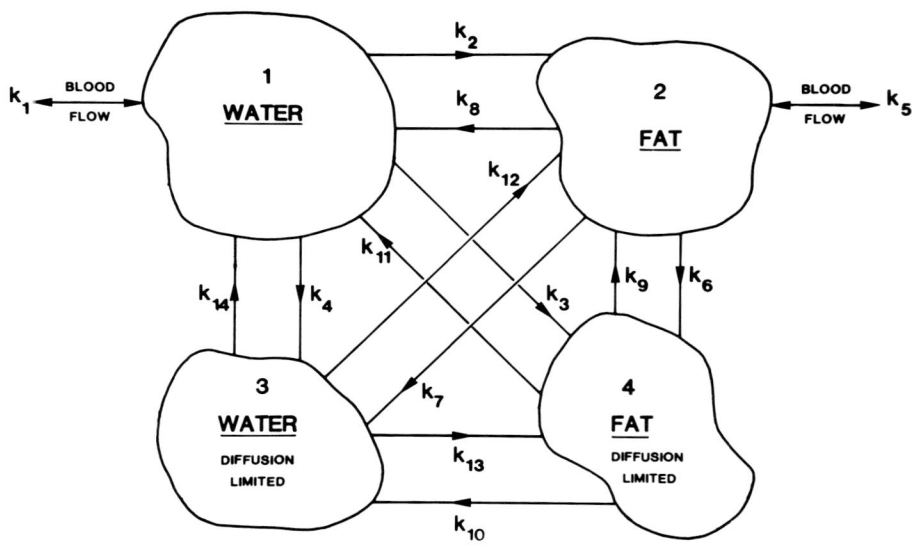

*Figure 17.8. Four-compartment tissue for dissolved gas, model A.*

There are 12 intercompartmental parameters and 2 input parameters. Thus a total of 14 different (and as yet still unknown) tissue parameters are required to describe the overall response to a change in the environmental inert gas partial pressure.

The model proposed by Schreiner (1971) is probably the nearest conventional (i.e., perfusion limited) equivalent: four levels of solubility and perfusion generate a set of 16 non-interacting compartments. The overall response specified by 14–15 parameters (one 'fast' compartment is ignored) can be made mathematically equivalent to the 4-compartment model above. However the perfusion–diffusion interaction is absent, an important limitation in that the change in the diffusion response for another inert gas cannot be represented.

If this tissue complex contains undissolved gas as well, then model A may reasonably be simplified to a two-compartment model (B), as shown in Figure 17.9. If the undissolved gas is intravascular, then the perfusion limited compartments will become essentially diffusion limited. The original diffusion-limited compartments are lumped together to form a single diffusion-limited compartment. The undissolved gas, although probably distributed throughout the tissue, as shown in the lower right half of Figure 17.9, is lumped into a single gas-filled compartment (shown in the lower left half) at approximately

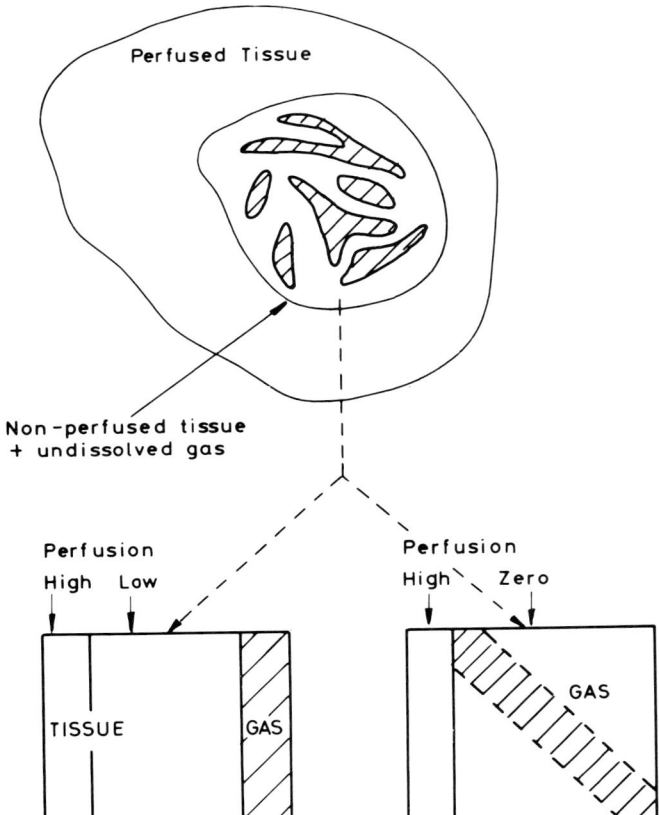

*Figure 17.9. Two-compartment tissue for undissolved gas, model B.*

ambient pressure (corrected to allow for tissue elasticity, surface tension and the presence of other tissue gases in the gas phase).

This model, which requires only 2 parameters, has been used successfully to describe the volume of the undissolved gas which begins to form in, for example, a joint, at some stage during a reduction in environmental pressure (Hennessy, 1980).

The 4-compartment dissolved gas model is also a realistic macroscopic model of the joint which consists of several interacting perfusion and diffusion limited tissues such as tendon, cartilage, bone, joint capsule, synovial fluid, fat pads, etc.

In fact the joint is the most commonly affected tissue in mild bends so it is reasonable to assume that just one such compound tissue (i.e., model A) is involved in the avoidance of an excess of undissolved gas on an experimental series. However, the results of the trial will provide only an indirect estimate of the dissolved gas content and effectively of only one of the four compartments in model A. The reason for this is that the estimate depends directly on any undissolved content in one of the compartments in model B. Any indirect data will increase the uncertainty in the model parameters.

Furthermore, if it was actually possible to measure directly the partial pressure of an inert gas component in one of the top (perfusion limited) compartments in Figure 17.8 such information would still not be sufficient to solve all the coefficients of the model. According to a theorem in compartmental analysis (Jacquez, 1972), only two compartments may be determined from data arising in one compartment and thus the 14+2 parameters cannot be solved unless new data can be obtained.

In fact any model which has a physiological basis requires new data to solve it. Such data might be extracted independently from another compartment (e.g., by using a radioactive tracer), or by an experiment using other gases such as helium or neon to obtain an estimate in the same compartment saturation time. For example, by using different gases on the same experiment, the parameters can in principle be solved because they are interrelated—the parameters for nitrogen and helium are related according to their partition coefficients and the square root of the ratio of their molecular weights. The best-fit set of parameters could then be found. So it is necessary to either try and re-interpret relevant experiments from the literature or design new ones to provide this information.

In any event this modelling approach has a greater physiological and biophysical basis than any numerically equivalent model consisting of a necessarily larger number of non-interacting elementary compartments.

However, if the parameters cannot be solved from the available data and new data is unobtainable, then there is no alternative but to simplify model A. Two compartments must be discarded to reduce the problem to a determinate one. It seems reasonable to drop the two diffusion limited compartments in Figure 17.8, leaving two interacting perfusion limited compartments described by 4 parameters, and each compartment will therefore have a double exponential response. Unfortunately, by omitting a diffusion limited response the ability to convert a data curve from one gas to another is lost unless the new gas has an identical diffusion coefficient and solubility partition coefficient.

## 4. Other difficulties in modelling

There are severe difficulties in developing a suitable gas exchange model because valid calibration data cannot be obtained directly. Such data can only be extracted by the application of yet another model called the safe ascent criterion to the results of a series of decompression trials and associated incidence of bends. The safe ascent criterion is the maximum tolerable reduction in environmental pressure so as to be on the threshold of mild bends. This is effectively equivalent to that critical volume of gas that comes out of solution in the tissue after a rapid pressure reduction (Hennessy and Hempleman, 1977). Thus before the inert gas partial pressure in a tissue can be deduced from the result of a decompression trial, the allowable pressure reduction as the gas comes out of solution must be known.

There is no reliable independent method of monitoring dissolved or undissolved gas in tissue (Hempleman *et al.*, 1984). In fact the end-point in determining the outcome of a decompression trial and the associated safe ascent criterion still depends on whether decompression sickness in the form of mild limb bends occurs or not! The bends end-point is unfortunately highly subjective and imprecise because it depends on the interaction between diver (or aviator) and the attending medical officer.

Of course successful recompression therapy is usually taken as confirmation of a diagnosis of bends. But even here the medical attendant has to depend totally on the subjective feelings of the recompressed subjects.

## Interpretation of bends

The apparently precise end-point of decompression sickness as noted by an attending medical officer conceals a large number of possibilities for misinterpretation. A trial may have one of the following outcomes: (*a*) a bend may actually occur and be correctly reported, promptly; (*b*) be delayed to a shallow depth; or (*c*) go unreported (a false positive); (*d*) a bend may not occur, but be falsely reported (a false alarm); or (*e*) not be reported, i.e., no event.

Consider the sequence of events during a decompression in a hyper-baric chamber. All the subjects are strongly influenced by previous events, by each other and by their own perceptions of what is likely to happen to undissolved gas in their body.

If earlier trials have been uneventful there is a tendency for the subjects to ignore minor symptoms of undissolved gas (niggles). On the other hand if an earlier trial contained a more serious case of neurological decompression sickness then the subjects may (perhaps unconsciously) exaggerate any niggles in order to ensure a recompression therapy (in case something more sinister should happen were the decompression to be continued).

There is no doubt that, when several subjects all undergo a single chamber dive together (where the above effect is absent), the outcome can be quite different from the case where the trials are performed sequentially (using say pairs of subjects), despite the fact that statistically these have been treated as being identical (Wethersby *et al.*, 1984).

On a chamber dive the subjects strongly influence each other. There is considerable pressure not to be the first to report a bend and so niggles may go unreported for some time until severe pain forces the victim to complain. Unfortunately the depth at which this occurs is likely to be quite shallow. The consequence is that on the next trial, the shallower portions of a profile will be slowed (possible unnecessarily) rather than the deeper portions. This has profound implications in the subsequent modification of the model because the profiles still to be tested are recomputed and information on the tolerance in the model is lost.

There is less opportunity to conceal or stall the reporting of symptoms in open sea diving. Indeed, it is a remarkable fact that a decompression table developed in a chamber invariably fails when tested in the open sea. This phenomenon is usually attributed to the arduous working conditions and cold water offshore. However, a more plausible but unrecognized explanation is that the reassuring and socially interactive conditions surrounding a chamber dive allow marginally unsafe procedures to be accepted.

On the other hand concealment is unlikely on frankly unsafe schedules. When one subject reports a genuine event, it is common for the others to decide to report problems although they may well have been holding back for some time.

There are problems when the victim is recompressed. If all subjects are recompressed with the victim there is strong pressure on him to report relief so that he and his companions are not delayed unnecessarily in returning to the surface. In fact this is the prime reason for not reporting problems—the desire to terminate a long experiment on time (for private reasons, e.g., to take up a weekend pass). On the other hand extra time in the chamber may bring extra pay and there have been cases where bends have been falsely reported for this reason. In any case, if all subjects are recompressed merely as a precaution vital data will be lost.

Some subjects report a bend when in fact they are uncertain of their symptoms. On recompression the subject may detect no improvement and may by then have convinced himself that he was imagining a problem. However, to conceal his nervousness from his companions and sometimes to cheer up the concerned medical attendant he may report that the recompression brought relief!

Finally the approach of the medical attendant has an influence on the outcome of a trial. In the early days mild niggles were not considered a sufficient reason to halt a decompression. Consequently, even if reported, such events were not acted upon and were not always recorded in the dive log. Thus some of the less frequently performed schedules have survived even though they are marginal. Nowadays the climate has changed and in some laboratories the most trivial effect is promptly reported and may be immediately acted upon, carrying out therapeutic recompression or oxygen or both. These incidents will force a slowing up of the gas exchange parameters or reduction in the safe ascent criterion.

Clearly the interpretation of a series of decompression trials is a minefield for the unwary and any misjudgement will have a profound influence on the way a particular schedule and the safe ascent criterion is modified. This in turn affects the way in which the parameters of the generating model are adjusted to obtain a fit to the apparent data.

## 5. Conclusions

If the dissolved and undissolved gas content of a tissue cannot be independently measured directly or indirectly (with or without bends) then the safe maximum limits relative to the environmental pressure cannot be accurately determined by doing decompression trials. In turn it will not be possible to systematically develop a comprehensive biophysical model for gas exchange.

In any case a more objective end-point is needed if there is to be any prospect of improving a particular decompression model. There is no justification for choosing a multi-parameter or multi-tissue model until new experimental data demands it. In fact this is the prime reason for following Occam's Razor in using the smallest number of physiologically meaningful compartments and parameters.

It is proposed that a best-fit double compartment model be used for dissolved gas (model A) and a single compartment model for undissolved gas (model B), involving only 4+2 parameters. These are the simplest models consistent with the available data.

However, these reduced models do not establish the method of converting a nitrogen response to that of helium or vice versa and new decompression data should be sought. The first experiment might be a series of no-stop helium-oxygen dives where all nitrogen is eliminated from the body prior to doing a dive. This experiment has never been done before—all earlier dives on helium-oxygen were contaminated with a variable level of nitrogen at various parts of the exposure.

Because of these limitations an experimental series must be very carefully designed, with the aim of revealing the tolerance and biophysical basis of the above models and parameters. Without an objective end-point and a satisfactory biophysical model, the mere accumulation of statistics, despite the attractions of a purely empirical approach, is largely superficial and unlikely to produce any advance on its own.

# References

Boycott, A.E., Damant, G.C.C. and Haldane, J.S., 1908, The prevention of compressed air illness. *J. Hyg.*, **8**, 342−443.

Buhlmann, A.A., 1983, *Dekompression—Dekompressionskrankheit* (Berlin: Springer).

Des Grange, M., 1956, *Standard Air Decompression Tables*, Research Report 5-57, Washington, DC: US Navy Experimental Diving Unit.

Dwyer, J.V., 1956, *Calculation of Repetitive Diving Decompression Tables*, Research Report 1-57, Washington, DC: US Navy Experimental Diving Unit.

Hamilton, R.W., Kenyon, D.J., Freitag M. and Schreiner, H.R., 1973, NOAA Ops I and II. *Formulation of Excursion Procedures for Shallow Undersea Habitats*, Tech. Memo UCRI No. 731. Tarrytown, NY: Union Carbide Corporation.

Hempleman, H.V., 1952, *Investigation Into the Diving Tables. Report III, Part A: A New Theoretical Basis for the Calculation of Decompression Tables*, London: Medical Research Council, RN Personnel Research Committee, UPS 131.

Hempleman, H.V., 1963, Tissue inert gas exchange and decompression sickness. In *Proceedings of the Second Symposium on Underwater Physiology*, edited by C.J. Lambertsen and L.J. Greenbaum (Washington, DC: National Academy of Sciences), pp. 6−13.

Hempleman, H.V., Florio, J.T., Garrard, M.P., Harris, D.J., Hayes, P.A., Hennessy, T.R., Nichols, G., Torok, Z. and Winsborough, M.M., 1984, UK deep diving trials. In *Diving and Life at High Pressures*, edited by W.D.M. Paton, D.H. Elliott and E.B. Smith. *Phil. Trans. R. Soc. Lond.*, **B 304**, 119−141.

Hennessy, T.R., 1974, The interaction of diffusion and perfusion in homogeneous tissue. *Bull. Math. Biol.*, **36**, 505−526.

Hennessy, T.R., 1980, Decompression aspects. In *Human Physiological Studies at 43 bar*, edited by H.V. Hempleman. Report No AMTE(E) R80-40, pp. 217−240 (Gosport: AMTE, Physiological Laboratory).

Hennessy, T.R. and Hempleman, H.V., 1977, An examination of the critical released gas volume concept in decompression sickness. *Proc. R. Soc. Lond.*, **B 197**, 299−313.

Jacquez, J.A., 1972, *Compartmental Analysis in Biology and Medicine* (Amsterdam: Elsevier).

Miller, J.W. (editor), 1976, *Vertical Excursions Breathing Air from Nitrogen-Oxygen or Air Saturation Exposures* (Washington, DC: National Oceanic and Atmospheric Administration).

Perl, W., 1963, An extension of the diffusion equation to include clearance by capillary blood flow. *Ann. N.Y. Acad. Sci.*, **108**, 92−105.

Schreiner, H.R., 1971, A pragmatic view of decompression. In *Proceedings of the Fourth Symposium on Underwater Physiology*, edited by C.J. Lambertsen (New York: Academic Press), pp. 205–219.

Thalmann, E.D., 1986, Personal communication.

Vann, R.D., 1986, Personal communication.

Weathersby, P.K., Homer, L.D. and Flynn, E.T., 1984, On the likelihood of decompression sickness. *J. Appl. Physiol.: Respirat. Environ. Exercise Physiol.*, **57**, 815–825.

Weaver, R.S., Kuehn, L.A. and Stubbs, R.A., 1968, *Decompression Calculations: Analogue and Digital Methods*, Research Paper 703, Toronto, Defence and Civil Institute for Environmental Medicine (DCIEM, formerly DRET).

Workman, R.D., 1965, *Calculation of Decompression Schedules for Nitrogen-Oxygen and Helium-Oxygen Dives*. Report No. 6-65, Washington DC: US Navy Experimental Diving Unit.

Zuntz, N., 1897, Zur Pathogenese und Therapie der durch rasche Luftdruckänderungen erzeugten Krankheiten. *Fortschr. d. Med.*, **15**, 632–639.

# 18. Respiratory Load Imposed by the Diver's Protective Equipment

**Gunnar O. Dahlbäck**

## 1. Introduction

When designing personal protective equipment (PE) it is essential to identify the design factors that could reduce performance. For a complete PE including suit, helmet and breathing apparatus the following factors should be considered:

Flow-volume characteristics
Hydrostatic and elastic load
Dead space
Inhaled gas temperature and humidity
Suit system insulation
Partial pressure of inhaled gases
Mass of the system

The above factors will cause a reduced physical performance via four main physiological pathways:

Reduced or limited ventilation and gas exchange
Thermal imbalance
Increased demands on the metabolism
Unphysiological levels of inhaled gases

During diving ventilation is a very critical factor as it will be influenced by the increased gas density, hydrostatic loading and loading imposed by the breathing apparatus and the suit and harness system.
In this chapter the total load on the diver's respiratory system will be discussed.

## 2. Lung mechanics

To be able to assess this load a lung mechanical technique can be used. As shown in Figure 18.1, two variables are measured, lung volume changes with a spirometer and pressure changes around the lung (pleural pressure) with an oesophageal balloon. For further details see Dahlbäck *et al.* (1984).

Sitting dry with relaxed respiratory muscles and without any PE the inward recoiling force of the lung is balanced by the outward recoiling force of the chest and diaphragm. Thus there is a slight negative pleural pressure, the so-called relaxation pressure at this functional residual capacity (FRC) level. Figure 18.2 shows the volume—pressure curves of the lung and of the chest in two different diving situations.

If the lung volume is changed from maximal inhalation to maximal exhalation the pressure around the lungs is measured at several inter-mediate stops, when the flow is zero, a so called static pressure—volume curve is achieved (the broken line in Figure 18.2). When the diver inhales for instance 1 litre from the FRC level an extra pressure drop

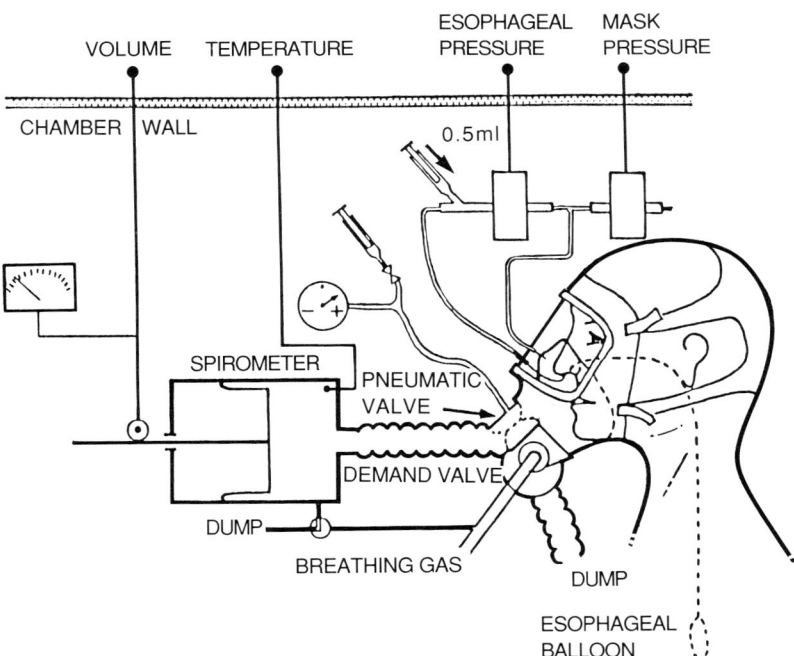

*Figure 18.1. The set up for measuring lung mechanics in a diving situation. For further details, see Dahlbäck et al. (1984).*

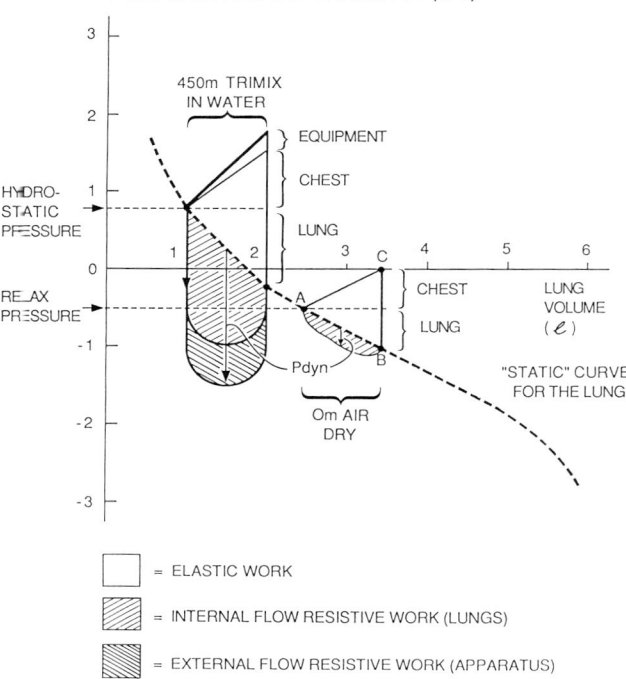

PRESSURE AROUND THE LUNGS (kPa)

☐ = ELASTIC WORK

▨ = INTERNAL FLOW RESISTIVE WORK (LUNGS)

▧ = EXTERNAL FLOW RESISTIVE WORK (APPARATUS)

*Figure 18.2.    Volume–pressure curves for the lung and chest with static and dynamic pressure changes during inhalation indicated for two diving situations.*

($P_{dyn}$) below the static curve is recorded (the thin line from $A$ to $B$) due to the flow resistance.

If at point $B$ after inhaling 1 litre the diver is asked to relax his respiratory muscles against a closed mouthpiece, the pressure will increase from $B$ to $C$. This pressure increase is a measure of the elastic force in the lung, chest and diaphragm taken up by the respiratory muscles during inhalation.

## 3. Power of breathing

Thus we can calculate the total work required by the inspiratory muscles during this inhalation, both the flow resistive work and the total elastic work.

It is essential that inspiration and expiration are dealt with separately as they involve separate muscle groups.

## Flow resistance power of breathing

The flow resistive work ($W_{res}$) during inhalation is the shaded area between the thin dynamic line and the broken static line in Figure 18.2. It is calculated by the following integration:

$$W_{res} = \int_A^B P_{dyn} \cdot d\text{Volume}$$

and the flow resistive power of breathing during inhalation is

$$\dot{W}_{res} = W_{res}/(\text{time for the inhalation})$$

By using the functional lung resistance values from a saturation dive (Örnhagen, 1984) where a lung mechanical study with different breathing gases was performed at different depths (Dahlbäck *et al.*, 1984), the internal power of breathing during inhalation can be calculated for the sitting dry situation at the surface. The formula used is (sinusoidal flow assumed):

$$\dot{W}_{res} = \dot{V}_E^2 \, R_t(l) \, \frac{\pi^2}{2}$$

Where $\dot{V}_E$ is minute ventilation and $R_f(l)$ is functional lung resistance, which is calculated from the actual measured work of breathing.

For further explanation see the above report and Dahlbäck (1985). Thus when $\dot{V}_E = 0{\cdot}67 \, \mathrm{l \, s^{-1}}$ (40 l min$^{-1}$) and $R_f(l) = 0{\cdot}13 \, \mathrm{kPa \, l^{-1} \, s}$ then $\dot{W}_{res}$ during inhalation will be 0·28 W.

## Elastic power of breathing

The elastic work is derived from the expression

$$W_{elast} = \text{inhaled volume } (V_T) \text{ pressure } (B-C)/2$$

The elasticity of the different parts of the respiratory system is expressed as compliance ($C$) with the unit (l kPa$^{-1}$), which within certain volume changes can be assumed to be constant. Thus total compliance is:

$$C_{tot} = 1/(1/C_{lung} + 1/C_{chest} + 1/C_{app}) \, \mathrm{l \, kPa^{-1}}$$

and

$$W_{elast} = V_T^2/(2\ C_{tot})$$

and as

$$V_T = \dot{V}_E\ T_{tot}$$

where $T_{tot}$ is the time for one breathing cycle ($T_{tot} = T_{in}$) the elastic power of breathing during inhalation will be

$$\dot{W}_{elast} = W_{elast}/T_{in} = \dot{V}_E^2\ (T_{tot}/C_{tot})$$

In the sitting dry situation both $C_{lung}$ and $C_{chest}$ could be approximated to $2\,l\,kPa^{-1}$ and thus $C_{tot} = 1\,l\,kPa^{-1}$. A breathing frequency of $20\ min^{-1}$ will then give a $T_{tot}$ of 3 s, thus $\dot{W}_{elast}$ will be 1·33 W.

## Total power of breathing at the surface

The total power of breathing during inhalation in the sitting dry situation at 0 m breathing 40 l air $min^{-1}$ will then be about 1·6 W, as shown in Figure 18.3, where the elastic work is about 80% of the total power.

## Total power of breathing at 450 m

In an extreme diving situation at 450 m, which could be reality in the near future, we will have a dramatically different situation.

The following is assumed. The diver is working ($\dot{V}_E = 40\ l\ min^{-1}$) in the water at 450 m in an upright position with the demand valve at his mouth level. This will, as shown in Figure 18.4, induce a hydrostatically imbalanced system equal to a negative pressure breathing situation of about $-2$ to $-3$ kPa. This imbalance will reduce the lung volume by about 1·5 litres, increase functional lung resistance by about 30% and lung compliance will almost be halved (Dahlbäck, 1978).

When calculating the power of breathing in this situation we have extrapolated the data from Dahlbäck *et al.* (1984) for a Trimix with 94% He, 5% $N_2$ and 1% $O_2$.

This will give a functional lung resistance of 1·3 kPa $l^{-1}$ s in the dry situation which due to the hydrostatical imbalance in our example will be 30% higher, i.e., 1·7 kPa $l^{-1}$ s. Thus $\dot{W}_{res}$ will be 3·73 W.

The flow resistive power required by the breathing apparatus ($\dot{W}_{res}(app)$) is assumed to be according to Morrison's and Reimer's (1982) tolerance criteria

$$\dot{W}_{res}(app) = 0·5\ \dot{V}_E + 2·4\ \dot{V}_{E^2}\ (W) = 1·40\ W$$

which is equal to a function apparatus resistance of 0·64 kPa $l^{-1}$ s.

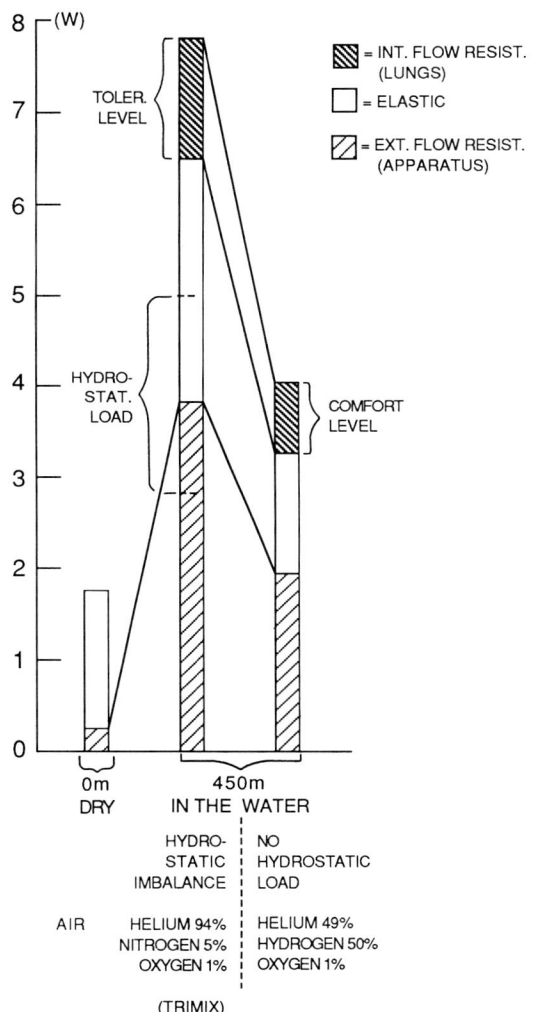

**CALCULATED POWER OF BREATHING
DURING INHALATION**

*Figure 18.3.* *Total power of breathing calculated for three situations. For details see the text.*

The elastic power of breathing is calculated from the assumptions that $C_{lung}$ is 1 l kPa$^{-1}$ and $C_{chest}$ is reduced to about 1·5 l kPa$^{-1}$ and that the suit and harness will strap the chest, adding a compliance of 3 l kPa$^{-1}$. This will give a $C_{tot}$ of 0·5 l kPa$^{-1}$ and $\dot{W}_{elast}$ will be 2·67 W. Half of the elastic power is due to the hydrostatic imbalance and the strapping of the chest.

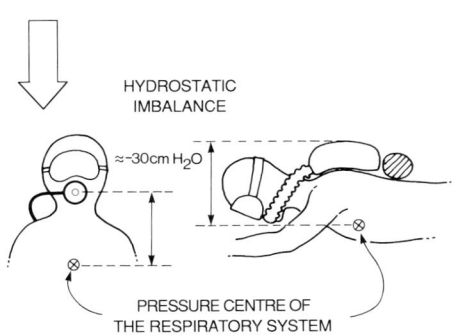

HYDROSTATIC
IMBALANCE

≈-30cm H₂O

PRESSURE CENTRE OF
THE RESPIRATORY SYSTEM

*Figure 18.4.   Hydrostatic imbalance in two diving situations with a demand valve in the mouth in the upright position and in the prone position with the breathing bag on the diver's back.*

The total power requirement in this diving situation in water at 450 m with a Trimix and protective equipment that is not optimally designed will be 7·8 W, as shown in Figure 18.3. The elastic power is now 34% of the total power requirement.

## 4. Techniques for reducing respiratory load

Several improvements can be made to the above high total load on the respiratory system. These are using a lighter gas mixture with hydrogen, balancing the breathing circuit hydrostatically, reducing the strapping of the chest and designing the breathing apparatus so it will fulfill Morrison's and Reimer's (1982) comfort criteria.

The use of hydrogen at 450 m in a mixture with 49% He, 50% $H_2$ and 1% $O_2$ will reduce functional lung resistance to about 0·9 kPa $l^{-1}$ s using the same extrapolation technique as before. Thus internal flow resistive power of breathing during inhalation when breathing at 40 l $min^{-1}$ with a hydrostatically balanced breathing circuit will be 1·97 W, a reduction of almost 50% compared with the unfavourable imbalanced Trimix situation.

The use of the comfort criterion (Morrison and Reimer, 1982)

$$\dot{W}_{res}(app) = 0.5 \, \dot{V}_E + 1.2 \, \dot{V}_E^2$$

will give an external flow resistive power of 0·88 W. This criterion is of course easier to fulfil with a lighter gas mixture.

The elastic power will also be halved to 1·33 W, thus the total power of breathing during inhalation in this 'optimally' designed diving situation will be 4·2 W, a reduction of almost 50%.

Much further work is required to quantify the effects of different gas mixtures at different depths and workloads, different levels of hydrostatic imbalance and different suit and harness designs.

# 5. Summary

—Inhalation and exhalation load should be treated separately, as different muscle groups are utilized.
—The elastic work is a major part of the load.
—The design of suit and harness may add to the elastic work.
—Hydrostatic imbalance of $-2$ to $-3$ kPa will increase flow resistive load by about 30% and elastic load by about 100%.
—During deep diving, a lighter gas (hydrogen) can reduce the flow resistive load.

## References

Dahlbäck, G.O., Warkander, D. and Örnhagen, H., 1984, Lung function during hydrox breathing at 1·3 MPa. In *Hydrogen-Oxygen (Hydrox) Breathing at 1·3 MPa*, edited by H. Örnhagen. National Defence Research Institute, Stockholm, FOA Report C 58015-H1.

Dahlbäck, G.O., 1985, Hydrogen in breathing gas reduces work of breathing. Paper presented at Undersea Medical Society workshop, *The Use of Hydrogen as a Component of Breathing Gas Mixture for Deep Diving*, Wilmington, NC, 3—5 October.

Morrison, J.B. and Reimers, S.D., 1982, Design principles of underwater breathing apparatus. In *The Physiology and Medicine of Diving*, 3rd edition, edited by P.B. Bennett and D.H. Elliott (London: Balliere-Tindall), pp. 55—98.

Örnhagen, H. (editor), 1984, *Hydrogen-Oxygen (Hydrox) Breathing at 1·3 MPa*. National Defence Research Institute, Stockholm, FOA Report C 58015-H1.

# 19. Development of Ergonomic Design Standards for Underwater Breathing Apparatus

## James B. Morrison

## 1. Introduction

Although recent advances in the knowledge of diving physiology have resulted in exposure of humans to pressures equivalent to 650 m of seawater within hyperbaric chambers, attempts to work underwater have been restricted to shallower depths. The transfer of diving technology from the simulated environment of the hyperbaric chamber to the ocean floor is complicated by the remoteness and hostility of the worksite and the requirement of adequate life support systems for the diver operating outside the diving bell or habitat. The design and performance of underwater breathing apparatus in particular is now recognized as a critical factor in underwater work. In order to provide breathing equipment which satisfies the diver's needs, the respiratory requirements of the diver must be understood and the effects of respiratory loading on physiological status must be carefully controlled. In the development of ergonomic design data for breathing apparatus, factors which should be specified include gas temperature, hydrostatic breathing pressure, work of breathing, ventilation, respiratory gas mixture and physical work capacity.

## 2. Ventilation and gas exchange

As the gas mixtures breathed by divers normally provide a greater oxygen pressure than air at sea level, it should be possible for divers to achieve their maximum aerobic capacity when working underwater. Several factors in the working environment can, however, interfere with work capacity. Ventilatory response to exercise is reduced by a change in pulmonary airway resistance in response to increased gas

340

density, leading to a state of hypercapnia (Lanphier, 1963; Hesser *et al.*, 1968). This effect can be further exacerbated by imbalance of hydro-static pressures between the apparatus and the lung (Ting *et al.*, 1960; Agostoni *et al.*, 1966), or by additional external respiratory resistance imposed by the apparatus (Morrison *et al.*, 1976).

Hypercapnia in divers is a problem which has been well documented, although there are aspects which still remain unresolved. Figure 19.1 shows the effect of gas density and breathing apparatus on a diver's ventilation and gas exchange when working underwater (Morrison and

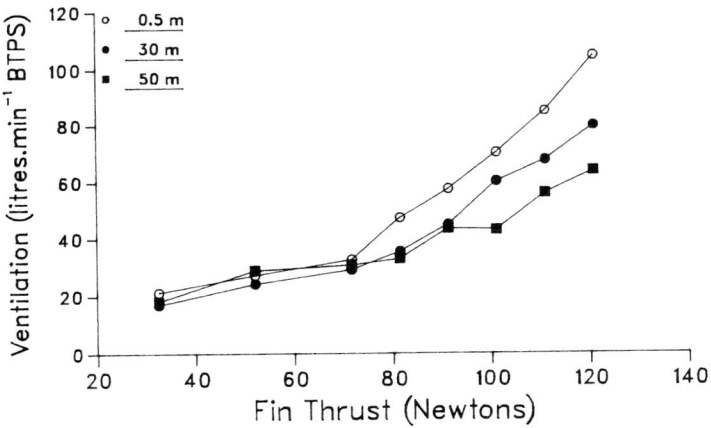

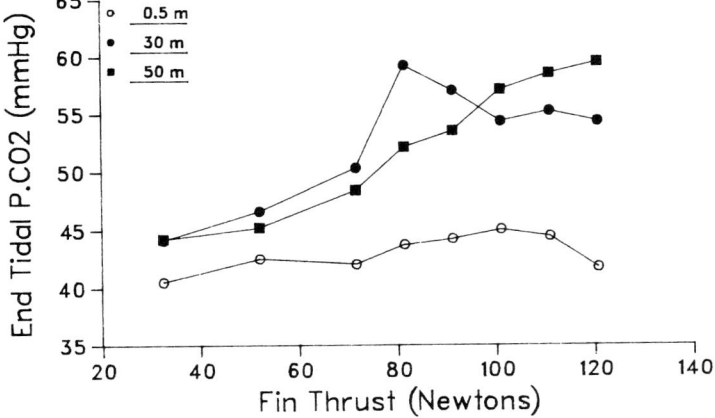

*Figure 19.1.* *Relationship of ventilation and end-tidal $PCO_2$ to work load when breathing air at 0·5, 30 and 50 metres seawater. Results of one diver breathing from open circuit demand U.B.A. and swimming in an ergometer. Work load is measured as fin thrust (N).*

Wood, 1986). The top diagram shows ventilatory response to work rate, and the lower diagram shows the corresponding end tidal carbon dioxide partial pressure $P_{ET}CO_2$. The effect of reduced ventilatory response at raised ambient pressures is to increase $P_{ET}CO_2$, which should normally remain below 45 mm Hg, to approximately 50−60 mm Hg during hard work.

Elevated $P_{ET}CO_2$ can cause headache, nausea and impaired judgement. The synergystic action of carbon dioxide and nitrogen in promoting narcosis was demonstrated by Case and Haldane (1941) in a series of exposures to 10 ata (91 m) in a dry compression chamber. The susceptibility of individuals to the combined effects of $CO_2$ and $N_2$ narcosis varied considerably. Carbon dioxide retention has also been implicated in the onset of panic and loss of consciousness underwater (Morrison *et al.*, 1978, 1981) and in the early onset of oxygen convulsions when breathing hyperoxic gas mixtures (Lanphier, 1955; Lambertsen *et al.*, 1959; Marshall and Lambertsen, 1961). In view of these findings it is preferable that $P_{ET}CO_2$ is held below 50 mm Hg in $O_2-N_2$ diving and it should certainly not exceed 60 mm Hg.

The physiological strain imposed upon divers (in the form of hypercapnia) is directly dependent on the stress applied by the underwater breathing apparatus (in the form of airflow resistance, gas delivery pressure and breathing gas mixture). For this reason it is tempting to use $P_{ET}CO_2$ as a criterion in the design and testing of breathing apparatus: for example, any breathing apparatus which elevates $P_{ET}CO_2$ above 50 mm Hg should be discarded. A problem in this approach is the natural variation of ventilatory response to work rate amongst divers. There is a trend amongst divers to adapt to elevated $PCO_2$ (Schaefer, 1955; Goff and Bartlett, 1957; Lalley *et al.*, 1974; Florio *et al.*, 1979) to the point where $P_{ET}CO_2$ levels of 50 mm Hg occur during work, even at the surface (Lanphier, 1955). Figure 19.2 shows four divers with a normal ventilatory response to work rate and one diver with a low response (Morrison *et al.*, 1981). Whereas $P_{ET}CO_2$ increases to approximately 50 mm Hg in four divers, it reaches 70 mm Hg in the diver with a low response. Thus, the use of $P_{ET}CO_2$ as a measure for testing the design and performance of breathing apparatus is complicated by subject selection and open to misinterpretation.

A more consistent assessment of a particular apparatus can be achieved by direct physical measures of the breathing characteristics using a respiratory simulator. A data base obtained from human experimentation is still necessary, however, as the physiological strain resulting from external respiratory stress must be known in order to set reference standards for mechanical evaluation.

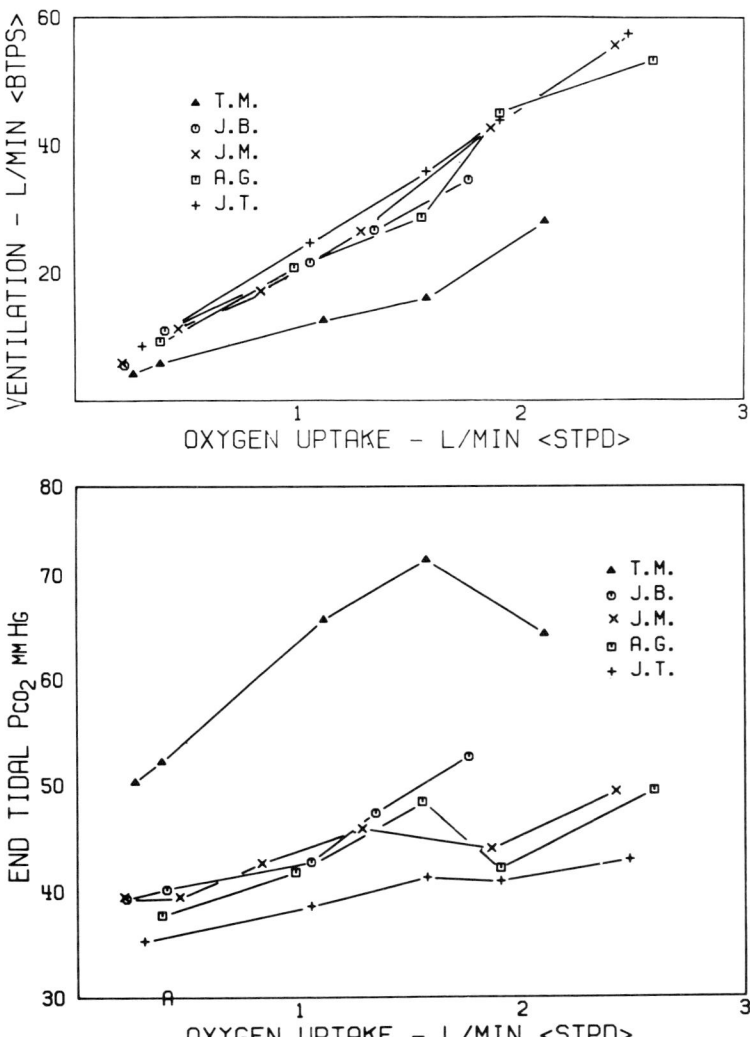

*Figure 19.2.*    *Response of ventilation and end-tidal PCO₂ to work rate in 5 divers at 50 m.*

## 3. Inspired carbon dioxide pressures

An additional source of hypercapnia when working underwater is the presence of $CO_2$ in the inspired breathing gas. Inspired carbon dioxide $(P_ICO_2)$ can result from either inefficient design of the $CO_2$ absorbent canister in the recirculating apparatus or from excessive external respiratory dead space in open circuit apparatus. In response to $P_ICO_2$ a diver can either increase ventilation in order to maintain alveolar $PCO_2$ at previous levels or tolerate a higher alveolar $PCO_2$ in order to achieve the same gas exchange rate. For example, in order to maintain normal gas exchange a $P_ICO_2$ of 10 mm HG (1·3 kPa) would necessitate either an increase of 33% in alveolar ventilation or a corresponding increase of 10 mm Hg in alveolar $PCO_2$ (Lanphier, 1969). During strenuous work both solutions would impose a severe burden on a diver. In practice a compromise involving increases of both ventilation and $PCO_2$ generally results.

It has been suggested that $P_ICO_2$ should not exceed 0·5 kPa (Morrison and Reimers, 1982). This goal is attainable in most types of underwater breathing apparatus. In recirculating apparatus an efficient design of $CO_2$ absorbent canister will maintain a downstream $PCO_2 < 0·5$ kPa for a prolonged period followed by an exponential rise of $PCO_2$ as the absorbent nears exhaustion. An absorbent breakthrough of 0·5 kPa is commonly used to determine the safe life of the apparatus (Middleton, 1979; Middleton and Thalmann, 1981; Morrison, 1981b; Segedal and Morrison, 1985). In this type of apparatus ergonomic design is important, as the performance and life of the absorbent is sensitive to gas temperature, humidity, air flow distribution, dwell time of expired gas and the thermal properties of the absorbent canister (Smith, 1973, 1974; Wang, 1975, 1976; Paulsen and Jarvi, 1977; Middleton, 1979). Poor design can result in both inefficient absorption of $CO_2$ under normal conditions, and a loss of up to 75% in the safe life of the scrubber when diving in cold water (Bentz, 1976; Middleton, 1979).

In open circuit demand apparatus the provision of a clean air supply to the diver ensures that $P_ICO_2$ will not exceed 0·2 kPa when air diving to 50 m, and is negligible in $O_2$-He or Trimix diving. A secondary source of $P_ICO_2$ arises from the respiratory dead space within the mouthpiece or helmet of the diver. The effect of apparatus dead space is dependent on the inspired volume of the diver. When breathing at a tidal volume of 2·5 l a dead space of 200 ml would contribute a mean $PCO_2$ of 0·4 kPa to the inspired gas.

A design standard of $PCO_2 < 0·5$ kPa poses a problem in the design of open circuit free flow breathing apparatus. In this type of equipment a more lenient requirement of $PCO_2 < 2·0$ kPa (15 mm Hg) is normally accepted (US Navy, 1975). As discussed earlier, this standard imposes a considerable physiological stress upon the diver. With improved aerodynamic design it is possible that the respired $PCO_2$ can be reduced below 1·0 kPa in this type of apparatus (O'Neill, personal communication), and values greater than this should no longer be accepted.

# 4. Respiratory heat losses

In normobaric conditions respiration does not present a significant source of heat loss from the body, except at extremely low temperatures. When working underwater, the increased density of the breathing medium results in a corresponding rise in respiratory heat loss. The magnitude of respiratory heat loss is generally tolerable within the normal range of air diving (<50 m). When diving beyond 100 m in cold water, heating of inspired gas is essential to maintain comfort and thermal homeostasis. At 200 m, respiratory heat losses can approach metabolic heat production, and can result in rapid core cooling and bronchial congestion sufficient to terminate a dive or incapacitate the diver. As the mass of gas respired by the diver increases with ambient pressure, inspired gas temperatures must also rise in order to limit heat loss and presumably maintain airway temperatures at a constant level. Vaernes *et al.* (1983) found that at 500 m comfortable oxygen–helium gas temperatures were restricted to a range of a few degrees ($\pm 3°C$) and that in one instance when gas heating failed, shivering thermogenesis began within 20 s even though skin temperatures were unaltered and inspired gas temperature did not drop below 14°C.

A limit of inspired gas temperature was adopted by the US Navy (1975) which equated allowable respiratory heat losses to spontaneous thermogenesis. This limit, shown in Figure 19.3, was designed to maintain thermal equilibrium provided there were no other avenues of heat loss. Later work proved this limit to be unsuitable due to local cooling of the airways which can produce serious problems of bronchial congestion before a point of general core cooling is reached (Hayes *et al.*, 1981). For this reason, Piantadosi (1980) and Hayes (1980) suggested a more conservative curve, representing the minimum safe inspired gas temperature. Higher temperature limits were proposed by Pasche and Tonjum (1983), to ensure respiratory comfort based on experimental data from saturation dives to 500 msw (Figure 19.3).

# 5. Hydrostatic pressure differences

Peterson and Wright (1976) have reported divers working towards maximum effort at extreme gas densities of 25 g $l^{-1}$ (which is equivalent to 1 500 m depth when breathing helium). In these circumstances the divers remained in fairly good condition. However, these experiments were undertaken in dry conditions. In underwater work Morrison (1973) and Thalmann *et al.* (1979) both noted respiratory problems in divers breathing oxygen–nitrogen mixtures at high gas densities and

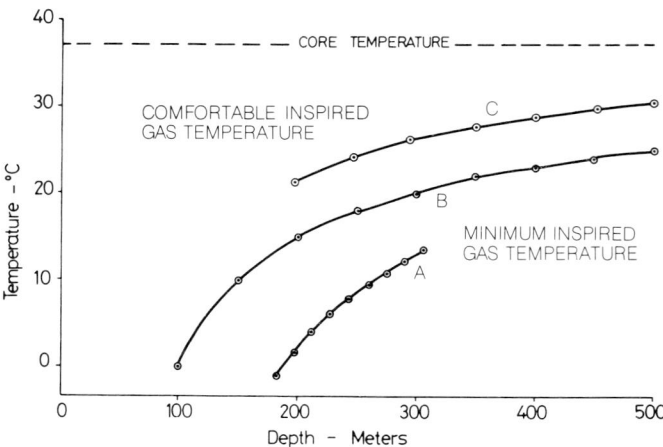

Figure 19.3.   *Proposed limits of inspired gas temperature when breathing $O_2$-He mixtures: Curve A, limits of tolerance, USN Diving Manual (1975); Curve B, minimum safe inspired temperature, Piantadosi (1980); Curve C, comfortable inspired gas temperature. From Pasche and Tonjum (1984) (see text for details).*

at maximum workloads. In contrast are the results of Spaur *et al.* (1977) and Dwyer *et al.* (1977) whose divers experienced unexpected and incapacitating breathlessness in oxygen-helium dives of 400–500 m. Although this involved similar gas densities to the above two experiments, breathlessness resulted at only moderate workloads.

The apparent differences between these results remain unresolved but it is possible that breathlessness may be caused by hydrostatic imbalance between the lung centroid pressure (i.e., mean hydrostatic pressure acting on the thorax) and the breathing gas supply pressure. In experiments in which divers exercised in a prone position, Thalmann *et al.* (1979) found that dyspnoea occurred in the presence of a hydrostatic imbalance of ±20 cm $H_2O$ whereas less dyspnoea was reported when the imbalance was within the range of ±10 cm $H_2O$. In general the severity of dyspnoea experienced increased with the degree of hydrostatic imbalance.

When working in an upright position and breathing from a diving helmet, hydrostatic imbalance is approximately −20 to −30 cm $H_2O$ depending on apparatus design and the locus of lung centroid. In these conditions (i.e., a negative hydrostatic imbalance) vital capacity, forced expired volume ($FEV_{1.0}$), maximum voluntary ventilation and lung relaxation volume are decreased (Morrison, 1983). The latter effect will cause an increase in air flow resistance (Agostoni *et al.*, 1966) due to a narrowing of the airways, and an increase in elastic work (Hong *et al.*, 1969) due to a lower respiratory compliance at the reduced relaxation volume. By compensating the breathing gas delivery pressure of the apparatus to balance the external thoracic pressure, lung relaxation volume can be restored to normal and hence the work of breathing substantially reduced (Taylor, 1987). In order to accomplish this the mean external thoracic pressure (lung centroid pressure) must be determined and the breathing gas delivered must be at or close to that pressure. In addition, to avoid discomfort facial pressure must also be balanced with breathing gas delivery pressure.

The lung centroid pressure has been estimated by several investigators (Hong *et al.*, 1969; Jarrett, 1965; Craig and Dvorak, 1975) and its anatomical location is generally defined relative to the suprasternal notch or 7th cervical vertebra. The original study by Jarrett (1965) reported the centre of pressure to be located 7 cm posterior and 19 cm inferior to the sternal notch. Although the value has been widely accepted by designers, it is based on only three subjects, who displayed somewhat unusual pulmonary compliance relationships. A recent study of 20 divers immersed in upright and prone positions locates lung centroid pressure 13·6 cm inferior and 7·0 cm posterior to the sternal notch (Taylor, 1987), and offers better agreement with others (Agostoni *et al.*, 1966; Hong *et al.*, 1969). The use of the lung centroid pressure as a design goal is impractical, however, unless face pressure is suitably compensated for gas delivery pressure. At present there is no apparatus, with the possible exception of the traditional free-flow helmet used in standard diving dress, and the AGA ACSC recirculating apparatus, which can approach the goal of hydrostatic imbalance <10 cm $H_2O$. Both of these systems are limited by other design factors to a working range of approximately 0−50 m.

## 6. *Work of breathing*

A major factor affecting respiratory effort is the resistance to airflow imposed by the apparatus. There are several studies of airway resistance of the lungs (Ferris *et al.*, 1964; Agostoni *et al.*, 1966), and of external resistance to airflow (Silverman *et al.*, 1945; Bentley *et al.*, 1973). Although measurement of airflow resistance is informative to the designer it is not very helpful as an ergonomic design standard. Resistance is the driving pressure divided by the resultant flow rate (usually reported in units of cm $H_2O$ $l^{-1}$ $s^{-1}$ or kPa $l^{-1}$ $s^{-1}$). It generally varies as a non-linear function of flow rate throughout the breathing cycle and therefore any values are only applicable to a specific flow.

A more useful quantity in assessing the respiratory effort is the work of breathing from the apparatus. This quantity is computed from the integral of pressure with respect to volume and can be represented as the area of a pressure–volume diagram. Figure 19.4 shows a typical pressure–volume diagram for an open circuit demand regulator. Area

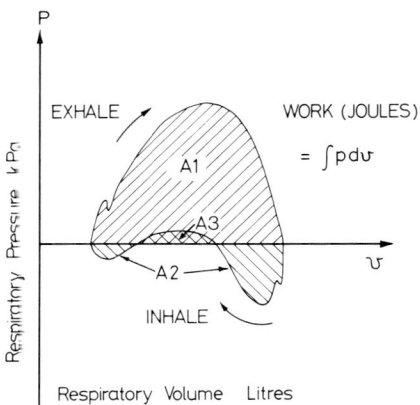

*Figure 19.4.    Pressure volume diagram for open circuit demand regulators (see text for details).*

A1 represents expiratory work and area A2 represents inspiratory work. During inspiration (A3) an area of negative work may exist (that is, the regulator is actively assisting inspiration). Once muscular work has been expended it cannot be restored, thus the area A3 cannot be subtracted from the expiratory work, which is correctly represented as the area A1 between the pressure–volume curve and the volume axis. As the phase of negative inspiratory work represents assisted breathing and will reduce internal respiratory work during that time, the net additional inspiratory work can be represented as area A2−A3. Hence, in order to fully describe the characteristics of breathing apparatus, the positive, negative and net work of both inspiration and expiration and net total work should be reported. An example of breathing apparatus demonstrating these characteristics is shown in Figure 19.5. The practice of reporting the area enclosed within the pressure volume loop as the external work of breathing is erroneous for any apparatus having a negative work component, and can be highly misleading in terms of breathing characteristics. As the respiratory work rate (i.e., power output) of the diver is also dependent on the frequency of breathing, data describing respiratory work are generally presented either as power (watts) or as work per unit ventilation (joules litre$^{-1}$ or kPa).

Most of the earlier studies of external respiratory work were concerned with design of mine rescue equipment. Silverman and co-workers (1945) studied the tolerance of workers to an externally imposed breathing resistance. Based on these measures it was suggested that external respiratory power should not exceed 0·6% of total body work rate. Later, Cooper (1960) used his own data and that of Silverman to define a tolerance limit for external respiratory work of 2·5 J l$^{-1}$, but also suggested that the ideal (or comfort) limit would be half

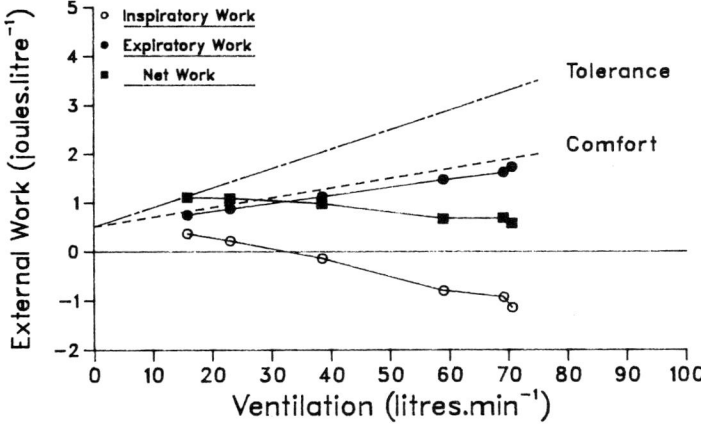

*Figure 19.5.    Acceptance standards for work of breathing from underwater breathing apparatus. The upper line denotes maximum acceptable values, the lower line denotes preferred values. The inspiratory, expiratory and net work of breathing air from an open circuit demand regulator at 50 m is also shown (USD Royal). Negative inspiratory work denotes ventilation assisted by the regulator.*

this value. Since then a number of other limits, standards or 'goals' for work of breathing have been suggested by various authors.

A value of $1 \cdot 7 \, J \, l^{-1}$ was suggested by Bentley *et al.* (1973) for mine safety apparatus and adopted by Reimers (1974) for underwater breathing apparatus. This limit was based entirely on inspiratory resistance with no expiratory resistance and no hydrostatic pressure effects. Another limit (or 'goal') was suggested by Middleton (1980) for open circuit demand regulators of $1 \cdot 4 \, J \, l^{-1}$. The value was based on the performance of the best available apparatus rather than physiological data or tolerance of respiratory work. Of the apparatus tested by Middleton, only a few could meet the recommended goal, and many were far in excess of the desired limit.

A slightly different approach was taken by Sennek (1962), who suggested that the relationship between respiratory power and minute ventilation is a third order polynomial function. Sennek therefore proposed a non-linear standard based on the experimental data of several investigators. The standard of Sennek may overestimate the increase of respiratory power (watts) with ventilation as Cooper (1961) has suggested that a second order relationship is more appropriate.

More recently Morrison (1981a,b,c) proposed a standard for external work of breathing which provides a linear relationship between unit work in $J \, l^{-1}$ and ventilation in $l \, min^{-1}$

$$W_T = 0 \cdot 5 + 0 \cdot 04 \, V \, J \, l^{-1}$$

where $W_T$ = external work of breathing measured in J $l^{-1}$ ventilation; and $V$ = ventilation in l $min^{-1}$.

This standard, shown in Figure 19.5, is based on the principle that the external respiratory work should be limited to a constant fraction of internal respiratory work, irrespective of ventilatory rate. The equation chosen to define the relationship is based on physiological data, and subjective tolerance reported by previous investigators. In addition to an acceptance limit based on tolerance, Morrison also proposed a more stringent design recommendation based on comfort:

$$W_E = 0.5 + 0.02 \ V \ (\text{J } l^{-1}) \qquad (2)$$

When diving deeper than 400 m, or breathing gases denser than 7.8 g $l^{-1}$, which is equivalent to air at 50 m, it is recommended that the lower (comfort) limit should be adhered to.

The principle proposed by Morrison was later demonstrated in the work of Sedagal *et al.* (1984), shown in Figure 19.6. Data were measured for a diver working at 400 m simulated depth. The open circles represent the total work of breathing (both internal and external) estimated by oesophageal pressure measurements. The triangles represent the external work of breathing measured at the mouth. The total work of breathing (per litre ventilation) is approximately a linear function of ventilation. The external work of breathing from this particular apparatus represents approximately 20–30% of total measured respiratory work and falls within the limit of comfort proposed in Figure 19.5. Measurement of respiratory work by oesophageal pressure changes excludes work done to overcome tissue resistance of structures external to the pleural space. This work is estimated to be approximately 25% of the total internal work of breathing (Otis *et al.*, 1950; Ferris *et al.*, 1964). Hence, if the internal work of Figure 19.6 were adjusted to include this component, the external respiratory work in this case would be more accurately represented as approximately 20% of total respiratory work (or 25% of the internal work of breathing).

## 7. *Implementation of standards*

Until recently, there were no detailed standards for the design of underwater breathing apparatus. In 1981 a set of performance goals and test procedures were reported by Middleton and Thalmann. These goals were adopted by the US Navy in the selection of new apparatus. The goals reported are designed to select the best apparatus in each category rather than to prevent the manufacture or use of inadequate equipment. Thus the limitations applied to each type of apparatus vary

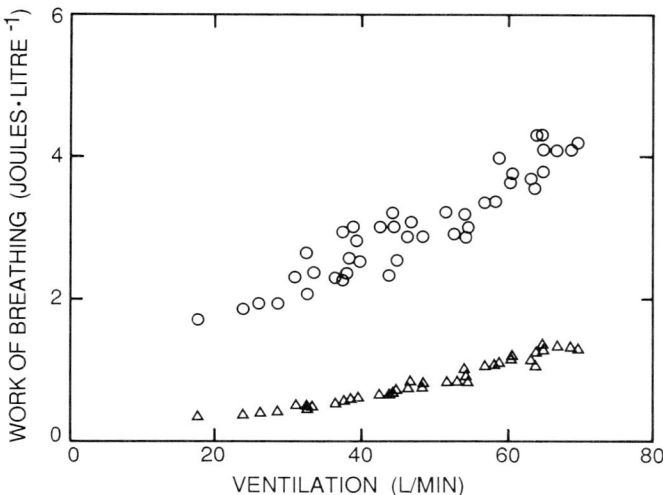

*Figure 19.6.   Work of breathing at a simulated depth of 400 m. The triangles represent the external work of breathing imposed by breathing apparatus and the circles represent 'total' work of breathing as measured by oesophageal pressure changes. Adapted from Segadal et al. (1984).*

according to technological development rather than reflecting the physiological needs of the diver. Recommendations are limited to work of breathing and do not address questions of temperature, hydrostatic balance or breathing gas mixture.

More recently the British and Norwegian Authorities published Draft Guidelines for performance requirements of breathing apparatus (Norweigian Petroleum Directorate, 1984) to be used in North Sea commercial diving. This document is based on the recommendations of Morrison (1981a,b), and includes the suggestions of Middleton and Thalmann (1981) on test procedures, and Pasche and Tonjum (1983) on respiratory gas temperatures. Credit must also be given to other investigators mentioned in this review on whose data the specifications also rely. The guidelines are written in a general format applicable to all types of apparatus, with the objective of providing quantified advice on design parameters to manufacturers. Specifications include limits for external work of breathing, respiratory pressures, hydrostatic pressure imbalance, lung overpressure in event of equipment failure, inspired $CO_2$ pressure and inspired gas temperatures. Also included are a comprehensive set of test procedures to enable evaluation and comparison of equipment. The principle features of these performance requirements and test procedures are summarized in Tables 19.1, 19.2 and 19.3.

Table 19.1.  *Performance requirements of breathing apparatus.*

| | |
|---|---|
| External respiratory work | |
| maximum | $0.5 \pm 0.04\ V\,\mathrm{J}\,\mathrm{l}^{-1}$ |
| preferred | $0.5 \pm 0.02\ V\,\mathrm{J}\,\mathrm{l}^{-1}$ |
| Range of ventilation | $10-75\ 1\ \mathrm{min}^{-1}$ |
| External respiratory pressures | |
| maximum | $\pm 2.5$ kPa |
| preferred | $<\pm 1.5$ kPa |
| Hydrostatic imbalance, $H$ | |
| maximum | $-3.5 < H < +2.0$ kPa |
| preferred | $-2.0 < H < +1.0$ kPa |
| Maximum lung over and under pressure (in event of apparatus failure) | $\pm 6.0$ kPa |
| Inspired $PCO_2$ (measured as inspired volume average) | |
| maximum | $2.0$ kPa |
| preferred | $<1.0$ kPa |

Table 19.2.  *Test procedures for underwater breathing apparatus. Breathing simulator respiratory wave form sinusoidal: ± 5% tolerance on amplitude and timing.*

| Respiratory minute volume, $V$ (1 min) | Frequency, $f$ $(\mathrm{min}^{-1})$ | Stroke volume, $VT$ (1) | $CO_2$ injection rates $(1\ \mathrm{min}^{-1})$ |
|---|---|---|---|
| 15 | 15 | 1.0 | 0.6 |
| 22.5 | 15 | 1.5 | 0.9 |
| 40 | 20 | 2.0 | 1.6 |
| 62.5 | 25 | 2.5 | 2.5 |
| 75 | 30 | 2.5 | 3.0 |

| Depth (msw) | Inspiratory gas temperature (°C) |
|---|---|
| 150 | $20 \pm 2$ |
| 200 | $20 \pm 2$ |
| 250 | $25 \pm 2$ |
| 300 | $25 \pm 2$ |
| 350 | $30 \pm 2$ |
| 400 | $30 \pm 2$ |

Although these standards are published as guidelines rather than enforced as regulations, the response from industry has been positive, and companies are requiring that breathing equipment comply with the recommendations. In deeper diving operations, the preferred rather than the acceptance limit has been sought. The guidelines have also

*Table 19.3.    Test procedures for a $CO_2$ scrubber in recirculatory apparatus.*

Expired gas conditions at scrubber inlet

| | |
|---|---|
| temperature | 30°C |
| humidity | 80% RH |
| $PCO_2$ | 4.0 kPa |
| water temperature | 5 ± 1°C |

Test cycle: to be repeated until breakthrough of $PCO_2 = 1\cdot0$ kPA

| Time (min) | $VCO_2$ (l min$^{-1}$) | RMV (l min$^{-1}$) | $VT$ (l) | $f$ (min$^{-1}$) |
|---|---|---|---|---|
| 4 | 0·9 | 23·0 | 2·0 | 11·5 |
| 6 | 2·0 | 50 | 2·0 | 25 |

been used as a basis for the development of acceptance standards for other forms of apparatus such as bail out systems, welding masks and chamber emergency breathing systems (Segadal and Morrison, 1985).

## 8. *Evaluation of breathing apparatus*

Despite the above advances in knowledge and design capabilities, most nations do not require underwater breathing apparatus to meet ergonomic standards. This is true of both commercial and sport diving activities. In the field of sport diving, where a great majority of apparatus is sold, few manufacturers provide performance data or depth limits for their apparatus. Thus the buyer must rely for information on 'hearsay' reports, and the relatively vague and unsubstantiated claims of the manufacturer such as 'low resistance'. In a study of seven open circuit Scuba regulators used in Canada, only two were capable of meeting the work of breathing requirements shown in Figure 19.5 at a depth of 50 m (Morrison *et al.*, 1986). The remaining apparatus failed to meet the acceptance standard at depths greater than 30 m.

Segadal *et al.* (1984) studied the performance of commercial breathing apparatus at simulated depths of 200–400 m. They concluded that a number of breathing apparatus could meet the preferred limit (equation 2) for work of breathing at 200 m, and one apparatus could meet this limit at 300 m. Based on physiological data and unmanned testing of apparatus Segadal *et al.* (1984) concluded (*a*) that there was no justification for using equipment which fell outside the limits of acceptance given in equation 1 (i.e., the tolerance limit); and (*b*) that the preferred limit (of comfort) given in equation 2 was a more realistic requirement of current design technology, particularly for deep diving.

## References

Agostoni, E., Gurtner, G., Torrie, G. and Rahn, H., 1966, Respiratory mechanics during submersion and negative-pressure breathing. *J. Appl. Physiol.*, **21**(1), 251–258.

Bentley, R.A., Griffin, O.G., Love, R.G., Muir, D.C.F. and Sweetland, K.F., 1973, Acceptable levels for breathing resistance of respiratory apparatus. *Arch. Environ. Health*, **27**, 273–280.

Bentz, R.L., 1976, Some design considerations for hyperbaric $CO_2$ scrubbers. In *Proceedings of the Divers' Gas Purity Symposium* (Columbus, OH: Batelle Columbus Laboratories) (Report No. 7–76), pp. 9.1–9.10.

Case, E.M. and Haldane, J.B.S., 1941, Human physiology under high pressure. *J. Hygiene*, **41**, 225–249.

Cooper, E.A., 1960, Suggested methods of testing and standards of resistance for respiratory protective devices. *J. Appl. Physiol.*, **15**, 1053–1061.

Cooper, E.A., 1961, *Behaviour of Respiratory Apparatus*, Medical Research Memorandum 2, UK National Coal Board.

Craig, A.B. and Dvorak, M., 1975, Expiratory reserve volume and vital capacity of the lungs during immersion in water. *J. Appl. Physiol.*, **38**, 5–9.

Dwyer, J., Saltzman, H.A. and O'Bryan, R., 1977, Maximal physical-work capacity of man at 43·4 ATA. *Undersea Biomed. Res.*, **4**, 359–372.

Ferris, B.G., Mead, J. and Opie, L.H., 1964, Partitioning of respiratory flow resistance in man. *J. Appl. Physiol.*, **19**, 653–658.

Florio, J.T., Morrison, J.B. and Butt, W.S., 1979, Breathing pattern and ventilatory response to carbon dioxide in divers. *J. Appl. Physiol.*, **46**, 1076–1080.

Goff, L.G. and Bartlett, R.G., 1957, Elevated end tidal $CO_2$ in trained underwater swimmers. *J. Appl. Physiol.*, **10**, 203–206.

Hayes, P.A., 1980, Hazards of diving—cold and heat. In *Oceanology International 80*, Brighton, UK, Technical Session G. Diving Operations, pp. 33–40.

Hayes, P.A., Padbury, E.H., Florio, J.T. and Fyfield, T.P., 1981, *Respiratory Heat Transfer in Cold Water and During Rewarming*, Admiralty Marine Technology Establishment Report (E) 80–401 (London: HMSO).

Hesser, C.M., Fagraeus, L. and Linnarsson, D., 1968, *Cardio-respiratory Responses to Exercise in Hyperbaric Environment*, Report, Laboratory of Aviation and Naval Medicine (Stockholm: Karolinska Institutet).

Hong, S.K., Cerretelli, P., Cruz, J.C. and Rahn, H., 1969, Mechanics of respiration during submersion in water. *J. Appl. Physiol.*, **27**(4), 535–538.

Jarrett, A.S., 1965, Effect of immersion on intrapulmonary pressure. *J. Appl. Physiol.*, **20**, 1261–1266.

Lalley, D.A., Zechman, F.W. and Tracy, R.A., 1974, Ventilatory responses to exercise in divers and non-divers. *Resp. Physiol.*, **20**, 117–129.

Lambertsen, C.J., Owen, S.G., Wenndel, H., Stroud, M.W., Lurie, A.A., Lochner, W. and Clark, G.F., 1959, Respiratory and cerebral circulatory control during exercise at 0·21 and 2·0 atmospheres inspired $PO_2$. *J. Appl. Physiol.*, **14**, 966–982.

Lanphier, E.H., 1955, Use of nitrogen-oxygen mixtures in diving. In *Proceedings of the 1st Symposium on Underwater Physiology*, edited by L.G. Goff, National Research Council Publication 377 (Washington, DC: National Academy of Sciences), pp.74–78.

Lanphier, E.H., 1963, Influence of increased ambient pressure upon alveolar ventilation. In *Proceedings of the 2nd Symposium on Underwater Physiology*, edited by C.J. Lambertsen and L.J. Greenbaum, Jr, National Research Council Publication 1181 (Washington, DC: National Academy of Sciences), pp.124–133.

Lanphier, E.H., 1969, Pulmonary function. In *The Physiology and Medicine of Diving and Compressed Air Work*, 1st ed., edited by P.B. Bennett and D.H. Elliott (London: Bailliere, Tindall and Cassell), pp. 58–112.

Marshall, J. and Lambertsen, C.J., 1961, Interactions of increased $PO_2$ and $PCO_2$ effects in producing convulsions and death in mice. *J. Appl. Physiol.*, **16**, 1–7.

Middleton, J.R., 1979, *Unmanned Evaluation of US Navy UBA Ex–16 Prototype Closed Circuit Rebreather*, Navy Experimental Diving Unit Report 11–79 (Panama City, FL: Dept of the Navy).

Middleton, J.R., 1980, *Evaluation of Commercially Available Open Circuit SCUBA Regulators*, Navy Experimental Diving Unit Report 2–80 (Panama City, FL: Dept of the Navy).

Middleton, J.R. and Thalmann, E.D., 1981, *Standardized NEDU Unmanned UBA Test Procedures and Performance Goals*, Navy Experimental Diving Unit Report 3–81 (Panama City, FL: Dept of the Navy).

Morrison, J.B., 1973, Oxygen uptake studies of divers when fin swimming with maximum effort at depths of 6–176 feet. *Aerospace Med.*, **44**, 1120–1129.

Morrison, J.B., 1981a, *Physiological Acceptance Criteria for Underwater Breathing Apparatus* (London: Petroleum Production Division of Department of Energy).

Morrison, J.B., 1981b, *Unmanned Test Procedures for Underwater Breathing Apparatus* (London: Petroleum Production Division of Department of Energy).

Morrison, J.B., 1981c, Unmanned test procedures and physiological acceptance criteria. In *Divetech 81*, Proceedings of an International Conference (London: Society for Underwater Technology).

Morrison, J.B., 1983, The influence of breathing system design on the capacity of divers to work at depth. In *Offshore Goteborg, '83, Underwater Technology and Diving*, Session B3a (Goteborg, Sweden: Svensk Massan Stiftelse).

Morrison, J.B. and Reimers, S.D., 1982, Design principles of underwater breathing apparatus. In *The Physiology and Medicine of Diving and Compressed Air Work*, 2nd ed., edited by P.B. Bennett and D.H. Elliott (London: Bailliere, Tindall), Chapter 5.

Morrison, J.B. and Wood, I., 1986, Ventilatory responses and perceived respiratory effort of divers working at maximal physical capacity. *Ann. Physiol. Anthrop.*, **5**, 169.

Morrison, J.B., Butt, W.S., Florio, J.T. and Mayo, I.C., 1976, Effects of increased $O_2-N_2$ pressure and breathing apparatus on respiratory function. *Undersea Biomed. Res.*, **3**, 217–234.

Morrison, J.B., Florio, J.T. and Butt, W.S., 1978, Observations after loss of consciousness under water. *Undersea Biomed. Res.*, **5**, 179–187.

Morrison, J.B., Florio, J.T. and Butt, W.S., 1981, Effects of $CO_2$ insensitivity and respiratory pattern on respiration in divers. *Undersea Biomed. Res.*, **8**(4), 209–217.

Morrison, J.B., Morariu, G. and Wood, I., 1986, Performance characteristics of open circuit demand regulators from 0·5 to 50 metres seawater. *Ann. Physiol. Anthrop.*, **5**, 169.

Norwegian Petroleum Directorate, 1984, *Draft Guidelines for Minimum Performance Requirements and Standard Unmanned Test Procedures for Underwater Breathing Apparatus.* Stavanger, Norway.

Otis, A.B., Fenn, W.O. and Rahn, H., 1950, Mechanics of breathing in man. *J. Appl. Physiol.*, **2**, 592–607.

Pasche, A. and Tonjum, S., 1983, Body heat balance during operational offshore diving. In *Offshore Goteborg, '83, Underwater Technology and Diving*, Session B3a (Goteborg, Sweden: Svenska Massan Stiftelse).

Paulsen, H.N. and Jarvi, R.E., 1977, *Swimmer Life Support System (SLSS Mk1) Technical Evaluation*, Navy Experimental Diving Unit Report 14–76 (Panama City, FL: Dept of the Navy).

Peterson, R.E. and Wright, W.G., 1976, Pulmonary mechanical function in man breathing dense gas mixtures at high ambient pressure—predictive studies III. *Proceedings of the 5th Symposium on Underwater Physiology*, edited by E.J. Lambertsen (Bethesda, MD: FASEB), pp. 67–77.

Piantadosi, C.A., 1980, *Respiratory Heat Loss Limits in Helium Oxygen Saturation Diving*, Navy Experimental Diving Unit Report 10–80 (Washington, DC: Dept of the Navy).

Reimers, S.D., 1974, *Proposed Standards for the Evaluation of the Breathing Resistance of Underwater Breathing Apparatus*, Navy Experimental Diving Unit Report 19–73 (Washington, DC: Dept of the Navy).

Schaefer, K.E., 1955, The role of carbon dioxide in the physiology of human diving. In *Proceedings of the 1st Symposium on Underwater Physiology*, edited by L.G. Goff, National Research Council Publication 377 (Washington, DC: National Academy of Sciences), pp.131–139.

Segadal, K., 1984, Breathing resistance—keeping the requirements realistic. In *Divetech '84*, Proceedings of an International Conference (London: Society for Underwater Technology).

Segadal, K. and Morrison, J.B. 1985, *Acceptance Criteria and Unmanned Test Procedures for Underwater Breathing Apparatus*, Report for Norsk Hydro A/S. OTP W.O. 335. (Bergen: NUTEC).

Segadal, K., Furevik, M. and Myrseth, E., 1984, Breathing resistance—keeping the requirements realistic. In *Divtech '84*, Proceedings of an International Conference (London: Society for Underwater Technology).

Senneck, C.R., 1962, Breathing apparatus for use in mines. In *Design and Use of Respirators*, edited by C.N. Davies (Oxford: Pergamon Press), pp. 143–159.

Silverman, L., Lee, G., Yancey, A.R., Amory, L., Barney, L.J. and Lee,

R.C., 1945, *Report No. 5339* (Washington, DC: Office of Scientific Research and Development).

Spaur, W.H., Raymond, L.W., Knott, M.M., Crothers, J.C., Braithwaite, W.R., Thalmann, E.D. and Uddin, D.F., 1977, Dyspnea in divers at 49·5 ATA: mechanical not chemical in origin. *Undersea Biomed. Res.*, **4**(2), 183–198.

Smith, J.G., 1973, *Low Temperature Performance of $CO_2$ Systems* (Atlanta, GA: W.R. Grace).

Smith, J.G., 1974, *Comparison of Sodasorb and Baralyme Performance* (Atlanta, GA: W.R. Grace).

Taylor, N., 1987, Effect of breathing gas pressure on respiratory mechanics of immersed man. PhD Thesis, Simon Fraser University, Burnaby, Canada.

Thalmann, E.D., Sponholtz, D.K. and Lundgren, C.E.G., 1979, Effects of immersion and static lung loading on submerged exercise at depth. *Undersea Biomed. Res.*, **6**(3), 259–290.

Ting, E.Y., Hong, S.K. and Rahn, H., 1960, Lung volumes, lung compliance and airway resistance during negative–pressure breathing. *J. Appl. Physiol.*, **15**, 554–556.

US Navy, *U.S. Navy Diving Manual*, 1975, NAVSEA 0994 LP 001 9010 (California: Best Bookbinders).

Vaernes, R., Pasche, A., Tonjum. S. and Peterson, R., 1983, Working in water at 500 msw breathing heliox: An analysis of diver performance as a function of HPNS and body temperature. In *Proceedings of the 8th Symposium on Underwater Physiology*, edited by A.J. Bachrach and M.M. Matzen (Bethesda, MD: Undersea Medical Society).

Wang, T.C., 1975, *Temperature Effects on Baralyme, Sodasorb and Lithium Hydroxide* (Harbour Branch Foundation).

Wang, T.C., 1976, *$CO_2$ Absorbent Comparison Analyses* (Harbour Branch Foundation).

# PART VI
# THE HYPOBARIC
# ENVIRONMENT

# 20. Prediction of Severe Body Cooling in Hypobaric Environments

### Yasunobu Nishi

## 1. Introduction

Metabolic heat production is balanced by radiative, convective and evaporative heat loss from the skin surface, and also by respiratory evaporative and convective heat losses. Among these heat transfer processes, significant effects of barometric pressure may be theoretically predicted for convection, skin evaporation and also for pulmonary heat loss.

## 2. Convective heat transfer

Convective heat transfer from the body surface may be generally expressed in terms of the convective heat transfer coefficient by

$$C = hc \ (T_{cl} - T_a) \ (\text{kcal m}^{-2} \text{ h}^{-1})  \tag{1}$$

Where $C$ = rate of convective heat transfer in kcal m$^{-2}$ h$^{-1}$; $hc$ = convective heat transfer coefficient in kcal m$^{-2}$ h$^{-1}$ °C$^{-1}$; $T_{cl}$ = clothing surface temperature in °C; and $T_a$ = ambient air temperature in °C. In a hypobaric environment, the convective heat transfer coefficient ($hc'$) is affected by barometric pressure due to a change in air density (Gagge and Nishi, 1977; ASHRAE, 1985).

$$hc' = (P_B/760)^{0.55} h_c (\text{kcal m}^{-2} \text{ h}^{-1} \text{ °C}^{-1})  \tag{2}$$

where $P_B$ = barometric pressure in Torr; and 760 = sea level barometric pressure in Torr. Based on this equation, on the summit of Mt Everest, where the pressure is approximately one-third of an atmosphere, the value of $hc'$ decreases by about 50%. Thus on high mountains, provided the temperature gradient $T_{cl} - T_a$ remains the same, humans may lose less heat by convective heat transfer.

## 3. Evaporative heat transfer

In a similar manner as shown in equation (2), the rate of evaporative heat transfer can be described in terms of the evaporative heat transfer coefficient and the water vapour pressure difference between the skin surface and the air:

$$E = w \, he \, (P_{sk} - P_a) \; (\text{kcal m}^{-2} \, \text{h}^{-1}) \tag{3}$$

where $E$ = rate of evaporative heat loss in kcal m$^{-2}$ h$^{-1}$ °C; $w$ = skin wettedness in N D; $he$ = evaporative heat transfer coefficient in kcal m$^{-2}$ h$^{-1}$ Torr$^{-1}$; $P_{sk}$ = saturated water vapour pressure at mean skin temperature in Torr; and $P_a$ = ambient water vapour pressure in Torr.

In the equation above skin wettedness ($w$) represents a fraction of the skin surface area made wet by sweating (Gagge and Nishi, 1977; ASHRAE, 1985). The evaporative heat transfer coefficient ($he$) may be derived by applying the heat and mass transfer analogy as before and is given in terms of the convective heat transfer coefficient ($hc$) by

$$he = 2 \cdot 2 \, hc \; (\text{kcal m}^{-2} \, \text{h}^{-1} \, \text{Torr}^{-1}) \tag{4}$$

In the equation the factor 2·2 (°C Torr$^{-1}$) is known as the modified Lewis Relation (LR); this factor is also inversely proportional to the barometric pressure ($P_B$) (Gagge and Nishi, 1977; Nishi and Gagge, 1977; ASHRAE, 1985).

$$LR = 2 \cdot 2 \, (760/P_B) \; (°C \, \text{Torr}^{-1}) \tag{5}$$

Finally, in a hypobaric environment $he'$ can be described in terms of ($P_3$) and the $hc$ value for sea level by the following equation:

$$he' = 2 \cdot 2 \, (760/P_B)^{0 \cdot 45} h_c (\text{kcal m}^{-2} \, \text{h}^{-1} \, \text{Torr}^{-1}) \tag{6}$$

Using this equation at one-third of an atmosphere, or on the summit of Mt Everest, evaporative capacity of the environment is predicted to increase by as much as 60%. That is to say, in hypobaric environments evaporative cooling on the skin surface becomes quite effective to dissipate body heat to the environment.

## 4. Water loss through pulmonary ventilation

If the exhaled air is saturated with water vapour at body temperature, 37°C, water loss ($m$) from the lungs may be calculated by:

$$m = (60/1000) \, V_{btps} \, (1/vs) \, (x_E - x_I) \; (\text{kg h}^{-1}) \tag{7}$$

where $m$ = rate of water loss in kg h$^{-1}$; $V_{btps}$ = rate of pulmonary ventilation in 1 min$^{-1}$; $vs$ = specific volume of moist air in m$^3$ kg$^{-1}$ (D.A.); $x_E$ = humidity ratio of exhaled air in kg kg$^{-1}$ (D.A.); and $x_I$ = humidity ratio of inhaled air in kg kg$^{-1}$ (D.A.).
and

$$x_E = 0.622 \ P_E \ (P_B - P_E)^{-1} \ (\text{kg kg}^{-1} \ (\text{D.A.})) \tag{8}$$

where $P_E$ = water vapour pressure in exhaled air in Torr.

## 5. Enthalpy of moist air

The enthalpy ($i$) of moist air is measured in terms of the temperature ($T$) and humidity ratio ($x$):

$$i = 0.24 \ T + x \ (597.5 + 0.44 \ T) \ (\text{kcal kg}^{-1} \ (\text{D.A.})) \tag{9}$$

The first term is the energy content of the dry air itself and the second term is the energy content of the moisture. The constant factor 579.5 (kcal kg$^{-1}$) is the heat of vapourization of the water, and the two factors 0.24 and 0.44 (kcal kg$^{-1}$ °C$^{-1}$) are the specific heats of dry air and water vapour respectively.

Figure 20.1 shows how the enthalpy of the saturated air increases at high altitude. This is why in hypobaric environments the gaseous concentration of air itself is quite low, but a unit volume of air can contain a certain amount of the moisture which is only dependent on temperature.

## 6. Respiratory heat loss

The third avenue of body heat loss which will be significantly affected by the barometric pressure is pulmonary ventilation. Using enthalpy, the rate of respiratory heat loss at altitude may be predicted by:

$$RHL = (60/1000) \ V_{btps} \ (1/vs) \ (i_E - i_I \ (\text{kcal h}^{-1}) \tag{10}$$

where $V_{btps}$ is the rate of the pulmonary ventilation as described before; $vs$ is the specific volume of saturated exhaled air at 37°C; $i_E$ is the enthalpy of exhaled air which is saturated at 37°C; and $i_I$ is the enthalpy of inhaling ambient air. Mountain climbers who reach an altitude of 4 000 m (13 000 feet) sometimes suffer from mountain sickness due to the reduced oxygen concentration in the inhaled air. Thus mountain climbers who ascend more than 7 000 m usually carry compressed

*Environmental Ergonomics*

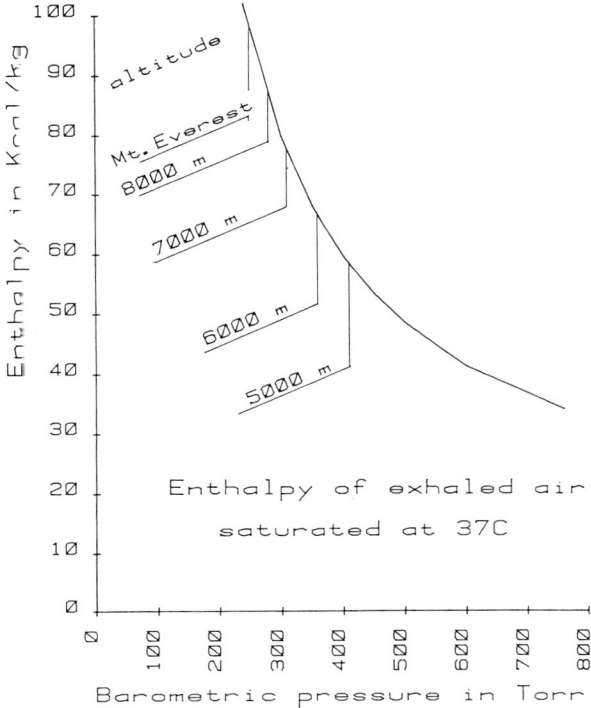

*Figure 20.1. Enthalpy of exhaled air saturated at body temperature at altitude.*

oxygen to prevent problems related with hypoxia. However, when a climber breathes thin ambient air without compressed oxygen, he will have to inhale a large volume of thin air to get the required amount of oxygen. This is known as hyperventilation.

Figure 20.2 shows the hyperventilation observed by Balke on a resting subject at simulated high altitude. The data show a sharp increase in ventilation rate at about 5 000 m (Balke, 1972). Also in the Figure, data observed by West and his associates on the summit of Mt Everest are indicated, where the ventilation rate increased to as much as $40\,l\,min^{-1}$ at rest, and substantially more during physical work (West *et al.*, 1983).

If the assumption is made that the air in the lungs is saturated at body temperature, and this may be true since water is supplied abundantly from the respiratory tract, heat loss through respiration becomes significant. Based on both increased ventilation rate and the increased enthalpy value at altitude the rate of respiratory heat loss may be predicted, as shown in Figure 20.3.

As indicated in Figure 20.3, at altitudes of 7 000–8 000 m humans may lose about 60 kcal of body heat in 1 hour. The amount of this heat

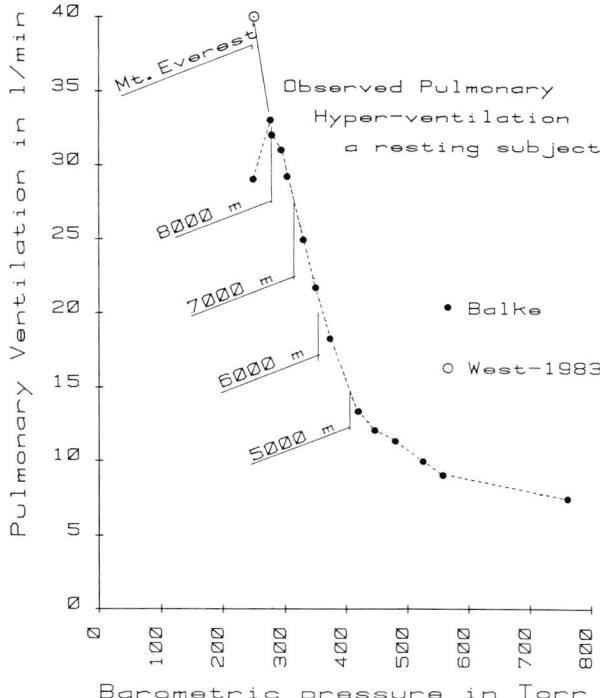

*Figure 20.2.* *Pulmonary hyperventilation observed at simulated altitude and at the summit of Mount Everest.*

loss almost corresponds to human resting metabolic energy. In other words humans may lose most of their metabolic energy solely through excessive pulmonary ventilation at high altitude.

In addition to this, in the cold, heat losses by convection, radiation and skin evaporation must be taken into account, and thus severe body cooling should be expected in cold hypobaric environments.

Figure 20.4 shows predicted water loss through pulmonary ventilation derived from data shown in Figure 20.2 and equation (7), and suggests possible dehydration at altitude.

## 7. *Conclusions*

In hypobaric environments, unique characteristics of the heat transfer processes are:

1. Convective heat transfer may be eased due to a decrease in air density.

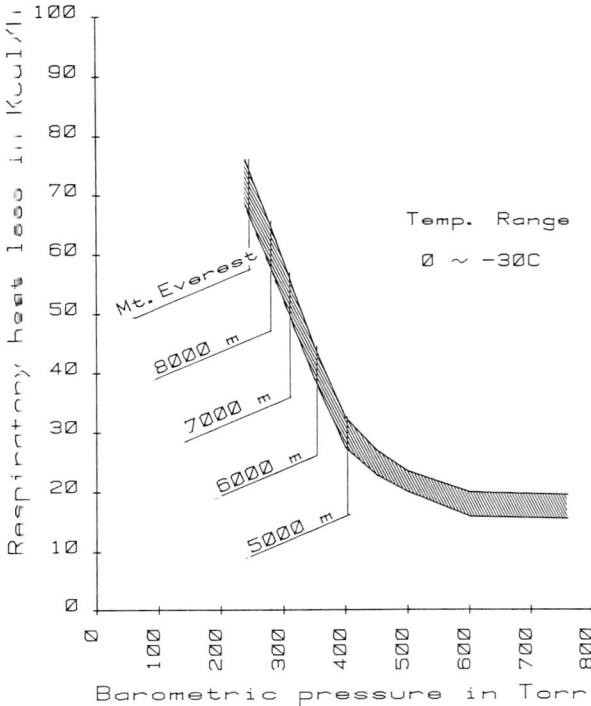

*Figure 20.3.  Predicted pulmonary heat loss at altitude.*

2. Evaporative heat transfer may be accelerated. Its ultimate mechanism is by vacuum vaporization or drying. However, on a cold, high mountain, regulatory sweating may not be expected, and thus the effect of evaporative cooling may be minimized.

3. Heat loss through pulmonary ventilation may increase significantly due to the hyperventilation and psychrometric characteristics of moist air at altitude. The amount of this heat loss can correspond almost to the resting metabolic energy.

A significant number of fatal accidents on high mountains are obviously attributable to failure of physiological adaptation related to hypoxia. However, severe body cooling resulting from hyperventilation, as predicted in this chapter, may also force even well-trained climbers to their physiological limits, even if they are well protected with heavily insulated garments.

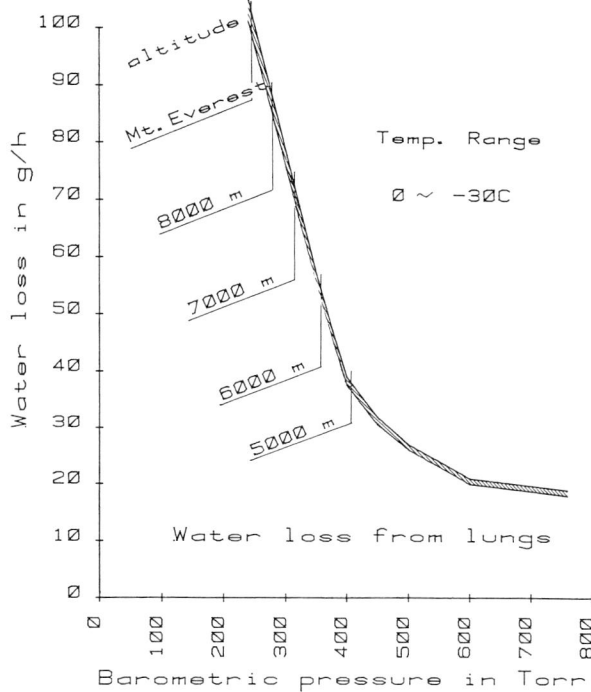

*Figure 20.4.　Predicted water loss from lungs at altitude.*

## References

ASHRAE, 1985, *Handbook of Fundamentals*, SI edition (New York: American Society of Heating, Refrigerating and Air-Conditioning Engineers).

Balke, B., 1972, Physiology of respiration at altitude. In *Physiological Adaptations—Desert and Mountain*, edited by M.K. Yousef *et al.* (New York: Academic Press), Chapter 14.

Gagge, A.P. and Nishi, Y., 1977, Heat exchange between human skin surface and thermal environment. In *Handbook of Physiology. Reaction to Environmental Agents*, edited by D.H.K. Lee (Bethesda, MD: American Physiological Society), Section 9, Chapter 5, pp. 69–92.

Nishi, Y. and Gagge, A.P., 1977, Effective temperature scale useful for hypo- and hyperbaric environments. *Aviat. Space Environ. Med.*, Feb, 97–107.

West, J.B., Hackett, P.H., Maret, K.H., Milledge, J.S., Peters, J.R., Pizzo, C.J. and Winslow, R.M., 1983, Pulmonary gas exchange on the summit of Mount Everest. *J. Appl. Physiol.: Respirat. Environ. Exercise Physiol.*, **55**(3), 678–687.

# 21. The Lower Critical Temperature at Acute Hypoxia

**Masahiko Sato and Kazuya Matsuda**

## 1. Introduction

The lower critical temperature (LCT) has been regarded as a sensitive index for estimating cold adaptability of homeotherms since Scholander *et al.* (1950) suggested it to be near 27°C in most tropical species and with a range from −50°C to 15°C in arctic species. However, the LCT at high altitude, to our knowledge, has not been investigated.

Several authors have reported that acute hypoxia affects the increase in MR caused by cold exposure (Wezler and Frank, 1948; Bullard, 1961; Clegg *et al.*, 1970; Velasquez, 1970; Busch *et al.*, 1985). And it has been postulated that peripheral vasodilation is caused by direct effects of moderate hypoxia on peripheral vessels (Krogh, 1929; Abramson *et al.*, 1943; Anderson *et al.*, 1946; Black and Roddie, 1958; Goemoeri *et al.*, 1960). Therefore, acute moderate hypoxia may exert a marked influence on LCT. The present investigation of the effects of hypoxia on LCT is significant to our understanding of human adaptability to high altitude.

## 2. Methods

Seven young adult males, ranging in age from 21 to 25 years, all in good physical condition, volunteered for this study. The mean and standard deviation of their stature were 174·6 cm and 6·5 cm respectively and corresponding values of body weight were 64·8 kg and 9·8 kg.

The LCT of each subject was estimated at three different grades of simulated altitude, that is, 0 m, 2 500 m and 5 000 m, according to an indirect intersect method (Erikson *et al.*, 1956), using the metabolic rate

(MR) during pedalling an ergometer. The subjects quietly assumed a reclining position for 2 hours at 15°C in a climatic chamber. The relative humidity and air velocity were kept at about 50% and 25 m min$^{-1}$, respectively. They reclined with their legs fully extended to maximize the exposure. Then they moved to a bicycle ergometer in the same ambient conditions, and sat and pedalled for about 1 hour while maintaining constant rectal temperature ($T_r$) (Yoshimura and Yoshimura, 1969; Wilkerson *et al.*, 1972; Sato *et al.*, 1985). $T_r$ at a depth of 10 cm was continuously recorded with a thermistor. Before entering the chamber, all subjects rested at 28°C and 50% humidity in a recumbent position for more than 30 minutes. Every subject was measured at each altitude on separate days in April or May. The order of altitude exposure was randomized between the subjects.

Each exposure was performed with the subject in the postabsorptive state. Expired gas was collected with Douglas bags through low-resistance valves for 3 minutes at intervals of 30 minutes during the resting and pedalling exposures. The ventilatory volume was measured with a wet gas meter and aliquots of mixed expired gas were analysed for oxygen and carbon dioxide by a mass spectrometer. Mean skin temperature ($T_{sk}$) was calculated from continuous recordings of the skin temperature at the chest, upper arm, thigh and lower leg, according to the formula of Ramanathan (1964). All the subjects were wearing the same type of thin T-shirts, shorts, socks and shoes. An additional exposure to 28°C at 0 m was conducted for each subject following the other exposures to obtain control data in a thermoneutral condition.

Shivering activity was monitored visually and quantified by observers according to an arbitrary scale: 0 = no shivering activity, 1 = mild shivering in bursts, 2 = generalized but discontinuous shivering, 3 = very marked and continued shivering movements.

## 3. Results

The MR was increased by the cold exposure under all the altitude conditions. Figure 21.1 shows the equilibrated values of MR at the final stage of thermoneutral and cold exposures, plotted against the altitude from 0 to 5 000 m. It was observed that the cold-induced increase in MR was enhanced at high altitude, especially at 5 000 m. An analysis of variance confirmed that the altitude factor significantly ($P<0.005$) influenced the MR. The subjects exhibited obvious, though discontinuous, shivering movements, which became evident within 30 minutes after the onset of cold exposure, as shown in Figure 21.2. Shivering tended to become more vigorous at higher altitude. The increase in shivering movements with increase in altitude and exposure

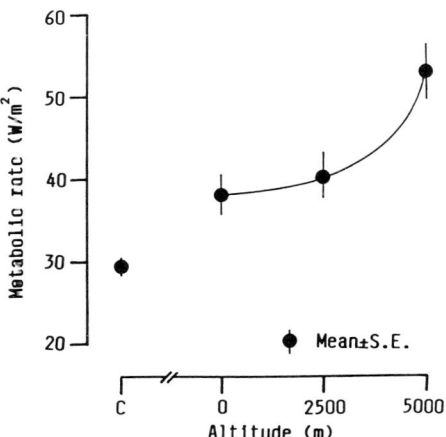

*Figure 21.1.* MR (mean±SE) in the final stage of control and cold exposures at different altitudes.

time was significant ($P<0.01$) according to a nonparametric t–test (Wilcoxon, 1945).

It was observed that $T_{sk}$ at 5 000 m was higher than that at the other altitudes (Figure 21.3). These differences were statistically significant ($P\leq0.005$).

As is evident from Figure 21.4, $T_r$ at 5 000 m was significantly lower ($P\leq0.005$) than during any other condition.

The LCT was estimated from the MR during cycle ergometry (Sato *et al.*, 1985). The mean and standard error of LCT at each altitude is shown in Figure 21.5. An analysis of variance confirmed that

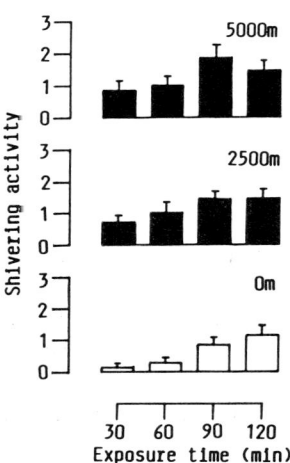

*Figure 21.2.* *Visible shivering intensity, recorded at intervals in arbitrary units (see text).*

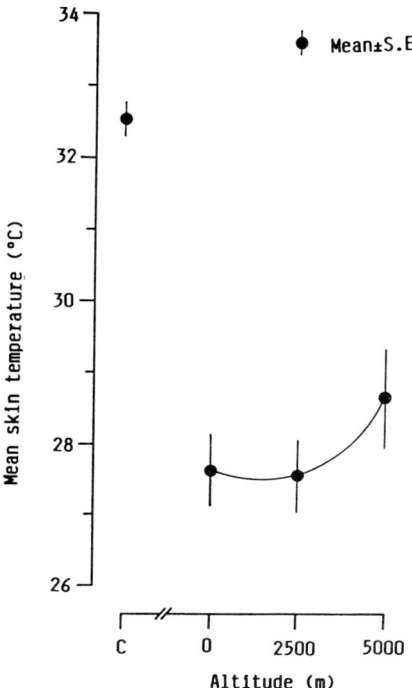

*Figure 21.3.* T$_{sk}$ *(mean±SE) in the final stage of control and cold exposures at different altitudes.*

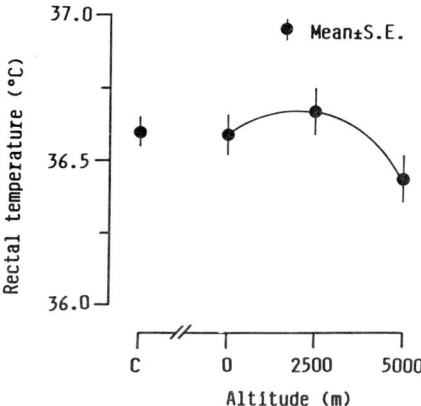

*Figure 21.4.* T$_r$ *(mean±SE) in the final stage of control and cold exposures at different altitudes.*

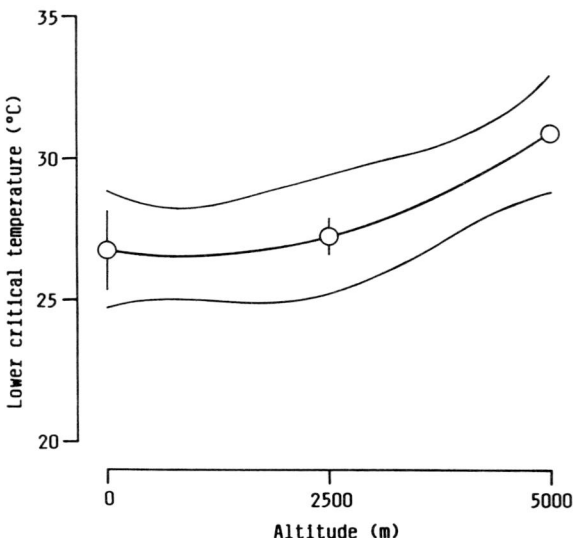

*Figure 21.5.    LCT (mean±SE) at three grades of simulated altitude and the regression curve of LCT on altitude and its 95% confidence limits.*

the altitude factor significantely influenced LCT ($P\leq0\cdot005$). The regression equation of LCT (°C) on altitude ($H$ km) was calculated as:

$$LCT = 26\cdot8 - 0\cdot422\ H + 0\cdot248\ H^2$$

The correlation coefficient between the LCT and altitude was $0\cdot61$ ($P\leq0\cdot01$).

   The thermal conductance of peripheral tissue was calculated by the method of Hardy (1961) and is shown in Figure 21.6. Thermal conductance under cold conditions became larger as the degree of hypoxia became more severe (($P\leq0\cdot005$). The value of this variable at 5 000 m was almost equal to that at 28°C under normoxic conditions.

## 4. Discussion

The effect of hypoxia on cold-induced increase in MR remains equivocal. Blatteis and Lutherer (1976) pointed out that conflicting results may be ascribable to different environmental conditions, experimental subjects, and protocols used in past studies. In the present study, the level of MR exhibited under hypoxia by the subjects was higher than that at 0 m. A similar enhancement of the calorigenic response to cold of

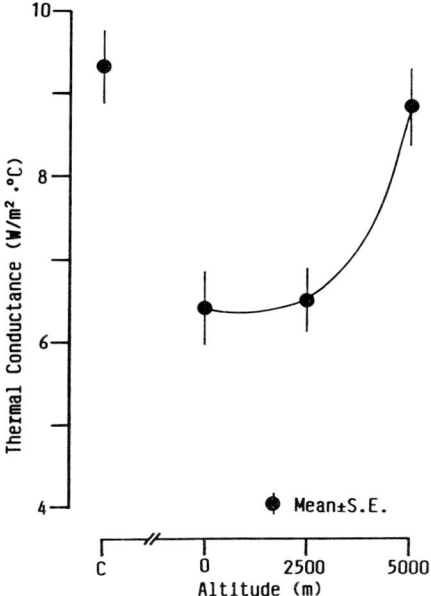

*Figure 21.6.    Thermal conductance (mean±SE) in the final stage of control and cold exposures at different altitudes.*

unacclimatized subjects has already been reported (Wezler and Frank, 1948; Bullard, 1961; Clegg *et al.*, 1970; Velasquez, 1970).

As shown above, visible shivering intensity was increased under hypoxic conditions in the subjects. This result is also in accordance with previous findings (Wezler and Frank, 1948; Bullard, 1961; Davis *et al.*, 1967; Blatteis and Lutherer, 1976). The intensity of shivering seems to be in agreement with the degree of increase in MR.

$T_{sk}$ at 5 000 m was higher than under other conditions of altitude. This kind of alleviation of the cold-induced lowering in skin temperature by hypoxia has been already indicated by others (Krogh, 1929; Abramson *et al.*, 1943; Anderson *et al.*, 1946; Black and Roddie, 1958; Goemoeri *et al.*, 1960; Davis *et al.*, 1967; Hanna, 1970; Cipriano and Goldman, 1975). The peripheral vasodilation accounting for higher $T_{sk}$ under conditions of relative hypoxia has been attributed to the accompanying hypocapnia (Black and Roddie, 1958). This phenomenon means that heat loss from the body surface is greater under conditions of acute hypoxia.

$T_r$ was different among the altitude conditions. The low $T_r$ and high MR observed at 5 000 m may imply that the drive, in the present subjects, for increased heat production in the cold was related to a lowering of core temperature.

The LCT of the present subject group at 0 m was estimated to be

26·8°C. This estimation agrees with previous determinations on Japanese males (Yoshimura and Yoshimura, 1969; Sato *et al.*, 1980; Sato and Yamasaki, 1985). LCT at 2 500 m was determined to be 27·3°C, and at 5 000 m to be 30·9°C. Thus, LCT was shown to rise as the altitude increases. The quadratic regression of LCT on altitude suggests that the rate of increase in LCT would be in direct proportion to the altitude.

LCT has been defined as the lowest temperature for homeothermic mammals to maintain their body temperature by only vasomotor control. As shown by the high skin temperature under hypoxic conditions, the high LCT at high altitude seems to be caused mainly from the peripheral vasodilation produced by the hypoxic effect.

Thermal conductance at 5 000 m was larger than that at the other altitudes. An analysis of variance confirmed that altitude had a significant effect on LCT ($P \leq 0.05$). The large thermal conductance of small resistance to heat loss at 5 000 m easily explains the high LCT at this altitude.

## Acknowledgements

The authors express their sincere thanks to their seven subjects and to the members of the Department of Ergonomics, Kyushu University of Design Sciences for their important assistance in this study.

## References

Abramson, B.I., Landt, H. and Benjamin, J.E., 1943, Peripheral vascular responses to acute anoxia. *Arch. Int. Med.*, **71**, 583−593.

Anderson, D.P., Allen, W.G., Barcroft, H., Edoholm, O.G. and Manning, G.W., 1946, Circulatory changes during fainting and coma caused by oxygen lack. *J. Physiol.*, **104**, 426−436.

Black, J.E. and Roddie, I.C., 1958, The mechanism of the changes in forearm vascular resistance during hypoxia. *J. Physiol.*, **143**, 265−235.

Blatteis, C.M. and Lutherer, L.O., 1976, Effects of altitude exposure on thermoregulatory response of man to cold. *J. Appl. Physiol.*, **41**, 848−858.

Bullard, R.W., 1961, Effects of hypoxia on shivering in man. *Aerosp. Med.*, **32**, 1143−1147.

Busch, M.A., Tucker, A. and Robertshaw, D., 1985, Interaction between cold and altitude exposure on pulmonary circulation of cattle. *J. Appl. Physiol.*, **58**, 948−953.

Cipriano, L.F. and Goldman, R.F., 1975, Thermal responses of unclothed men exposed to both cold temperatures and high altitude (3475 m). *J. Appl. Physiol.*, **39**, 796−800.

Clegg, E.J., Harrison, G.A. and Baker, P.T., 1970, The impact of high altitude on human populations. *Human Biol.*, **42**, 486−518.

Davis, T.R.A., Nayer, H.S., Sinha, K.C., Nisbith, S.D. and Rai, R.M., 1967, The effect of altitude on the cold responses of low altitude acclimatized Jats, high altitude acclimatized Jats and Tibetans. In *Biometeorology 2*, edited by S.W. Tromp and W.H. Weihe (Oxford: Pergamon Press), pp. 191–198.

Erikson, H., Krog, J., Andersen, K.L. and Scholander, P.F., 1956, The critical temperature in naked man. *Acta Physiol. Scand.*, **37**, 35–39.

Goemoeri, P., Kovack, A.G.B., Takacs, L., Foldi, M., Szabo, G., Nagy, Z., Wiltner, W. and Kallay, K., 1960, The regulation of cardiac output in hypoxia. *Acta Med. Hung.*, **16**, 93–98.

Hanna, J.M., 1970, A comparison of laboratory and field studies of cold responses. *Amer. J. Phys. Anthropol.*, **32**, 227–232.

Hardy, J.D., 1961, Physiology of temperature regulation. *Physiol. Rev.*, **41**, 521–606.

Krogh, A., 1929, *The Anatomy and Physiology of Capillaries*, revised ed. (New Haven, CT: Yale University Press), p. 133.

Ramanathan, N.L., 1964, A new weighting system for mean surface temperature of human body. *J. Appl. Physiol.*, **19**, 531–533.

Sato, M. and Yamasaki, K., 1985, The lower critical temperature and native place. *Ann. Physiol. Anthropol.*, **4**, 181–182.

Sato, M., Takasaki, Y. and Yamasaki, K., 1980, The lower and upper critical temperatures for oxygen intake at rest. *J. Anthropol. Soc. Nippon*, **88**, 133–140.

Sato, M., Watanuki, S., Iwanaga, K. and Shinozaki, F., 1985, The influence of clothing ensembles on the lower critical temperature. *Eur. J. Appl. Physiol.*, **54**, 7–11.

Scholander, P.F., Hock, R., Walters, V., Johnson, F. and Irving, L., 1950, Heat regulation in some arctic and tropical mammals and birds. *Biol. Bull.*, **99**, 237–258.

Velasquez, T., 1970, Aspects of physical activity in high altitude natives. *Amer. J. Phys. Anthropol.*, **32**, 251–258.

Wezler, K. and Frank, E., 1948, Chemische Warmeregulation gegen Kalte und Hitze in Sauerstoffmangel. *Pflugers Arch.*, **250**, 439–464.

Wilcoxon, F., 1945, Individual comparisons by ranking method. *Biometrics Bull.*, **1**, 80–83.

Wilkerson, J.E., Raven, P.B. and Horvath, S.M., 1972, Critical temperature of unacclimatized male Caucasians. *J. Appl. Physiol.*, **33**, 451–455.

Yoshimura, M. and Yoshimura, H., 1969, Cold tolerance and critical temperatures of the Japanese. *Int. J. Biometeorol.*, **13**, 163–172.

# 22. Changes in Pulmonary Diffusing Capacity at Simulated High Altitudes Under Different Ambient Temperatures

**Akira Yasukouchi, Kaoru Inoue and Masahiko Sato**

## 1. Introduction

There have been problems in assessing changes in pulmonary diffusing capacity ($D_L$) and pulmonary capillary blood volume ($V_c$) at high altitude. One of these is due to a thermal effect: although ambient temperature decreases with increased altitude, the effect of temperature on changes in $D_L$ and $V_c$ has not been investigated. In addition, $V_c$ has usually been estimated by the method of Roughton and Forster (1957). This method makes the assumption that $V_c$ is not changed by a change in oxygen tension. This assumption seems reasonable in normoxia and hyperoxia but it may be invalid under hypoxic conditions because there is a possibility of capillary recruitment (Capen *et al.*, 1981; Capen and Wagner, 1982). These factors may be the cause of the conflicting reports on $D_L$ and $V_c$ at high altitude (Dempsey *et al.*, 1971; Kreuzer and Campagne, 1965; Weiskopf and Severinghaus, 1972). $D_L$ should be measured at the same partial $O_2$ pressure as that in ambient air at a given high altitude to assess hypoxic effect on $D_L$ and $V_c$.

The purpose of this study was to examine whether ambient temperature affects a change in $D_L$ caused by acute exposure to simulated high altitude when differences in the rate of the CO reaction with haemoglobin at different altitudes is taken into consideration.

## 2. Method

Subjects were five young male adults. Climatic conditions were produced in a biotron with ambient temperatures of 16, 20, 24 and 28°C

(RH 50% in all cases) combined with barometric pressures corresponding to sea level and altitudes of 2 000 and 4 000 m as the test conditions. The subjects, who wore thin running shirts and shorts, were each exposed to a total of 12 conditions, in random order, for 2 h 30 min.

$D_L$, oxygen intake and heart rate were measured at the 10th, 60th and 120th minute of exposure at rest in a sitting position. After a rest period, subjects performed exercise with a cycle ergometer at a work rate of 50 W for about 7 min and then the power was increased to 100 W and they continued to perform the exercise for a further 7 min. The same measurements were repeated at the 4th minute of the exercise period at each work rate.

$D_L$ was measured by a breath-holding method at almost the same partial $O_2$ pressure as that of room air under each condition. The CO and Ne contents of the inspired and the expired gas samples were determined by a gas chromatograph technique (Smith and Hamilton, 1962). Oxygen intake was measured by the Douglas bag method and the expired $O_2$ and $CO_2$ concentrations were measured by an electromagnetic gas analyser previously calibrated by the micro-Scholander method (Scholander, 1947). Heart rate was determined from electrocardiograms.

## 3. Results and discussion

The average $D_L$ was calculated from data on three measurements obtained during the 2 h rest because $D_L$ did not change with exposure time. Figure 22.1(a) shows the % ratio of average $D_L$ to that obtained at rest at 28°C at sea level. There was no significant change in $D_L$ between temperatures at sea level. However, the temperature effect was significant at 2 000 m ($p \leq 0.05$) and a temperature effect was also observed at 4 000 m ($p \leq 0.005$). This figure shows that $D_L$ increases with altitude and that the increase in $D_L$ caused by hypoxia becomes larger in a cool environment. Figure 22.1(b) shows the change in $D_L$ during exercise. The same tendency was observed during mild exercise (at 50 W), but when work rate increased the temperature effect was not as much as at rest.

Figure 22.2 shows the relationship of $D_L$ to oxygen intake. At sea level and 2 000 m the slope and the elevation of these regression lines were statistically constant between temperatures (at sea level, $F=0.115$ for slope, $F=0.583$ for elevation; at 2 000 m, $F=1.159$ for slope, $F=1.858$ for elevation). However, the elevation was significantly different at 4 000 m ($F=3.557$, $p \leq 0.05$), being higher in a cool environment than at 28°C although the slope was the same ($F=0.784$).

Figures 22.1 and 22.2 show that an increase in $D_L$ caused by hypoxia may not be explained solely by acceleration of the reaction rate between

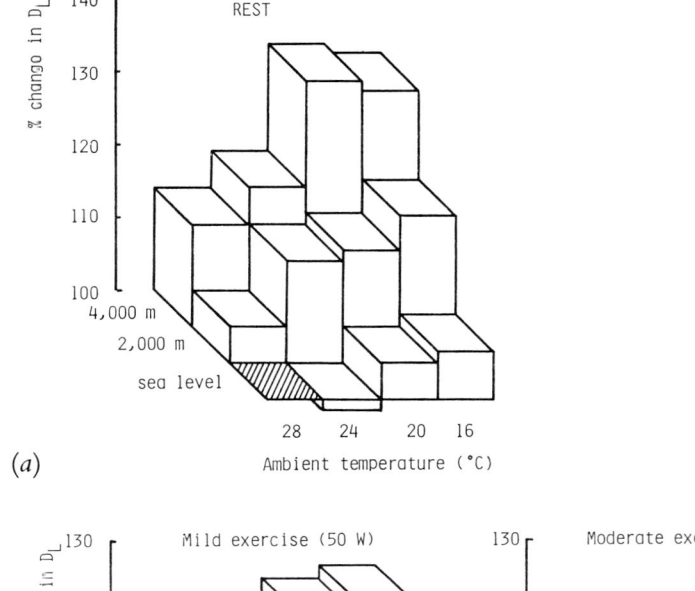

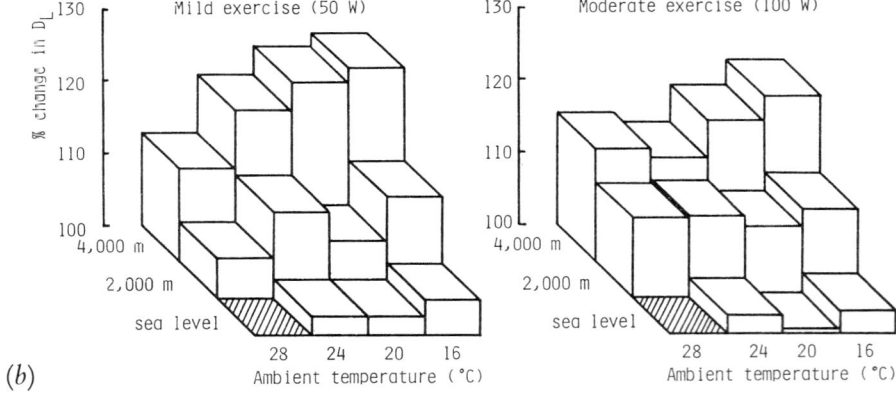

*Figure 22.1.* Hypoxic temperature effects on pulmonary diffusing capacity (DL) (a) at rest and (b) during exercise. Pulmonary diffusing capacity is expressed as a percentage ratio to that in an ambient temperature of 28°C at sea level.

CO and haemoglobin. To clarify this suggestion, it is necessary to examine whether capillary volume, and/or membrane components of diffusion, change at high altitude. Since the method of Roughton and Forster (1957) to estimate diffusion components may be invalid under hypoxic condition (Capen *et al.*, 1981; Capen and Wagner, 1982), we attempted to compare measured $D_L$ with predicted $D_L$ at high altitude assuming that the diffusion components were the same as those at sea level, shown in Figure 22.3. Roughton and Forster (1957) derived a mathematical formula relating diffusing capacity of the whole lung

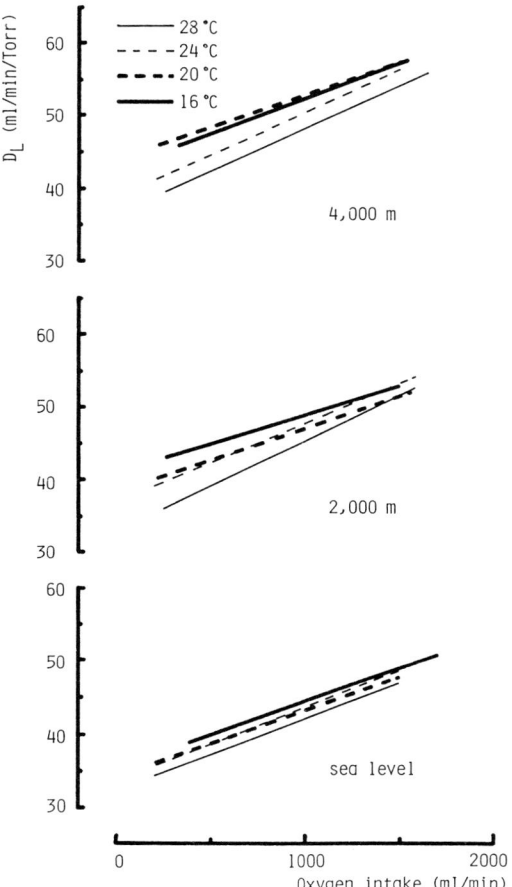

*Figure 22.2. The relationship between pulmonary diffusing capacity (DL) and oxygen intake under each condition.*

($D_L$) to the diffusion components—the functional capacity of the capillary bed ($V_c$) and the resistance to diffusion through the alveolocapillary membrane ($D_m$). Here, $\theta$ represents the reaction rate between CO and haemoglobin which is calculated from alveolar oxygen tension (Bates *et al.*, 1960). Since the mathematical formula shows that $1/D_L$ is a function of $1/\theta$, the slope of the line drawn by the formula represents $1/V_c$ and the *y*-intercept represents $1/D_m$. If predicted $D_L$ at high altitude consists of the same $V_c$ and $D_m$ as those at sea level, then $1/D_L$ must be on the line drawn at sea level and the $D_L$ value can be found at the crossing point between the line and a vertical line corresponding to $\theta$, or alveolar oxygen tension measured at a given high altitude. The line to predict

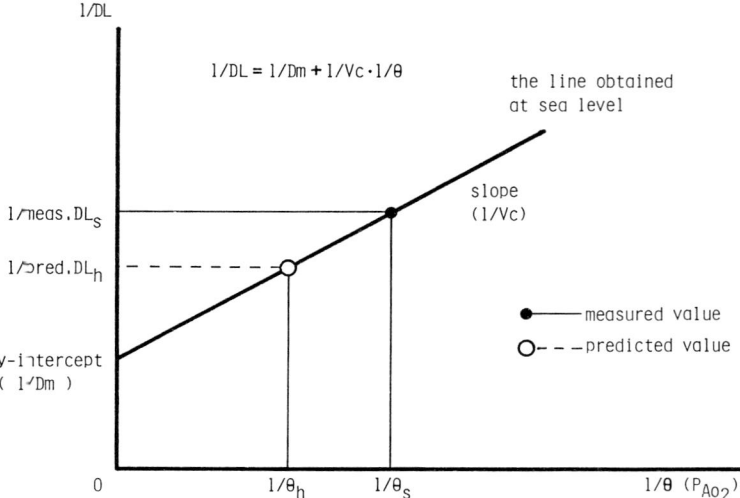

*Figure 22.3* The schematic relationship of the inverse of pulmonary diffusing capacity $(1/D_L)$ to the inverse of the reaction rate between CO and haemoglobin $(1/O)$. $D_{Ls}$ and $O_s$ represent the value obtained at sea level, while $D_{Lh}$ and $O_h$ represent the value obtained at high altitude.

$D_L$ may be drawn from data on $D_L$, $\theta$ and the slope obtained at sea level. However, there are no data on the slope because the method of Roughton and Forster (1957) was not used in this study. Therefore, this slope was estimated from equations in previous studies (Yasukouchi, 1984, 1985), in which $V_c$ has a very close relation with $D_L$ (Table 22.1). Thus, $D_L$ may be predicted from the line drawn at sea level and the alveolar oxygen tension measured at high altitude.

Figure 22.4(a) shows the result of comparing average measured $D_L$ with average predicted $D_L$ at each high altitude under the resting condition. At 2 000 m, in all temperature conditions, there was no significant difference between measured and predicted values. However, a significant increase in measured $D_L$ was observed in a cool environment at 4 000 m. This cannot be explained by the increase in $\theta$,

*Table 22.1.* The relationship of pulmonary capillary blood volume $(V_c)$ to pulmonary diffusing capacity $(D_L)$ at rest and during exercise.

|  | Equation | $r$ | Sy.x | Condition |
|---|---|---|---|---|
| Rest | $V_c = -44.7 + 3.50 D_L$ | 0.93 | 5.7 | Ambient temperature: 12, 20, 24 and 28°C |
| Work | $V_c = -24.4 + 2.88 D_L$ | 0.88 | 9.4 | Work load: 50, 70, 90, 110 and 130 W |

Abbreviations:
$r$ = the correlation coefficient.
Sy.x = the standard deviation for the errors in estimating the regression equation.

or by the same $V_c$ and $D_m$ as those at sea level. Figure 22.4(*b*) shows the case during exercise. The same tendency may be seen during mild exercise, but the temperature effect seems to become small with increasing work rate.

At 2 000 m on every occasion measured $D_L$ was the same as the predicted $D_L$ obtained from the line in Figure 22.3. This means that

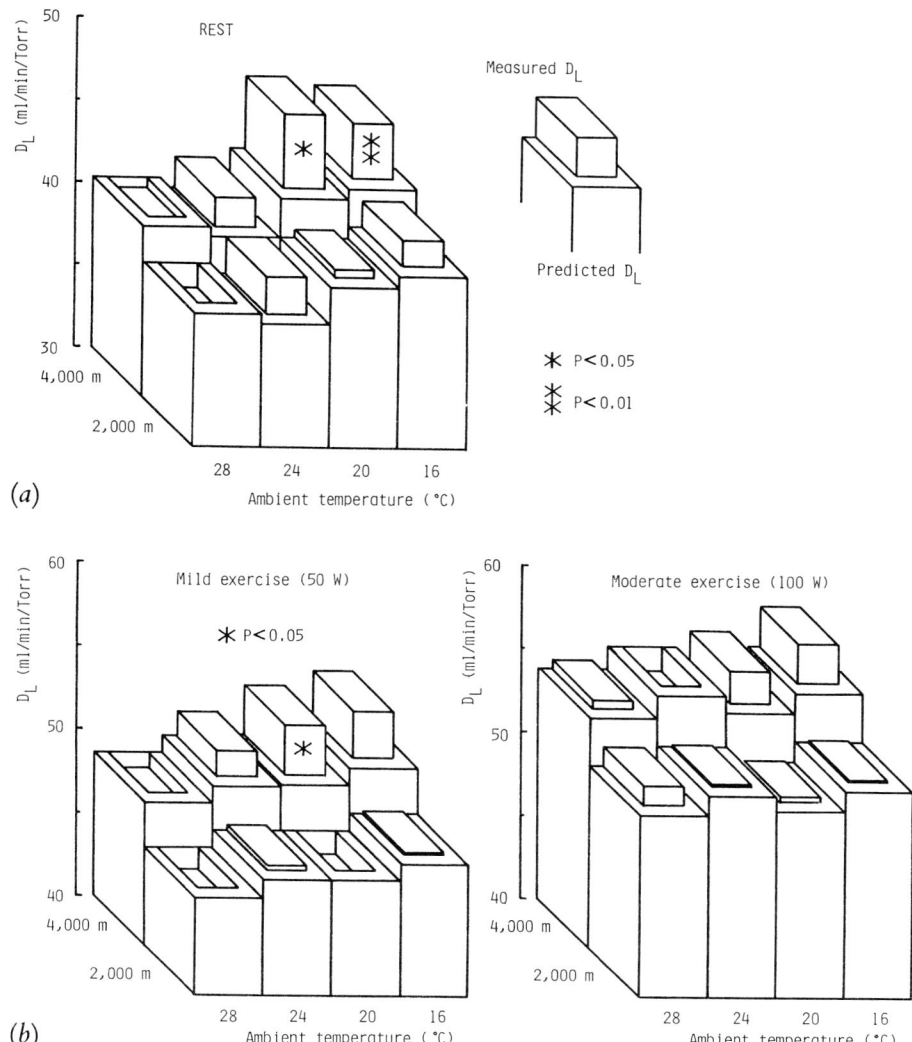

(*a*)

(*b*)

*Figure 22.4.  A comparison between the average measured and average predicted pulmonary diffusing capacity ($D_L$) at each high altitude (a) at rest and (b) during exercise. Smaller and larger square prisms represent measured and predicted diffusing capacity, respectively.*

*Environmental Ergonomics*

the increase in $D_L$ from sea level to 2 000 m was caused mainly by the increase in $\theta$ due to fewer $O_2$ molecules competing with CO for binding sites on the haemoglobin. On the other hand, at 4 000 m in a cool environment, except in the case of moderate exercise (at 100 W), measured $1/D_L$ was located below the line, as shown in Figure 22.5. In this case, two types of lines through this point may be assumed. Broken line A represents the case of the same slope and the lower $y$-intercept, that is, the same $V_c$ and an increased $D_m$ compared with those at sea level, and broken line B represents the case of the smaller slope and the same $y$-intercept, that is, an increased $V_c$ and the same $D_m$. In both cases, even if the reaction rate between CO and haemoglobin is adjusted to sea level, $D_L$ has a larger value in a cool environment at 4 000 m than at any other temperature at sea level. This may result from an increase in the surface area for gas exchange during hypoxia in a cool environment at 4 000 m because of an increase in pulmonary capillary blood volume and/or a change in distribution of capillary blood flow in the lungs. However, these effects were relatively reduced at ambient temperatures above 24°C or work rates above 50 W.

In conclusion, $D_L$ increases at altitude and this increase may not always be explained by an increase in the reaction rate between CO and haemoglobin. It is suggested that the extent of hypoxic effects on pulmonary diffusing capacity is closely related to ambient temperature and energy expenditure. The effects become larger in a cool environment at a higher altitude, especially under resting conditions. These

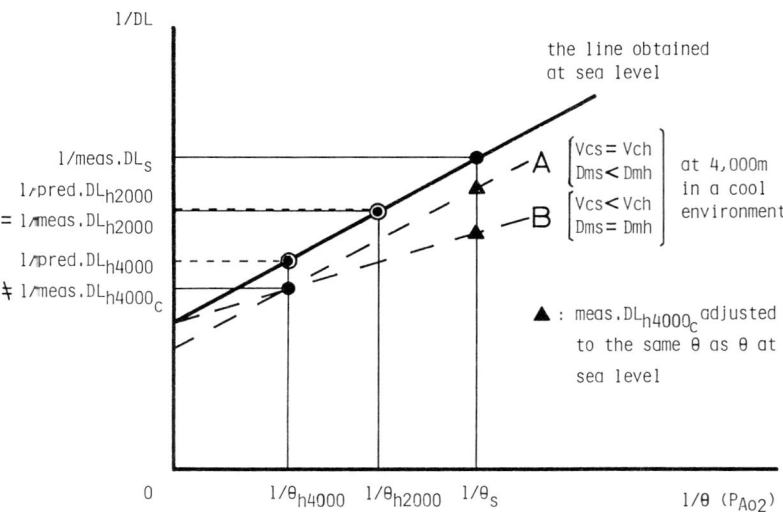

*Figure 22.5. A comparison between the measured and predicted pulmonary diffusing capacity ($D_L$) at high altitude as shown in Figure 22.3. Subscripts, h2000, h4000 and h4000c denote sea level, 2000 m, 4000 m and 4000 m in a cool environment, respectively.*

findings must be kept in mind when measuring pulmonary diffusing capacity under hypoxic conditions.

## References

Bates, D.V., Varis, C.J., Donevan, R.E. and Christie, R.V., 1960, Variations in pulmonary capillary blood volume and membrane diffusion component in health and disease. *J. Clin. Invest.*, **39**, 1401−1412.

Capen, R.L., Latham, L.P. and Wagner, W.W., Jr, 1981, Diffusing capacity of the lung during hypoxia: role of capillary recruitment. *J. Appl. Physiol.: Respirat. Environ. Exercise Physiol.*, **50**, 165−171.

Capen, R.L. and Wagner, W.W., Jr, 1982, Intrapulmonary blood flow redistribution during hypoxia increases gas exchange surface area. *J. Appl. Physiol.: Respirat. Environ. Exercise Physiol.*, **52**, 1575−1581.

Dempsey, J.A., Reddan, W.G., Birnbaum, M.L., Forster, H.V., Thoden, J.S., Grover, R.F. and Rankin, J., 1971, Effects of acute through life-long hypoxic exposure on exercise pulmonary gas exchange. *Respirat. Physiol.*, **13**, 62−69.

Kreuzer, F. and Campagne, P., 1965, Resting pulmonary diffusing capacity for CO and $O_2$ at high altitude. *J. Appl. Physiol.*, **20**, 519−524.

Roughton, F.J.W. and Forster, R.E., 1957, Relative importance of diffusion and chemical reaction rates in determining rate of exchange of gases in the human lung, with special reference to true diffusing capacity of pulmonary membrane and volume of blood in the lung capillaries. *J. Appl. Physiol.*, **11**, 290−302.

Scholander, P.F., 1947, Analyzer for accurate estimation of respiratory gases in one-half cubic centimeter samples. *J. Biol. Chem.*, **167**, 235−250.

Smith, J.R. and Hamilton, L.H., 1962, $D_{LCO}$ measurements with gas chromatography. *J. Appl. Physiol.*, **17**, 856−860.

Weiskopf, R.B. and Severinghaus, J.W., 1972, Diffusing capacity of the lung for CO in man during acute acclimation to 14,246 ft (4415 m). *J. Appl. Physiol.*, **32**, 285−289.

Yasukuochi, A., 1984, Uniformity of increase in pulmonary diffusing capacity during submaximal exercise in normal young adults. *Industr. Health*, **22**, 137−151.

Yasukuochi, A., 1985, Postural and thermal effects on pulmonary diffusing capacity at rest. *J. Anthrop. Soc. Nippon*, **93**, 87−95 (in Japanese, with English abstract).

# PART VII
# GRAVITATIONAL STRESS

# 23. Some Swedish Developments in Gravitational Physiology

## Hilding Bjurstedt

## 1. Introduction

Gravity, which is perhaps the most familiar force in nature, is capable of instigating compensatory reactions in the organism that are in several respects far from being explained. General interest in the effects of normal and increased gravitational stress on the human body has, however, a relatively short history. It is intimately interwoven with the development during the present century of aircraft capable of high speeds and manoeuvrability. Before the 1930s, accounts of the occurrence of pilot's blackout in curvilinear flight were mostly of an anecdotal nature. It was not until the rapid development of high performance aircraft during World War II that organized medical acceleration research gained impetus.

Gravitational stress assumes special significance as a physiological stimulus in that there is probably no influence, environmental or of other origin, that is capable of exerting such profound, yet reversible effects on the circulatory system. It is quite possible that centrifuge laboratories would exist today for research in gravitational physiology even if our civilization had not developed dive bombers or fighter aircraft. Experience has shown that the human centrifuge constitutes an excellent tool for basic cardiovascular research.

In the years before World War II the symptoms of blackout and loss of consciousness became increasingly familiar to pilots as manifestations of G stress during training in dive bombing and aerial dogfights. Although it was well known that these symptoms were caused by diminished blood supply to the eye and brain as a result of the action of the centrifugal force, there was scant information as to the physiological background to these hazards, and consequently the possibilities of protecting the pilots against them. For this reason, a number of human centrifuges were built during the war, and research in these laboratories helped to clarify the physiology of some of the more salient features of gravitational stress. Problems in this area increased in severity with the introduction of jet engines, and to permit continued physiological investigations, especially with regard to the effects of more long-lasting G stresses, several centrifuge laboratories have been built during the decades following World War II.

In the 1950s the Royal Swedish Air Force was in a build-up phase which had begun during the war. There emerged an increasing need for medical acceleration research reinforced by the presence of a rapidly expanding aircraft industry in the country. The Swedish centrifuge was constructed and built by ASEA and Saab and was installed at the Department of Aviation Medicine at the Karolinska Institute Medical School in Stockholm. It was ready for use in 1955 (Figure 23.1) and has sufficient performance capabilities to meet the demands of present-day requirements (Götzlinger and Helsing, 1955).

Below is a brief account of some of the centrifuge research conducted in selected areas of gravitational physiology. All centrifuge runs and other reference to inertial forces experienced in a head-to-foot direction will be referred to in terms of G rather than $+Gz$ as would be the standard terminology.

## 2. G-induced disturbances of pulmonary gas exchange

Because tolerance to increased G stress in the head-to-foot direction is limited by impaired blood and oxygen flow to the brain, with loss of consciousness rapidly ensuing at 4−6 G, the possibility that G stress might also seriously affect pulmonary function received little attention until the 1950s. At the time our centrifuge was ready for use, the susceptibility of the lung tissue and pulmonary circulation to the normal force of gravity had begun to be explored in many laboratories around the world; in these investigations a variety of methods were used to study how changes in the direction of the gravitational vector relative to the long axis of the lung affected the relative distribution of blood flow and ventilation in the lungs. For optimal gas exchange air and blood must be uniformly mixed in the lungs. In motionless standing, the action of normal gravity is sufficient to cause non-uniform distribution of blood flow through the lungs. When subjects were tilted from the supine to the upright position, we found that an alveolar dead space was created which was equivalent to a blocking of the blood perfusion to about 6% of the total number of lung alveoli (Bjurstedt *et al.*, 1962).

It is now well-recognized that gravity is the main factor determining the distribution of both blood and gas within the lungs under physiological conditions. However, in our early centrifuge work, it occurred

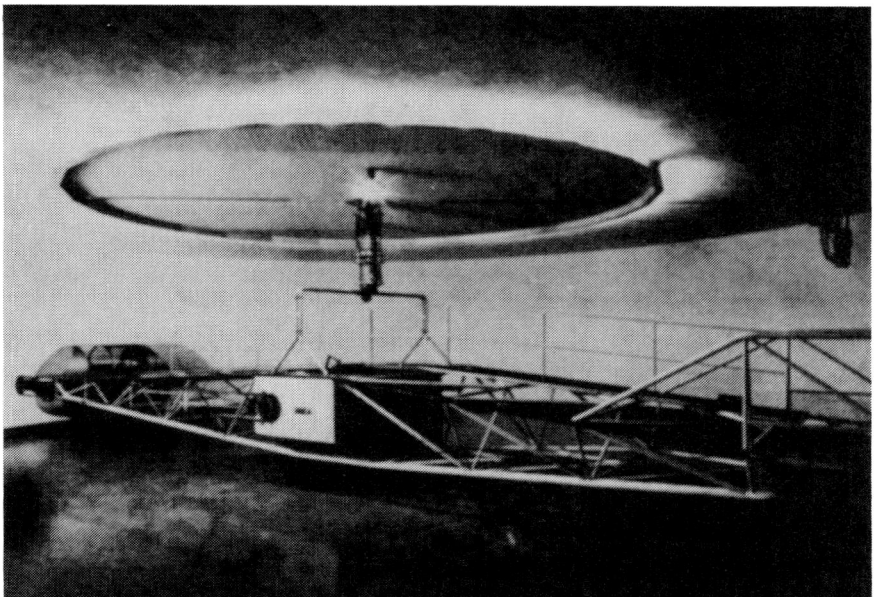

*Figure 23.1.   The human centrifuge (radius = 7·3 m) at the Karolinska Institute Medical School, Stockholm.*

to us that increased gravitational stress might exaggerate the non-uniformity of regional ventilation/perfusion ratios previously found to be caused by the action of normal gravity, and that with increased 'effective' weight of the blood a condition would eventually be reached in which all the blood in the lung would go to its base, and all ventilation would go to the upper regions. This would necessarily show in the degree of oxygenation of the blood. During the 1940s we had developed methods for continuous and simultaneous recording of the arterial $O_2$ saturation and hydrogen ion concentration in the streaming blood of anaesthetized dogs. Figure 23.2 shows a record from an experiment on the centrifuge in which an anaesthetized dog was exposed to the relatively low stress of 2·8 G with the inertial force acting in the head-to-tail direction. The development of progressive $O_2$ desaturation combined with hyperventilation and respiratory alkalosis in spite of the fact that the animal was breathing 100% $O_2$ demonstrated for the first time that increased gravitational stress is capable of inducing a large left-to-right shunt in the lungs; this can only be explained by a considerable amount of blood passing through collapsed portions of the lung (Barr *et al.*, 1959).

Barr (1963) extended the observations in anaesthetized dogs to man breathing air and exposed to 5 G for 2 min. In these experiments, one radial artery was catheterized with arterial blood being passed through

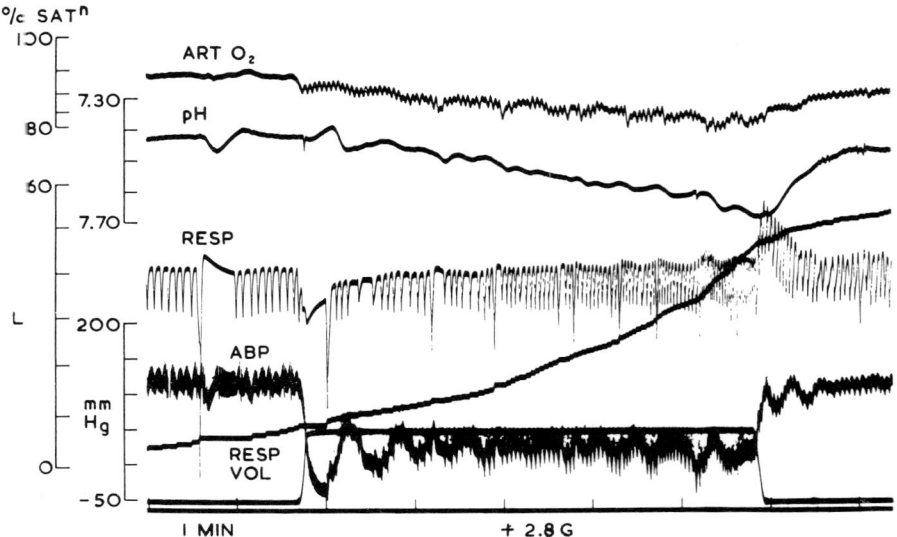

*Figure 23.2.*   *The effects of exposure to 2·8 G (head-to-tail) in a 17·5 kg dog under chloralose anaesthesia breathing 100% $O_2$. External counterpressure to abdomen. From above downwards: arterial $O_2$ saturation and pH; thoracic respiratory movements; gas meter tracing, slope of which increases with respiratory volume (calibration at extreme left in litres; accelerogram starting at and returning to zero G in the head-to-tail direction, plateau showing the magnitude and duration of G force as the centrifuge is rotating; 1 min between markings). Note hyperventilation with the development of progressive alkalosis and arterial $O_2$ desaturation, which demonstrates the development of a large pulmonary shunt. From Barr et al. (1959).*

sensors for hydrogen ion concentration and $O_2$ saturation as the centrifuge was spinning (Barr and Bjurstedt, 1959). In these 5 G experiments on nine subjects, the arterial $O_2$ saturation fell from an average of 96% to 87% and corresponded to a fall in calculated arterial $O_2$ tension from about 90 to 58 mm Hg, which suggests that a considerable shunting of blood must have occurred through collapsed lung units (Figure 23.3). Four of the nine test subjects showed values for $O_2$ tension which were lower than 50 mm Hg (Figure 23.4). The hazards to flight safety and efficiency due to this kind of hypoxaemia must be considered more serious than if a similar degree of hypoxaemia were induced by high altitude. This is because the oxygen flow to the brain is further reduced by the decrease in cerebral blood perfusion caused by G-induced hypotension at head level.

The possibility of preventing the occurrence of acceleration hypoxaemia by inhalation of 100% $O_2$ was investigated in a subsequent series of experiments (Barr et al., 1969). It was found that during exposure to 5 G for 3 min, arterial $O_2$ saturation was well maintained at 100% during the first minute, but subsequently could fall to 82% in

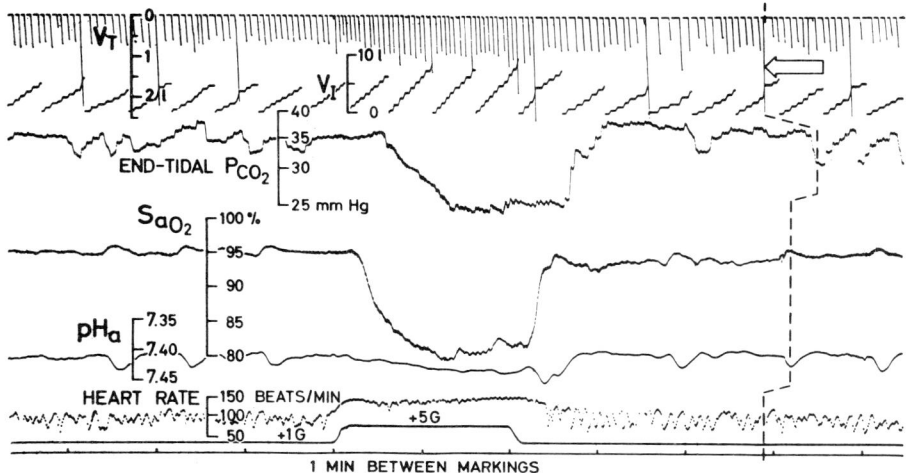

Figure 23.3. *The effects of exposure to 5 G of a subject wearing an anti-G suit and breathing air. Note the marked lowering of arterial O₂ saturation in spite of a concomitant increase in ventilation. The dashed line indicates lags after which pulmonary events are reflected in blood gas tracings. From Barr (1963).*

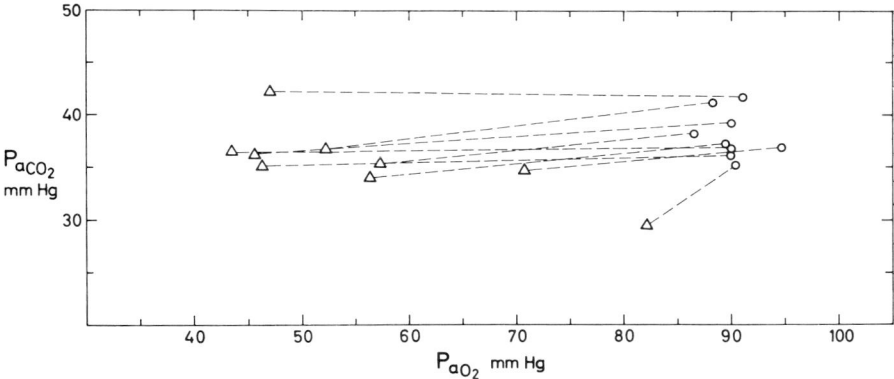

Figure 23.4. *Maximal individual changes in arterial O₂ and CO₂ tensions resulting from exposure to 5 G for 2 min, with subjects wearing anti-G suits and breathing air. Circles indicate pre-exposure values. Triangles indicate the lowest individual values during the course of the run. From Barr (1963).*

spite of concomitant hyperventilation. In these experiments, the breathing of oxygen probably caused alveolar collapse in the bases of the lungs to be more extensive than it would have been with air. Arterial pH showed a moderate acidotic shift, while end-tidal $CO_2$ pressure fell, indicating the development of a significant alveolar dead space.

As will be dealt with later, present-day high-performance combat aircraft are capable of producing considerably higher G levels for sustained periods of time than previously encountered. The human body possesses a number of defence mechanisms to cope with such extreme stress. However, gravitational stress interferes in various ways and at different levels with the transport of oxygen from the ambient air to the tissues (Bjurstedt, 1982); the lungs are the most susceptible to limitation by such stress. Possibilities of protection against G-induced lung collapse are at best marginal, and it is probable that the limit of man's resistance to high sustained G forces will ultimately be determined by the breakdown of his pulmonary gas exchange.

## 3. Anti-gravity action of the leg muscle pump

It is well known that prolonged motionless standing in a normal force environment leads to pooling of blood in dependent veins and progressive curtailment of cardiac output with eventual loss of consciousness. However, even mild leg exercise tends to reduce the transmural pressure and blood content in the dependent vascular beds, displacing blood from the legs towards the intrathoracic space and right heart and preventing dangerous loss of effective circulating blood volume. This effect implies that the leg muscle pump performs circulatory work; in fact, it is well established that the leg muscles are capable of contributing more than 30% of the energy required to circulate blood during running.

We wanted to investigate whether leg exercise might boost venous return to an extent that resistance to increased gravitational stress might be improved. Two series of experiments were carried out, in which the subjects performed graded leg exercise while exposed to 3 G for prolonged periods of time (Linnarsson and Rosenhamer, 1968; Rosenhamer, 1968). Figure 23.5 shows the position of the subject in the centrifuge gondola; an electrically braked cycle ergometer was used to provide the external work. It could be calculated that the average work load provided by the ergometer was essentially unaffected by the increase in gravitational load (Bjurstedt *et al.*, 1968).

Figure 23.6 shows that some of the subjects tolerated the exposure to 3 G in the resting condition for only a limited period of time, whereas all subjects were able to complete their runs when combined with exercise. The commencement of moderate-load exercise caused a rapid re-establishment of arterial pressure from a very low level at which the subject reported complete loss of vision, reflecting an instantaneous increase in stroke volume as exercise was started.

Additional series of experiments were performed to investigate more closely the influence of G stress on the adaptation of cardiovascular and

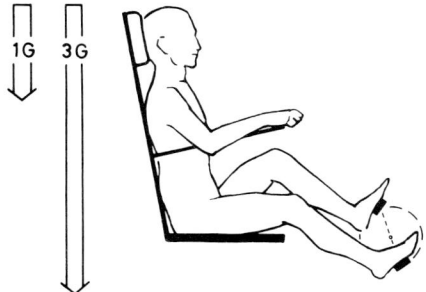

*Figure 23.5.  G forces and body contour with the subject seated in the centrifuge gondola, arrows indicating the magnitude of G force with the centrifuge standing still and running, respectively. From Rosenhamer (1968).*

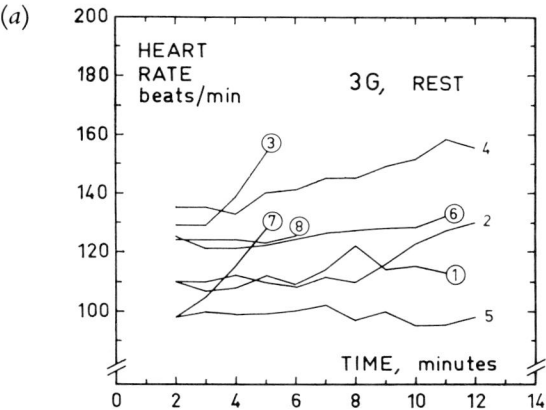

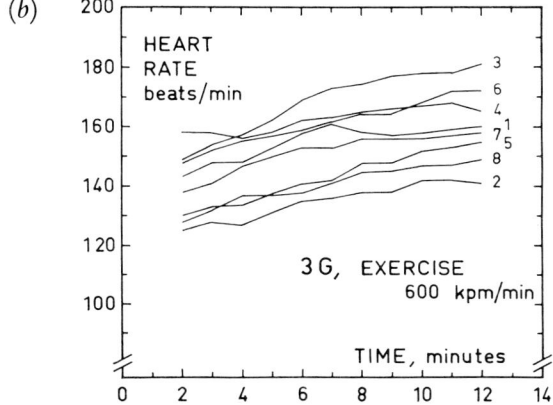

*Figure 23.6.  Heart rates in eight subjects at 3 G rest (a), and at 3 G exercise at 600 kpm min⁻¹ (b). Circles indicate the elective termination of experiment because of impending loss of consciousness. From Rosenhamer (1968).*

pulmonary function to leg exercise (Bjurstedt *et al.*, 1968, 1969; Rosen-hamer, 1968). One of the more striking observations was that stroke volume, as measured by the dye–dilution method, showed an immediate increase by no less than 80% as exercise at 3 G started, even with a mild load of 50 W. This illustrates, in an exaggerated fashion, the well-known difference in cardiac response to exercise in the supine and upright body positions: whereas in the supine position exercise increases cardiac output mainly through cardioacceleration, there is regularly a marked increase in stroke volume when exercise is started in the upright position.

The boost given to cardiac function by the leg muscle pump may also be studied during exercise in supine man with pooling in the leg vessels being induced by exposure of the lower portion of the body to subatmospheric pressure (lower body negative pressure, LBNP). In recent experiments, we found that with LBNP at $-50$ mm Hg, leg exercise caused a prompt increase in stroke volume by about 40%, with heart rate remaining remarkably unchanged or even showing a decrease (Eiken and Bjurstedt, 1985).

Bearing in mind the deleterious effects of gravitational stress on the pulmonary gas exchange, it may be questioned how such stress affects the oxygen transport to the muscles as work load is progressively increased. Figure 23.7 shows the relationship between the $O_2$ uptake in a group of subjects and the work load at normal gravity and at 3 G; individual data are also included. It can be seen that the G-induced increase of the work of postural and respiratory muscles caused an elevation of $O_2$ uptake at all loads. When the work load at 3 G was increased from 600 to 900 kpm min$^{-1}$, two subjects showed a levelling off of $O_2$ uptake and also considerably higher lactate values than the rest of the subjects, and one subject was unable to complete exercise at 900 kpm min$^{-1}$ at 3 G because of exhaustion. These observations together with other findings indicate that maximal $O_2$ uptake is lowered by G stress, and that the primary limitation imposed on the $O_2$ transport system by such stress occurs in the lungs. The conclusion can be drawn that the G-induced changes in pulmonary gas exchange present a greater handicap to $O_2$ transport and work capacity than do the changes in the systemic circulation (Bjurstedt *et al.*, 1968).

## 4. Autonomic heart blockade and G tolerance

By the use of β-adrenergic blocking agents and atropin, chemical autonomic blockade of the heart can be accomplished. In this way it is possible to investigate the influences of sympathetic drive and vagal restraint, respectively, on cardiac function during G stress. We have

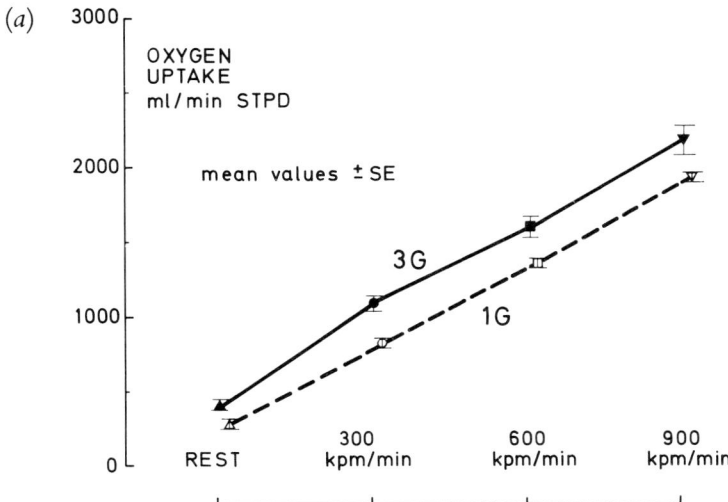

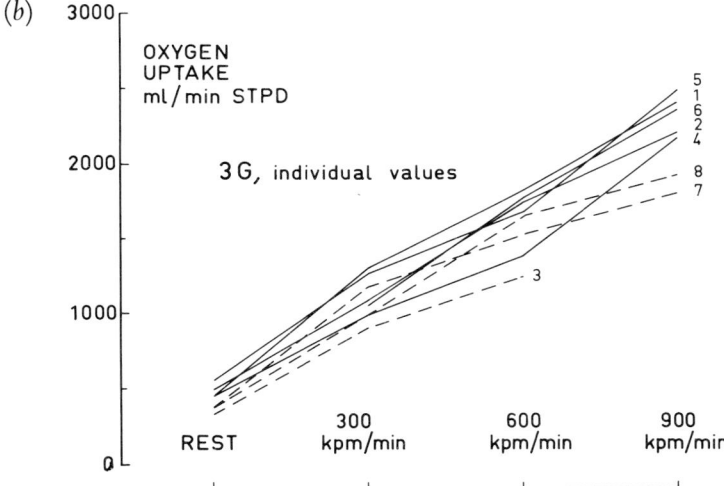

*Figure 23.7.* $O_2$ *uptake (ml min$^{-1}$) vs. external work load.* (a) *The effects of change in G level on group means for $O_2$ uptake at rest and during the 6th minute of exercise with increasing work load.* (b) *Individual data for the 3 G condition at rest and during exercise. Numbers 3, 7 and 8 showed the highest lactate levels during 3 G exercise at 900 kpm min$^{-1}$ (number 3 was not able to complete the 6 min experiment at this load). From Bjurstedt* et al. *(1968).*

studied the selective effects of the β-adrenergic blocking agent propranolol on the cardiovascular responses of human subjects exposed to G stress on the centrifuge (Bjurstedt *et al.*, 1974). In another series of experiments we have also used atropine in addition to propranolol to pharmacologically 'isolate' the heart from both sympathetic and vagal influence (Bjurstedt *et al.*, 1976). The latter approach to accomplish autonomic cardiac blockade was utilized to study the relative roles of cardiac and peripheral effector mechanisms in the circulatory defence against downward displacement of blood volume. This leads to central hypovolaemia, which is a salient effect of G stress in the head-to-foot direction. This condition was also studied with the body in the supine position by exposing its lower portion to subatmospheric pressure (Bjurstedt *et al.*, 1977), previously referred to as LBNP. This method may produce considerable pooling in dependent veins, mimicking the cardiovascular effects of rising from the supine position at normal gravity, or the effects of increased G stress, depending on the level of negative pressure applied.

In centrifuge experiments at 3 G under the influence of acute β-adrenergic blockage (0·25 mg kg$^{-1}$ b.wt., i.v.), the heart rate response averaged only 38% of that observed without blockade, which indicates that the increased heart rate during increased G stress is predominantly due to sympathetic stimulation. G-induced decline in cardiac output, measured by the dye–dilution method on the spinning centrifuge, was more marked after propranolol, yet subjective G tolerance was preserved, the arterial mean pressure being maintained by a strong compensatory 65% rise in systemic vascular resistance as against 43% without propranolol. Combined cardiac effector blockage, effected by the addition of atropine (0·04 mg kg$^{-1}$ i.v.), reduced the G-induced arterial pressure response, possibly by de-activating baroreflex control and/or by increased blood pooling in infracardiac vessels (Bjurstedt *et al.*, 1979). The negligible effect of β-blocking agents alone on G tolerance is of great practical importance in military aviation in view of the widespread use of such agents in the treatment of various cardiovascular disorders.

In LBNP experiments at −80 mm Hg, simulating gravitational stress considerably greater than normally experienced in the standing position, we found that tolerance to this condition was reduced by β-adrenergic blockade as judged from the arterial pressure response (Bjurstedt *et al.*, 1977). This is notable in view of our finding of a well preserved arterial pressure response in the centrifuge experiments. Figure 23.8 shows that LBNP during combined cardiac effector blockade reduced arterial pressure, but only to the same absolute level as obtained without blockade. Our experiments on the effects of autonomic heart blockade in conditions of true and simulated gravitational stress have been summarized by Tydén (1977).

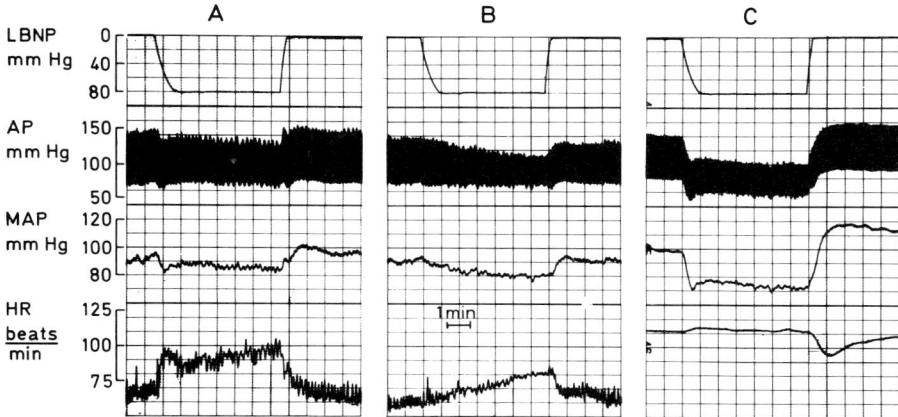

*Figure 23.8.* *The effects of LBNP on pressure in the radial artery (AP), mean arterial pressure (MAP), and heart rate before* (A) *and after beta-adrenergic* (B) *and combined beta-adrenergic and parasympathetic* (C) *blockade. From Bjurstedt* et al. *(1977).*

## 5. Protection against high sustained G forces

The latest generation of high-performance combat aircraft allow a manoeuvrability envelope which is much greater than found in yesterday's aircraft. As a result, high sustained G forces are created which impose increasing demands on the cardiovascular and pulmonary systems. The performance of certain aircraft in terms of magnitude and duration of G forces now easily exceeds the limits of human tolerance. That the potential for G-induced loss of consciousness and fatal accidents has shown an increase with the use of such aircraft is, therefore, not surprising.

For this reason, the need for more efficient countermeasures to prevent G-induced loss of consciousness is urgent. Up until recently, the best known protective measures consisted of an anti-G suit and the so-called M1 and L1 manoeuvres. Briefly, the anti-G suit, which was developed in the 1940s (Wood and Lambert, 1952), consists of a system of inflatable bladders covering portions of the legs and abdomen. The pressure in these bladders increases in rough proportion to the G force, through the action of a G sensitive valve, and by affording counter-pressure prevents excessive pooling in the high-capacitance veins below the heart. It also increases the resistance to flow in the arteries supplying the abdomen and legs and thus shunts blood flow away from these regions so that a larger portion of cardiac output may serve to perfuse the brain.

The M1 manoeuvre, which was originally devised at the Mayo

Clinic in the 1940s (Wood and Hallenbeck, 1945), consists of intermittent forceful exhalations against a partially closed glottis combined with tensing of the muscles of the legs, arms and abdomen. Through this manoeuvre, in principle consisting of frequent voluntary short-lasting strainings, the intrathoracic pressure is increased in an intermittent fashion. This pressure increase adds to that generated by the heart, thereby increasing the systemic arterial pressure and the perfusion of the brain.

In a remarkable series of experiments performed early in the 1970s at the School of Aerospace Medicine at Brooks AFB in Texas it was shown that the use of the M1 manoeuvre in combination with an anti-G suit in experienced selected subjects can permit exposures up to no less than 9 G for 45 s (Parkhurst *et al.*, 1972). It is evident, however, that the intensive use of M1 manoeuvres at such high G levels requires a straining pattern which causes severe fatigue. This may, in itself, increase the risk of loss of consciousness by weakening the strength of successive manoeuvres or otherwise interfering with the support of the systemic arterial pressure.

Additional methods for increasing individual resistance to high sustained G forces have therefore become necessary. Of methods recently tested the use of positive pressure breathing (PPB) and the introduction of physical conditioning programmes to improve strength straining have so far proved to be the most promising.

Experiments in our laboratory (Bjurstedt *et al.*, 1979) have shown that PPB at 30 mm Hg elevates the arterial pressure but reduces the driving pressure (arterial minus central venous pressure). The latter effect can be avoided by applying counterpressure to the body to prevent pooling in peripheral capacitance vessels. The beneficial effects of PPB in combination with an anti-G suit, which was first reported by Shubrooks (1973), have been further investigated in our laboratory. Balldin (1982) has used PPB at 30 mm Hg in combination with certain modifications of the anti-G suit including a more rapid inflation. He found a significant increase in the endurance time when his subjects were exposed to a sequence of 15 s periods at 1·5 and 7 G (Figure 23.9). Jointly with Burns (Burns and Balldin, 1983), he also investigated the protective effect of PPB with balancing external counterpressure applied on the thorax via an inflatable bladder ('assisted PPB'), and observed that such assisted PPB at 50 mm Hg combined with the M1 manoeuvre and the use of an anti-G suit permitted an increase of the endurance time at high G levels (varying between 5 and 9 G) by more than 100%. Also, Balldin, in co-operation with colleagues at the School of Aerospace Medicine at Brooks AFB, reported that PPB at 30 mm Hg may reduce G-induced pulmonary atelectasis (Haswell *et al.*, 1986). In Sweden, the effects on G tolerance of assisted PPB at 30 mm Hg in combination with a high-flow, ready-pressure anti-G

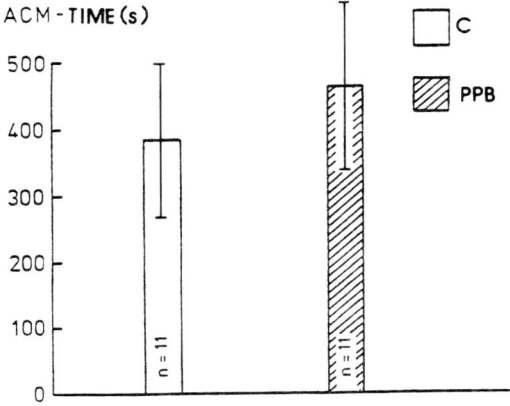

*Figure 23.9. G tolerance in the human centrifuge measured as the endurance time at successive 15 s periods at 1·5 and 7 G (ACM). C = subjects wearing a standard anti-G suit. PPB = subjects with positive pressure breathing at 30 mmHg, wearing a suit with faster filling and 'ready pressure'. The increase in ACM time in the latter subjects (19%) is statistically significant (p<0·005). From Balldin (1982).*

suit have been tested both in centrifuge runs up to 9 G and in a Saab Viggen aircraft during manoeuvres up to 6 G (Hjort and Balldin, 1986).

The straining pattern of the correct M1 breathing manoeuvre is somewhat similar to that used in weight-lifting. The strain breathing and concomitant muscle tensing requires practice, strength, stamina and endurance, and it is clear that this voluntary G-protective response can be exhausting in demanding air combat manoeuvres. We have investigated possible methods of preserving or increasing the efficacy of such G-protective strain breathing and muscle tensing. These investigations have in part been carried out in co-operation with Norwegian and Finnish researchers. The effects on G tolerance of an 11−12 weeks muscle strength training programme was studied in a group of 11 fighter pilots. It was found that mean G tolerance in terms of 'ACM endurance time' (exposure to 15 s alternating periods of 4·5 and 7 G until exhaustion) was increased by 39% (Tesch *et al.*, 1983). The high intra-abdominal pressures attained during strain breathing give rise to the question of whether isometric abdominal muscle training alone may be effective in increasing G tolerance. However, in a group of 10 experienced fighter pilots an 11-week abdominal muscle training programme was not sufficient to increase G tolerance in terms of ACM endurance time (Balldin *et al.*, 1985a). On the other hand, combined strength and aerobic training programmes, involving multiple muscle groups and conducted over a period of 12 months in a group of 20 fighter pilots, showed a significant increase in G endurance time over the first 6 months of training; no further improvement occurred during the remainder of the training period (Balldin *et al.*, 1985b). As a practical result of these and other investigations regarding the effects of strength training programmes, special weight exercise gymnasiums are currently in use for fighter pilots in the Royal Swedish Air Force.

Studies are under way in our laboratory to elucidate the mechanisms underlying changes in the tolerance to high sustained G. Histochemical studies of muscle fibre type and composition by needle biopsy have shown that strength training, which caused a substantial increase in muscular force production and G tolerance, was not associated with any changes in aerobic performance or in various histochemical indices (Tesch and Balldin, 1984). It seems more likely that neuromuscular factors are involved in the increased force production, such as improved innervation and additional recruitment of fast twitch motor units. Further studies are required to understand the mechanisms underlying the effects of strength training on G tolerance, so that optimal training programmes can be designed. An entirely different approach to the mechanisms underlying G tolerance is currently being used by Sporrong and co-workers (1986), who have studied relations between emotionally induced cardiorespiratory responses at normal gravity and G tolerance. It was observed that there was a positive relation between emotional responses of heart rate and arterial pressure induced by means of psychomotor performance testing on the one hand, and G tolerance on the other.

## 6. Conclusions

It must be emphasized that frequent exposures to high sustained G forces might conceivably cause considerable interference with the blood perfusion of internal organs. Of interest in this connection is that Balldin in collaboration with Norwegian researchers (Noddekind *et al.*, 1985) found significant amounts of protein and hyaline casts in urine samples from a group of fighter pilots after prolonged exposure to aerial combat simulation on the centrifuge (Figure 23.10), presumably a consequence of reduced renal blood flow. Further studies on the possible adverse effects of exposures to high sustained G forces on the function of different organ systems are required. It should be borne in mind that such effects may not only be the result of mechanical strain *per se*. One insidious effect is that of acceleration hypoxaemia, previously referred to, which in combination with reduced blood flow to a given organ, e.g., the brain, may result in severe functional disturbances. It is quite possible that a defective oxygen flow to vital organs secondary to pulmonary dysfunction will be the ultimate factor determining the limit of man's tolerance to high sustained G forces.

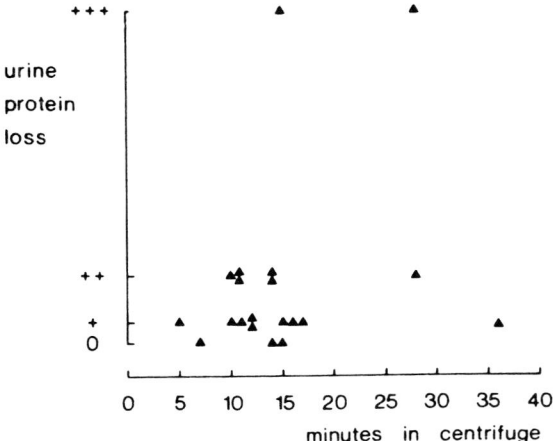

*Figure 23.10.   Urine protein loss (strip test) as a function of exposure time in centrifuge runs using a G profile of 15 s alternating plateaux at 3·5 and 5·5 G. Urine samples collected within 15 min of the end of individual exposure until exhaustion. From Noddeland et al. (1986).*

## References

Balldin, U.I., 1982, Positive pressure breathing and a faster filling ready pressure anti-G-suit: effects on +Gz tolerance. Preprint. Aerospace Medicine Association Meeting, pp. 16–17.

Balldin, U.I., Myhre, K., Tesch, P.A., Wilhelmsen, U. and Andersen, H.T., 1985a, Isometric abdominal muscle training and G-tolerance. *Aviat. Space Environ. Med.*, **56**, 120–124.

Balldin, U.I., Tesch, P.A., Rusko, H. and Kuronen, P., 1985b, G-tolerance after 12 months of combined strength and aerobic conditioning. Abstract. Aerospace Medicine Association Meeting. *Aviat. Space Environ. Med.*, **56**, 504.

Barr, P.O., 1963, Pulmonary gas exchange in man as affected by prolonged gravitational stress. *Acta Physiol. Scand.*, **58**, suppl., 207.

Barr, P.O. and Bjurstedt, H., 1959, Technique for continuous recording of blood pH and $O_2$ saturation in vivo in humans. *Rep. Lab. Aviat. Naval Med.*, **20**.

Barr, P.O., Bjurstedt, H. and Coleridge, J.C.G., 1959, Blood gas changes in the anesthetized dog during prolonged exposure to positive radical acceleration. *Acta Physiol. Scand.*, **47**, 16–17.

Barr, P.O., Brismar, J. and Rosenhamer, G., 1969, Pulmonary function and G-stress during inhalation of 100% oxygen. *Acta Physiol. Scand.*, **77**, 7–16.

Bjurstedt, H., 1982, Gravitational effects on oxygen transport. In *Oxygen Transport to Human Tissues*, edited by J.A. Loeppky and M.L. Riedesel (Amsterdam: North Holland).

Bjurstedt, H., Hesser, C.M., Liljestrand, G. and Matell, G., 1962, Effects of posture on alveolar-arterial $CO_2$ and $O_2$ differences and on alveolar dead space in man. *Acta Physiol. Scand.*, **54**, 65–82.

Bjurstedt, H., Rosenhamer, G. and Wigertz, O., 1968, High-G environment and responses to graded exercise. *J. Appl. Physiol.*, **25**, 713–719.

Bjurstedt, H., Rosenhamer, G. and Wigertz, O., 1969, Acceleration and muscular exercise. *Forsvarsmedicin*, **5**, 3–9.

Bjurstedt, H., Rosenhamer, G. and Tydén, G., 1974, Acceleration stress and effects of propranolol on cardiovascular responses. *Acta Physiol. Scand.*, **90**, 491–500.

Bjurstedt, H., Rosenhamer, G. and Tydén, G., 1976, Gravitational stress and autonomic cardiac blockade. *Acta Physiol. Scand.*, **96**, 526–531.

Bjurstedt, H., Rosenhamer, G. and Tydén, G., 1977, Lower body negative pressure and effects of autonomic heart blockade on cardiovascular responses. *Acta Physiol. Scand.*, **99**, 352–360.

Bjurstedt, H., Rosenhamer, G., Lindborg, B. and Hesser, C.M., 1979, Respiratory and circulatory responses to sustained positive-pressure breathing and exercise in man. *Acta Physiol. Scand.*, **105**, 204–214.

Burns, J.W. and Balldin, U.I., 1983, +Gz protection with assisted positive pressure breathing (PPB). Preprint. Aerospace Medicine Association Meeting, pp. 36–37.

Eiken, O. and Bjurstedt, H., 1985, Cardiac responses to lower body negative pressure and dynamic leg exercise. *Eur. J. Appl. Physiol.*, **54**, 451–455.

Gozzlinger, J. and Helsing, E., 1955, A human centrifuge for research into physiological flight stresses. *ASEA J.*, **28**, 1–9.

Haswell, M.S., Tacker, W.A., Balldin, U.I. and Burton, R.R., 1986, Influence of inspired oxygen concentration on acceleration atelectasis. *Aviat. Space Environ. Med.*, **57**, 432–437.

Hjert, H. and Balldin, U.I., 1986, Assisted positive pressure breathing and high flow ready pressure anti-G-suit during centrifuge and flight tests. Abstract. Aerospace Medicine Association Annual Scientific Meeting.

Linnarsson, D. and Rosenhamer, G., 1968, Exercise and arterial pressure during simulated increase of gravity. *Acta Physiol. Scand.*, **74**, 50–57.

Noddekind, H., Myhre, K., Balldin, U.I. and Andersen, H.T., 1986, Proteinuria in fighter pilots after high +Gz exposure. *Aviat. Space Environ. Med.*, **57**, 122–125.

Parkhurst, M.J., Leverett, S.D. and Shubrooks, S.J., Jr, 1972, Human tolerance to high, sustained +Gz acceleration. *Aerosp. Med.*, **43**, 708–712.

Rosenhamer, G., 1967, Influence of increased gravitational stress on the adaptation of cardiovascular and pulmonary function to exercise. *Acta Physiol. Scand.*, **68**, suppl., 276.

Rosenhamer, G., 1968, Antigravity effects of leg exercise. *Acta Physiol. Scand.*, **72**, 72–80.

Shubrooks, S.J., Jr, 1973, Positive-pressure breathing as a protective technique during +Gz acceleration. *J. Appl. Physiol.*, **35**, 294–298.

Sporrong, A., Baer, R., Balldin, U. and Nurmi, L., 1986, ACM G-tolerance and cardiovascular responsiveness to task-induced emotional stress at 1 G. Abstract. Aerospace Medicine Association Annual Scientific Meeting.

Tesch, P.A., Hjort, H. and Balldin, U.I., 1983, Effects of strength training on G tolerance. *Aviat. Space Environ. Med.*, **54**, 691—695.

Tesch, P.A. and Balldin, U.I., 1984, Muscle fiber type composition and G-tolerance. *Aviat. Space Environ. Med.*, **55**, 1000—1003.

Tydén, G., 1977, Aspects of cardiovascular reflex control in man. *Acta Physiol. Scand.*, suppl., 448.

Wood, E.H. and Hallenbeck, G.A., 1945, Voluntary (self-protective) maneuvers which can be used to increase man's tolerance to positive acceleration. *Fed. Proc.*, **4**, 78—79.

Wood, E.H. and Lambert, E.H., 1952, Some factors which influence the protection afforded by pneumatic anti-g suits. *Aerosp. Med.*, **23**, 218—228.

# Index

Mustang is proud to have been a corporate sponsor of the Second International Environmental Ergonomics Conference in association with the Kinesiology Department of Simon Fraser University.

We at Mustang are keenly aware of the valuable insight gained as a result of the research conducted by the Academic community in general and Simon Fraser University in particular. As designers and developers of critical personal life support products, the greater understanding of the mechanics of survival provided by this research has become a primary component of our product development process.

In particular this research is a key factor in the development of a range of Mustang products including Immersion/Exposure Suits for marine abandonment situations, Anti-Exposure Worksuits and Coveralls, buoyant and non-buoyant aviation protective clothing and inflatable life preservers.

Mustang looks forward to the next Ergonomics Conference and to continue to reinforce the spirit of mutual cooperation between Mustang Industries Inc. and Simon Fraser University.

**CANADA**

**MUSTANG INDUSTRIES INC.**
3810 Jacombs Road
Richmond, B.C., Canada
V6V 1Y6
Tel. (604) 270-8631

**U.S.A.**

**MUSTANG MFG. INC.**
2171 E. Bakerview Road
P.O. Box 5844
Bellingham, WA 98227
Tel. (206) 676-1782